Molecular Aspects of Insect–Plant Associations

Molecular Aspects of Insect–Plant Associations

Edited by

Lena B. Brattsten

Agricultural Products Department
E. I. DuPont de Nemours and Company, Inc.
Wilmington, Delaware

and

Sami Ahmad

Department of Biochemistry
University of Nevada
Reno, Nevada

PLENUM PRESS • NEW YORK AND LONDON

Library of Congress Cataloging in Publication Data

Molecular aspects of insect–plant associations.

Includes bibliographies and index.
1. Insect–plant relationships. 2. Allelopathic agents. 3. Insects—Physiology. I. Brattsten, Lena B. II. Ahmad, Sami.
QL495.M64 1986 582′.0233 87-2236

ISBN-13: 978-1-4612-9040-7 e-ISBN-13: 978-1-4613-1865-1
DOI: 10.1007/ 978-1-4613-1865-1

Softcover reprint of the hardcover 1st edition 1986
A Division of Plenum Publishing Corporation
233 Spring Street, New York, N.Y. 10013

PREFACE

Thanks to the meticulous and enthusiastic work of insect collectors and taxonomists over the past hundred years and more, we have today a large amount of information on the feeding habits and life styles of several hundred thousands of insect species. Insects that feed on plants during at least one of their life stages constitute about half of the three-quarters of a million described species. Their numbers both in terms of species and individuals together with their small but macroscopic sizes makes the insect-plant biological interface perhaps the most conspicuous, diverse and largest assemblage of intimate interspecies interactions in existence. It is also perhaps the most important biological interface because of the plants' role as primary producers upon which all other forms of earthly life depend, thereby bringing herbivorous insects occasionally into direct competition with human food and fiber production.

Early enthusiasm revealed many remarkable specializations and associations between insects and plants, and occasionally assigned chemical mediators for them. However, the modern practices of large scale crop protection by synthetic pesticides and their attendant problems, particularly with resistance in "pests" and destruction of natural enemies, have been in large measure responsible for drawing our attention to the mechanisms whereby plants control insect populations and insects adapt to the plants' defenses.

These practices have also brought home the importance of chemical mediators in practically all aspects of insect activities and, in particular, the importance of plant allelochemicals in maintaining and balancing insect-plant associations.

There are about 20,000 identified allelochemicals to date, estimated to be between ten and 50 percent of all possible such non-nutrient chemicals. Most of these have biological activity, mostly uncharacterized, when accumulated in sufficiently high concentrations. Many are acutely toxic to insects and other animals. A few have been used successfully for insect control and as drugs; more importantly, however, plant allelochemicals continue to provide model structures for the development of synthetic insecticides and drugs.

With the recent availability of high-resolution analytical instruments, natural history is rapidly turning into a modern, multidisciplinary biological science. The elucidation of the mechanisms which maintain the observed ecological equilibria in the insect-plant interface provide fascinating work and rich intellectual rewards. It is also necessary to understand these interactions for purposes of local food and fiber production, the premise being that understanding the molecular details of insect adaptations to plant allelochemicals may allow manipulations of those mechanisms for crop protection. At the same time, better insight in the basic mechanisms will likely enhance our ability to predict the ecological consequenses of such manipulations. Moreover, there is a compelling necessity to understand the factors that govern insect-plant associations for the conservation of an inhabitable global environment; the dynamic behavior of insect and plant populations is a sensitive indicator of environmental health, a canary in the coal mine.

Fortunately, the interest in studying insect-plant interactions is high. This is evident not least by the considerable number of books on the subject that have appeared in recent years many of which are symposia proceedings. Almost without exception, these publications emphasize the ecological and behavioral aspects of insect-plant associations, with a few contributions on population genetics: there are many more insect ecologists than insect biochemists at work. One of the functions of this book, which complements the existing ecological and behavioral information, is to stimulate interest in the basic mechanisms. Several other publications deal with the events leading to the finding and acceptance of a host plant. Here we concentrate on what happens after the long-distance search is over and the insect has contacted the plant. In doing so, we have not neglected the importance of behavior.

In order for an interaction to occur, the insect must have sensory modalities with which to recognize the plant and, specifically, its allelochemicals. Consequently, the first chapter in this book, by J. L. Frazier, describes the details of insect gustatory sensilla and discusses the mechanisms by which they transmit signals. The importance of seeking a correlation or common denominator between the in vitro electrophysiological recording and the actual feeding behavior of the insect is emphasized here. In the second chapter the focus is on the search for a connection between genetics and host plant specializations. This is done by feeding experiments with crosses and back-crosses of tiger swallowtail larvae obtained by hand mating so that the genetic constitution of the off-spring could be carefully controlled. These experiments are described and discussed by J. M. Scriber, and show how a seemingly very intricate complexity can be resolved into fairly controllable single variables.

Chapters three to seven delve deeply into the molecular mechanisms whereby insects defend themselves against toxic chemicals; the information available on specific insectan mechanisms was sometimes in short supply and some cases are illustrated with examples from vertebrates. The third chapter is a detailed account of the enzymes involved in the metabolism of xenobiotics, although not enough is known about the role of free radical-eliminating enzymes in insect-plant interactions for review now. The authors, S. Ahmad, L. B. Brattsten, C. A. Mullin and S. J. Yu, considered it important to include information on conjugating enzymes which may be essential in detoxifying polar plant allelochemicals; these enzymes get scant attention in most reviews of foreign compound metabolism because of the importance of cytochrome P-450 in the metabolism of lipophilic toxicants. However, the importance of cytochrome P-450 in the metabolism of lipophilic plant allelochemicals, notably terpenoids and phenylpropanes, can not be underestimated. The ability of this enzyme system as well as of glutathione transferases, epoxide hydrolases, and, perhaps, esterases to adapt to the chemical load of ingested plant tissues by induction is particularly interesting and potentially important. Therefore, in chapter four, S. J. Yu accounts for the construed consequenses of induction both in crop protection and insect-plant interactions.

In chapter five, C. A. Mullin provides a comprehensive comparison between adaptations in chewing insect herbivores and sucking ones such as aphids; the unique biochemical adaptations of aphids, most of which are highly specialized to one or a few related host plants, are pointed out and their potential importance vis a vis unavoidable plant allelochemicals is evaluated. L. B. Brattsten, in chapter six, complements the three preceding chapters by descriptions of the fate, mostly metabolic, of ingested plant allelochemicals, the possible influences of the major defense mechanisms on each other are discussed. Next, M. R. Berenbaum entreats us with the occurrence and evolutionary significance of adaptations in molecular target sites that render adapted insects resistant to several classes of otherwise highly insecticidal plant allelochemicals as well as to synthetic insecticides, in chapter seven.

D. W. Tallamy, in chapter eight, presents behavioral adaptations that appear to be prompted directly by the insect's encounter with toxic or impairing phytochemicals. Original research on the squash beetle - cucurbit interaction described in this chapter convincingly shows the complexity and sophistication of a behavioral defense evolved over perhaps millions of years in nature; this implies strongly the very limited importance of behavior as a defense against synthetic insecticides.

Finally, in chapter nine, K. F. Raffa discusses the possibilities and limitations in applying the results of basic research in insect-plant interactions to crop protection practices. He emphasizes the necessity of both basic and goal-directed research for achieving economically, socially, and, not least, biologically acceptable crop protection strategies, and the importance of candid communications between scientists involved in either kind of work.

Although the costs and benefits to the plant interactants are not the focus of any of the chapters in this book, this aspect of the interface is considered in many of the chapters. There seems to be a general paucity of botanists involved in collaborations with entomologists in studies of insect-plant interactions, even though it hardly seems realistic any longer to study plant evolution, ecology, genetics, and many other aspects without consideration of the effects of insects on plants. Another major function of this book is, thus, to stimulate interest in this area.

The full name of each insect species used to obtain the described information is included when the insect is first mentioned in each chapter, despite a certain hampering of the smooth flow of the reading. This is because of the importance of accurate identifications of the biological material under study. Together, the authors and editors succeeded in securing this information with a few exceptions. Nevertheless, the information provided remains valuable and, perhaps, important to other scholars even without full species identification; it would, however, be used with considerably more confidence and significance in interdisciplinary studies if the source was unequivocally stated.

The idea to put together this book was generated by a symposium sponsored by the Eastern Branch of the Entomological Society of America. Out of the ten symposium participants, eight contributed chapters all of which are updated and expanded in scope. We thank the authors for their enthusiastic and thoughtful cooperation in the gathering of this information into one volume. Their efforts should make the information easily accessible to students and investigators both at the bench and in the field.

We also thank all those who helped with reviews of the manuscripts and the staff at Plenum for advise and support. K. E. Kirkland helped substantially with the indexing. Most gratefully, we thank C. A. Cairo for her patient and careful work in transferring the text to camera-ready copy.

Wilmington and Reno L. B. Brattsten and S. Ahmad

CONTENTS

The Perception of Plant Allelochemicals that Inhibit Feeding 1
J. L. Frazier

1. Introduction 1
2. Insect Feeding Behavior 2
2.1. Generalized Sequence 2
2.2. Neuroethological Approach 3
3. The Contact Chemoreceptive System 7
3.1. Structure of Chemoreceptive Sensilla 7
3.2. Function of Chemosensilla 9
3.2.1. Contact Chemosensory Systems 9
3.2.2. Sensitivity of Contact Chemosensory Cells 10
3.2.3. Specificity of Contact Chemosensory Cells 11
3.2.4. Specificity of Chemoreceptor Sites 11
4. Inhibitory Plant Allelochemicals 13
4.1. Major Classes of Allelochemicals 13
4.2. Modes of Action on Chemosensory Cells 13
4.2.1. Potential Types of Action 13
4.2.2. Deterrent Chemosensory Cells 14
4.2.3. Effects of Alkaloids 28
4.2.4. Effects of Terpenes 29
4.2.5. Effects of Other Allelochemicals 31
4.3. Modes of Action on Other Excitable Cells 33
5. Summary 33
6. Acknowledgement 33
7. References 33

Allelochemicals and Alimentary Ecology: Heterosis in a Hybrid Zone? 43
J. M. Scriber

1. Introduction 43
1.1. A Case Study of Insect-Plant Interactions. The Insects 43
1.2. Estimation of Growth Characteristics 48
1.3. The Plant Allelochemicals 49
2. Hybridization Studies and Experimental Approach 49
3. Results 53
4. Discussion 57
5. Summary 59
6. Acknowledgement 60
7. References 60

Enzymes Involved in the Metabolism of Plant Allelochemicals 73
S. Ahmad, L. B. Brattsten, C. A. Mullin and S. J. Yu

1. Introduction 73
2. Cytochrome P-450 75
2.1. Occurrence 75
2.1.1. Occurrence in Species 75
2.1.2. Age-Related Distribution of Activity 77
2.1.3. Tissue Distribution and Subcellular Localization . 79
2.2. Biochemical Characteristics of Microsomal Cytochrome P-450 80
2.3. Apparent Reactions Catalyzed by Cytochrome P-450 83
2.3.1. CH-Hydroxylations 84
2.3.2. π-Bond Oxygenations 85
2.3.3. Oxygenation at an Unshared Electron Pair 88
2.4. Measurements of Cytochrome P-450-catalyzed Reactions in Insects 89
2.5. Inhibition of Cytochrome P-450 92
2.5.1. Endogenous Inhibitors 95
2.5.2. Synergists as Inhibitors 95
2.5.3. Plant Allelochemicals as Inhibitors 95
2.6. Induction of Cytochrome P-450 96
2.7. Regulation of Cytochrome P-450 Activity 96
3. Other Biological Oxidations 97
3.1. FAD-Monooxygenase 97
3.2. Dehydrogenases 98
4. Biological Reductions 98
4.1. Nitro-, Azo-, and Tertiary Amine Oxide Reduction 99
4.2. Carbonyl Reductases 99
5. Hydrolases 100
5.1. Esterases 100
5.1.1. Carboxylesterases 101
5.2. Glycosidases 114
5.3. S-Glycosyl Hydrolases 115
5.4. Ether Hydrolases 116
5.5. Other Hydrolases 117
6. Group Transfer Enzymes 118
6.1. Glutathione Transferases 118
6.2. Hexose Transferases 122
6.3. Amino Acid Conjugations 123
6.4. N-Acetylation 124
6.5. Phosphate Conjugation 124
6.6. Sulfotransferases 124
6.7. Thiosulfate Sulfur Transferase 125
7. Summary 126
8. Acknowledgement 127
9. References 127

Consequences of Induction of Foreign Compound-Metabolizing Enzymes in Insects 153
S. J. Yu

1. Introduction 153
2. Induction of Cytochrome P-450-Dependent Monooxygenases 154
3. Induction of Glutathione Transferases 158
4. Induction of Esterases 159
5. Induction of Epoxide Hydrolases 160
6. Consequences of Enzyme Induction 161
6.1. Toxicological Implications 161
6.2. Ecological Implications 165

7. Summary 168
8. Acknowledgement 169
9. References 169

Adaptive Divergence of Chewing and Sucking Arthropods to Plant Allelochemicals 175
C. A. Mullin

1. Introduction 175
2. Plant Resistance to Chewing and Sucking Insects 176
3. Susceptibility Differences to Toxicants 176
3.1. Susceptibility to Synthetic Toxicants 177
3.2. Susceptibility to Naturally-Derived Toxicants 178
4. The Aphid as a Model Phytophage of Phloem Chemicals 179
4.1. Phloem Toxicology 180
4.2. Chemical Ecology of Aphids 182
5. Adaptive Detoxification 183
5.1. Sequestration 183
5.2. Excretion 185
5.3. Metabolism of Plant Allelochemicals 187
5.4. Metabolism of Pesticides and Other Synthetic Toxicants . 188
5.5. Alternative Detoxification Strategies 193
6. Target Site Selectivities 195
7. Summary 195
8. Acknowledgement 196
9. References 196

Fate of Ingested Plant Allelochemicals in Herbivorous Insects . . . 211
L. B. Brattsten

1. Introduction 211
2. Terpenes 212
2.1. Monoterpenes 212
2.1.1. Bark Beetle Pheromones 212
2.1.2. Pyrethrins 214
2.1.3. Iridoid Glycosides 215
2.2. Sesquiterpenes 216
2.2.1. Juvenile Hormones 216
2.2.2. Sesquiterpene Antifeedants 217
2.3. Diterpenes 218
2.4. Triterpenes 218
3. Steroids 220
3.1. Phytosteroids 220
3.2. Ecdysones 222
3.3. Cardenolides 223
4. Phenylpropanoids 225
4.1. Precocenes 225
4.2. Furanocoumarins 227
4.3. Phenolics 228
5. Flavonoids 228
6. Amino acids 230
7. Cyanogenic Glycosides 231
8. Glucosinolates 233
9. Alkaloids 235
9.1. Tobacco Alkaloids 235
9.2. Pyrrolizidine Alkaloids 237
10. Summary 240
11. Acknowledgement 240
12. References 240

Target Site Insensitivity in Insect-Plant Interactions 257
M. R. Berenbaum

1. Introduction 257
2. TSI and Synthetic Insecticides 257
   2.1. The Sodium Gate 257
   2.2. The Synapse 259
3. TSI in Natural Systems 259
4. Examples of TSI in Insect-Plant Interactions 261
   4.1. Cardenolide Insensitivity 261
   4.2. Cyanide Insensitivity 262
   4.3. Canavanine Insensitivity 262
   4.4. Nicotine Insensitivity 263
5. Coevolutionary Consequences of TSI 263
   5.1. Consequences for the Insects 263
   5.2. Consequences for the Plants 265
6. Summary 266
7. Acknowledgement 267
8. References 267

Behavioral Adaptations in Insects to Plant Allelochemicals 273
D. W. Tallamy

1. Introduction 273
2. Avoiding Defenses in Time 274
3. Avoiding Defenses in Space 277
4. Mechanisms to Avoid Triggering Defenses 279
5. Overwhelming Plant Defenses 280
6. Blocking Plant Defenses 283
7. The Cost of Avoiding Plant Defenses 286
8. Summary 288
9. Acknowledgement 289
10. References 289

Devising Pest Management Tactics Based on Plant Defense
Mechanisms, Theoretical and Practical Considerations 301
K. F. Raffa

1. Introduction 301
2. General Strategies to Devising Pest Management Schemes 302
3. Key Processes in Plant-Insect Encounters 307
   3.1. Examples from a Model System 307
   3.2. A Generalized Model 310
4. Manipulating Plant-Insect Interactions 314
5. Can Pest Management Tactics Bases on Plant Defenses Favor
   Stability? 318
6. Summary 321
7. Acknowledgement 322
8. References 322

Index 329

THE PERCEPTION OF PLANT ALLELOCHEMICALS THAT INHIBIT FEEDING

James L. Frazier
E. I. du Pont de Nemours & Co., Inc.
Experimental Station, Building 402
Agricultural Products Department
Wilmington, Delaware 19898

1. INTRODUCTION

Plant-feeding insects have evolved an array of subtle discrimination capabilities involving their chemical senses. During the last 25 years an extensive list of plant chemicals that are detected by the contact and olfactory chemical senses of insects has accumulated. This list together with some 25-30 species of insects for which we have electrophysiological data on their chemical senses form the basis for our understanding of the chemical control of feeding. Many additional experiments at the behavioral level leave little doubt that the feeding of insects is regulated by some "umwelt" of plant chemicals, but the unique messages offered by host and nonhost plants have yet to be discovered (Dethier and Crnjar, 1982). The dynamic aspects of plant chemistry, as well as the many plant surface compounds that may have behavioral significance add to the complexity of this task (Fobes et al., 1985; Berenbaum, 1985; Gibson and Pickett, 1983; Woodhead, 1983). In addition, several types of physiological feedbacks that are important in regulating feeding have many details yet to be resolved (Shirahashi and Yano, 1984; Bernays and Simpson, 1982). These limitations in our understanding of the regulation of insect feeding come sharply into focus when we attempt to understand the importance of plant allelochemicals that inhibit feeding.

In this chapter I will discuss the progress that has been made in understanding how these chemicals exert their effects on feeding behavior through modifying chemosensory input, and indicate some factors that currently limit our efforts in this endeavor.

Since the mid 1960s, the terms feeding stimulant, incitant, and deterrent, have become an integral part of the literature. Their original definitions on a behavioral basis have served well when dealing at the whole organism level (Dethier et al., 1960). Now that the underlying neurophysiological bases of this behavior are within experimental grasp, the use of these terms in a mechanistic sense should be avoided. It may be convenient to classify the action of plant compounds as feeding stimulation or deterrency when dealing with ecological studies focused on multiple insect and plant species, but when considering the modes of action of plant compounds, one must focus on cell-specific mechanisms. Feeding deterrency can be produced by different cellular events and thus may be the result of different modes of action. The following sections will emphasize this rationale in more detail and indicate some of the cellular sites for plant allelochemical action.

2. INSECT FEEDING BEHAVIOR

2.1. Generalized Sequence

Several excellent reviews document our current understanding of the chemosensory regulation of insect feeding behavior (Bernays, 1985; Dethier, 1982; Hanson, 1983; Schoonhoven, 1982, 1986, Stadler, 1984; Miller and Strickler, 1984; Mustaparta, 1984). Although there has been much emphasis on understanding the chemosensory control of feeding, the importance of visual and mechanosensory inputs in regulating insect feeding are comparatively poorly studied and must be considered a major aspect of insect feeding about which we know very little (Miller and Strickler, 1984).

A generalized sequence of behavior that leads to feeding is shown in

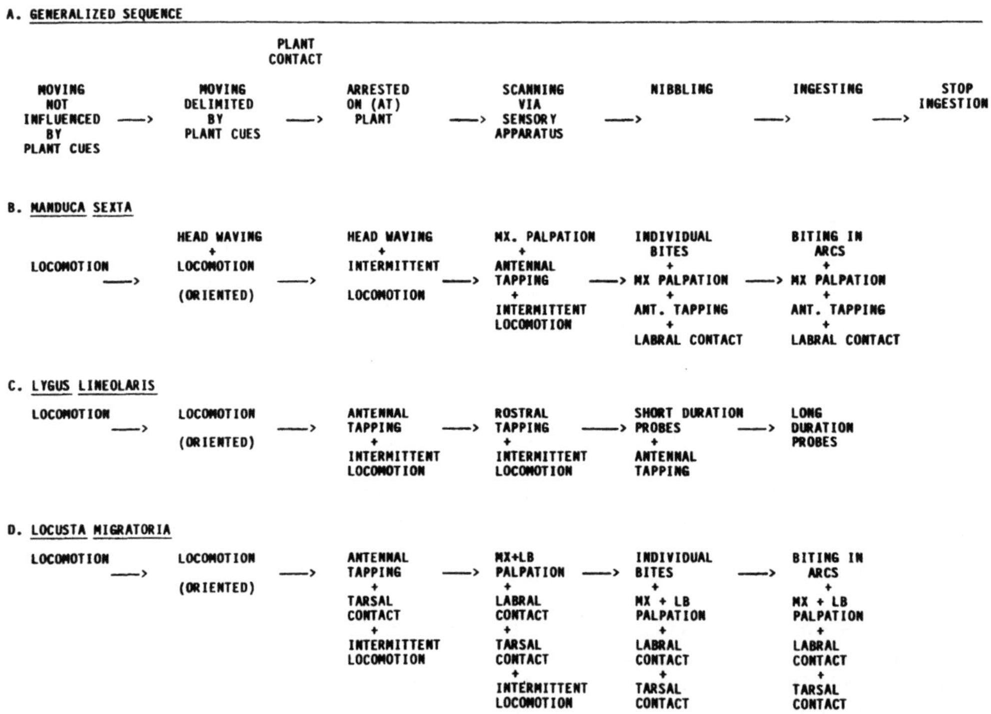

Fig. 1. A generalized sequence of feeding behavior in phytophagous insects with examples of the specific behavioral acts and their chemosensory inputs employed by three different species.
A. Generalized sequence modified from Miller and Strickler, 1984.
B. Oligophagous caterpillar, Manduca sexta, with chewing mouthparts (Frazier, unpublished).
C. Polyphagous true bug, Lygus lineolaris, (Pal. de Beauvois) (Hemiptera: Miridae) with piercing-sucking mouthparts (Hatfield et al., 1983).
D. Polyphagous grasshopper, Locusta migratoria, L. (Orthophera: Acrididae) with chewing mouthparts (Bernays 1985 and references therein).

Fig. 1 along with specific examples of the behavioral acts for three insect species. This generalized sequence is redrawn from Miller and Strickler (1984), and indicates that as the insect's behavior progresses sequentially from left to right, the number of sensory modalities stimulated increases, as does the number of sensory cues utilized. Although the authors caution against a strict interpretation of this sequence, and point to the danger in thinking that the essentials of insect foraging are captured accurately, the three examples included show that this generalized sequence is applicable to insects with greatly differing mouthparts and sensory complements. Each species passes successively through the behavioral states of the generalized sequence even though the specific acts of each are different.

The "hungry" insect moves randomly with respect to plant cues, until olfactory and probably visual signals release oriented locomotion toward the plant (Prokopy et al., 1983; Bernays and Wrubel, 1985). Upon contact with the plant, locomotion is arrested, and becomes intermittent as antennal contact, followed by mouthpart appendage contact with the plant increases in frequency and duration. During this stage of plant recognition, the exact roles of olfactory and contact chemosensing are difficult to separate. Olfactory information is, certainly, constantly incoming, while contact chemosensory information can vary with the frequency and duration of mouthpart appendage movements onto the plant. It is at this level of behavior that we can begin to appreciate the differences among species with respect to the number and locations of chemosensory sensilla. The movements of mouthparts can result in major differences in the chemosensory information made available to the CNS (Chapman 1982).

2.2. Neuroethological Approach

The marine mollusk *Aplysia* must stand at least a head ganglion above all other invertebrates in serving as a model system for the investigation of the neurophysiological basis of behavior. The major behaviors of *Aplysia*, including feeding, as well as the underlying neuromuscular circuits have been described. Serious inroads have been made in describing the functional capabilities of these circuits, even to the extent of addressing the cellular bases of habituation, motivation, and learning (Kandel, 1979; Barnes, 1986). Feeding behavior of phytophagous insects requires such an approach if we are to progress beyond our current level of understanding. The factors controlling feeding behavior in insects are understood almost entirely because of detailed work with a single insect, the black blow fly, *Phormia regina* (Meigen) (Diptera: Calliphoridae). Even in this insect many details at the cellular level are lacking (Shirahashi and Yano, 1984; Dethier, 1976). Clearly the prerequisite tools for a cellular approach to the understanding of insect feeding behavior are readily available. All that is required is their application to the problem.

Insects clearly offer many advantages for the cellular approach as evidenced by the analysis of acoustical behavior in crickets (Huber, 1984; Nolen and Hoy, 1984) and the genetic analysis in *Drosophila melanogaster* Meigen (Diptera: Drosophilidae) (Hall and Greenspan, 1979). Quantitative descriptions of the behavioral acts as fixed-action patterns are a necessary first step. From these descriptions, something about the functional connections of the underlying circuits can be inferred, offering the framework for further neural investigations. Quantitative measures of the chemosensory input regulating feeding in phytophagous insects have been underway for some time (Schoonhoven, 1986), but the sensory projections and interneuronal connections have only recently begun to be determined (Milburn, pers. comm. 1986). Functional studies at the cellular level can then complete the picture.

Recently the feeding behavior of the tobacco hornworm caterpillar Manduca sexta (L.) (Lepidoptera: Sphingidae) has been analyzed with respect to the chemosensory regulation of a single motor output (Frazier, unpublished). A single caterpillar is allowed to feed on a leaf disk, while a videocamera focused through a microscope records the movements of the mouthparts. A second video camera focused on an oscilloscope records the electromyogram (EMG) from the mandibular adductor muscle recorded through a wire implanted into the muscle. The signals from the two videocameras are mixed to form the composite picture in Fig. 2. Thus the live insect, tethered only by the EMG recording wires, feeds on the leaf and permits the simultaneous recording of the total behavior and a single motor output. Each bite of the leaf is accompanied by the EMG burst in the insert. With this system, the chemosensory input can be regulated experimentally, and the output of the EMG can be quantitatively described.

When a Manduca caterpillar is feeding on a glass fiber disk treated with a stimulatory mixture of α-D-glucose and myoinositol, it first contacts the disc, palpates, and exhibits a few single bites. It then begins to feed by biting in arcs of four to nine bites per arc. This first period of constant biting in arcs lasts for about 230 seconds with the stimulant mixture alone, but when the deterrent caffeine is mixed with the stimulus, this period of biting is decreased in a dose-dependent manner, Fig. 3. The same concentration range of caffeine mixed with the same stimulus also causes a dose-dependent reduction of feeding in no-choice tests on glass fiber disc. When the duration of biting in a single bout is regressed against the reduced consumption from the longer term feeding test for the same caffeine concentration range, a linear relationship is obtained, Fig. 4. The effect of caffeine is to reduce the duration of constant biting during each feeding bout and therefore the amount ingested over a period of several hours.

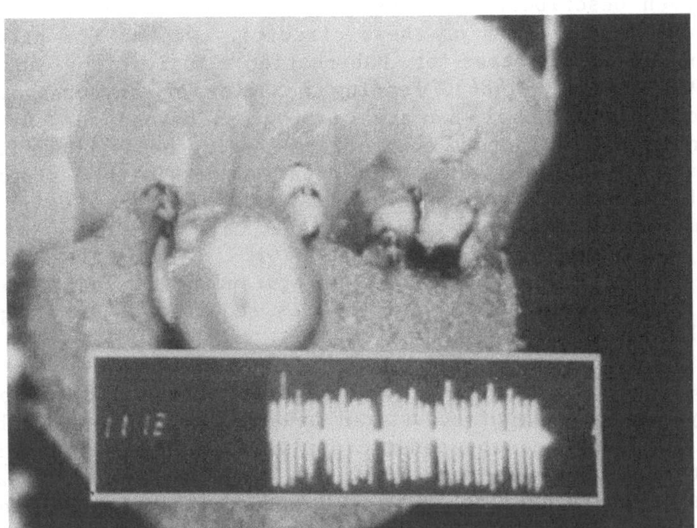

Fig. 2. A photograph of the simultaneous videorecording of the tobacco hornworm caterpillar feeding on a leaf disk and the electromyogram response (EMG) of the right mandibular muscle for each bite (insert at bottom). The gaps in the EMG record result from the time required to move the head dorsally to begin another biting arc (Frazier, 1986).

The information flow from chemosensory inputs to mandibular motor output is summarized in a conceptual model for a Manduca caterpillar in Fig. 5. Electrophysiological recordings from the chemosensory cells of Manduca have shown that both excitatory and inhibitory inputs are produced (Schoonhoven, 1972a; Dethier and Crnjar, 1982; DeBoer et al., 1977; Frazier, unpublished). The chemosensory cells on the maxilla and labrum project to the suboesophageal ganglion and apparently do not involve the frontal ganglion as in the caterpillar Pieris brassicae L. (Lepidoptera: Pieridae) (Ma, 1972). Manduca caterpillars with the frontal ganglion excised can feed and complete normal development (Bell, 1984). In the silkworm, Bombyx mori L. (Lepidoptera: Bombycidae), olfactory inputs from the basiconic sensilla on the maxillary palps either activate or inhibit mandibular biting activity by modifying a centrally generated motor program (Hirao et al., 1976). The motor neurons of the mandibular muscles appear to have reciprocal inhibitory connections typical of antagonistic muscle pairs (Seath, 1977). The model of information flow regulating mandibular movements in Manduca incorporates some of these features together with what is known about antennal olfactory and mouthpart contact chemosensory inputs. The central connections of this pathway are under investigation (Milburn pers. comm., 1986), and will greatly facilitate our understanding of the cellular control of this single motor output that is so crucial to ingestion. The resulting details of this pathway may offer additional cellular targets on which plant allelochemicals can act to reduce ingestion (see section 4.2.2).

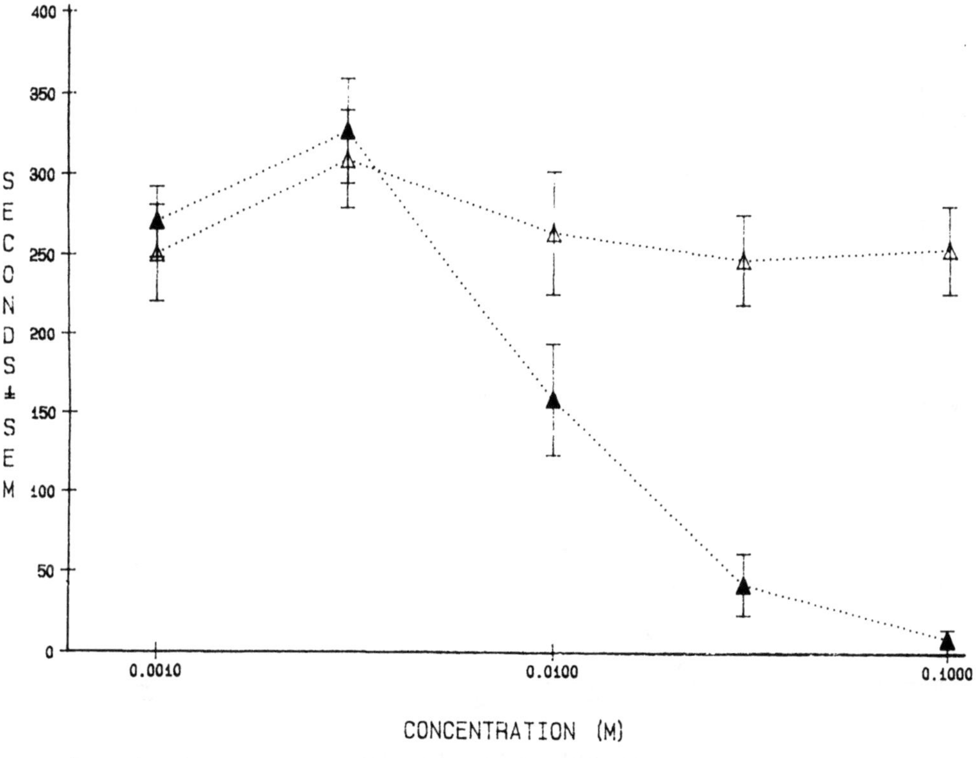

Fig. 3. The duration of the first biting phase (T1) by the tobacco hornworm caterpillar feeding on glass fiber disks. Control T1 on a disk treated with 0.3 M glucose + 0.1 M myoinositol is open triangles. Reduction in T1 as a function of added caffeine is closed triangles (Frazier, 1986).

The information available on insect chemosensory cells has been obtained by electrophysiological recordings from individual sensilla using the tip-recording technique (Hodgson et al., 1955). This technique has been used on Manduca to determine that the styloconica on the galea contain sensory cells responsive to glucose and to myoinositol (Schoonhoven, 1972b). Such studies use a strict time course of stimulus application and intertrial intervals to obtain reproducible responses of the cells. When one observes a caterpillar feeding, however, the maxillae move in rapid coordination with the mandibles (Frazier, unpublished). The maxillae tap the leaf surface so that the galeal styloconic sensilla contact the leaf surface and perhaps the cut edge of the leaf at a frequency above one per second for a duration of perhaps a few hundred milliseconds for each contact. Are the sensilla stimulated only during this brief contact, or do plant compounds constantly diffuse into the sensillar pores, producing a continual stimulation? The answer seems very important to our understanding of chemosensory input, but is currently unknown. Palpation reduces the rate of sensory adaptation in the migratory locust, Locusta migratoria L. (Orthoptera: Acrididae) (Blaney and Duckett, 1975); the extremely fast rate of disadaptation of contact chemosensory cells in some larval lepidopterans suggests that they are uniquely different from those of flies and perhaps other phytophagous insects (Ma, 1977; Wieczorek, 1976; Mitchell and Gregory, 1979; Frazier, unpublished). In the dark-sided cutworm, Euxoa messoria (Harris) (Lepidoptera: Noctuidae), the rate of palpal contact is much more rapid than in Manduca (Devitt and Smith, 1985). Clearly, such detailed descriptions of the fixed-action patterns

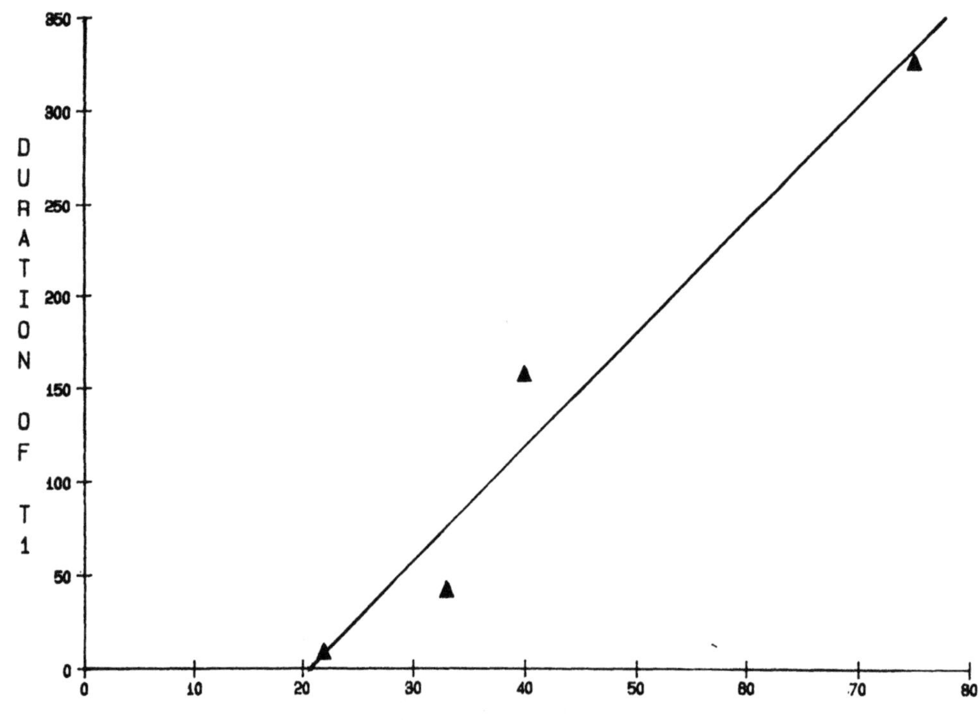

Fig. 4. The relationship of duration of a single biting phase (T1) to the amount eaten in a no choice feeding test of one hour duration for the same concentrations of caffeine and stimulus solution shown in Fig. 3 (r =0.95).

involved in insect feeding are needed as a framework for further neurophysiological studies to determine the function of the underlying circuits.

3. THE CONTACT CHEMORECEPTIVE SYSTEM

3.1. Structure of Chemoreceptive Sensilla

The surface of the insect body is richly supplied with sensilla of a variety of shapes and densities. Yet, regardless of the sensory modalities they contain, their structure is basically the same. The cuticular portion is hollow, and contains the dendrites of the sensory cells with various ultrastructural modifications (Zacharuk, 1985). If the sense cells are chemosensory, the cuticular portion has a single pore at the tip or multiple pores in the wall depending on whether the cells are principally contact or olfactory in function. Contact chemosensory cells have unbranched dendrites while olfactory chemosensory cells have branched or unbranched dendrites with some unique relationships to the wall pores (van der Wolk,

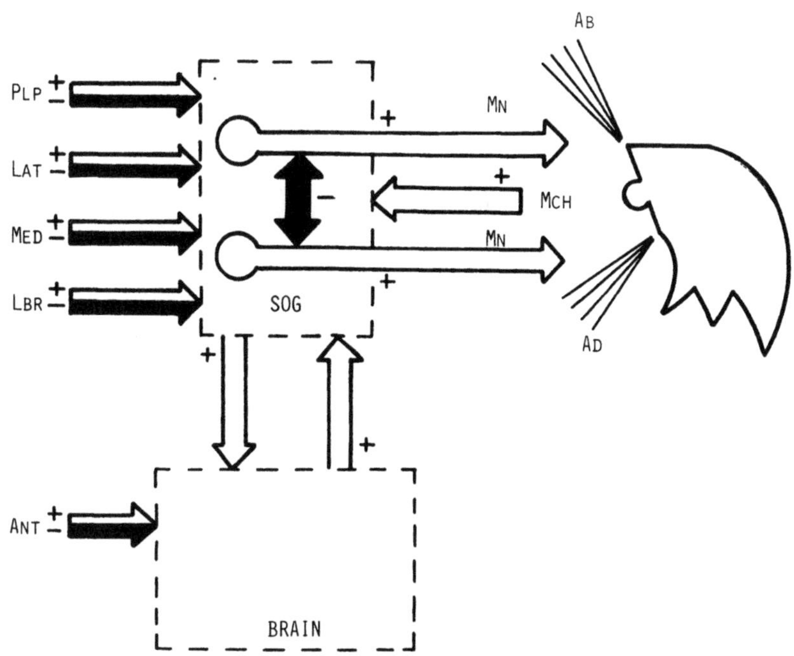

Fig. 5. A conceptual model of the information flow from chemosensory inputs to mandibular motor output in the tobacco hornworm caterpillar. A feedback loop from mechanosensory cells (MCH) to the suboesophageal ganglion (SOG) and a reciprocal inhibitory connection between motorneurons (MN) of the abductor (AB) and adductor (AD) muscles of the mandible are shown. Chemosensory cells from the maxillary palp tips (PLP), lateral styloconica (LAT), medial styloconica (MED), and labral epipharyngeal papillae receptor (LBR) project information into the suboesophageal ganglion. The antennal chemosensory cells (ANT) project their information into the olfactory lobe of the brain. Excitatory information flow is shown as (+) open arrows and inhibitory information flow is shown as (-) black arrows and are both defined relative to mandibular movement. Excitatory inputs increase the frequency and duration of biting, while inhibitory inputs decrease these parameters.

et al., 1984). The number of sense cells per sensillum commonly varies from one to ten with exceptions of over a 100 (Chapman, 1982). The sense cells within a single sensillum may be of different modalities, so that combinations of chemo-, mechano-, thermo-, and hygro-reception are known (Altner and Prillinger, 1980). Chemosensory cells have a constricted ciliary region that divides the dendrites into a distal and proximal region. The distal region is surrounded by a cuticular sheath that is shed at each molt. Usually three types of specialized sheath cells are found in association with each sensillum. These are known as the basal sheath cell, the outer sheath cell (tormogen), and the inner sheath cell (trichogen). The relationships of these cells to the sensory cells are shown diagrammatically in Fig. 6. These accessory cells have both tight and gap junctions among them and form an electrically insulating barrier between the extracellular space surrounding the dendrites and the hemolymph space below the epidermis (Kuppers and Thurm 1982). The outer sheath cells are densely covered with microvilli that extend into the extra

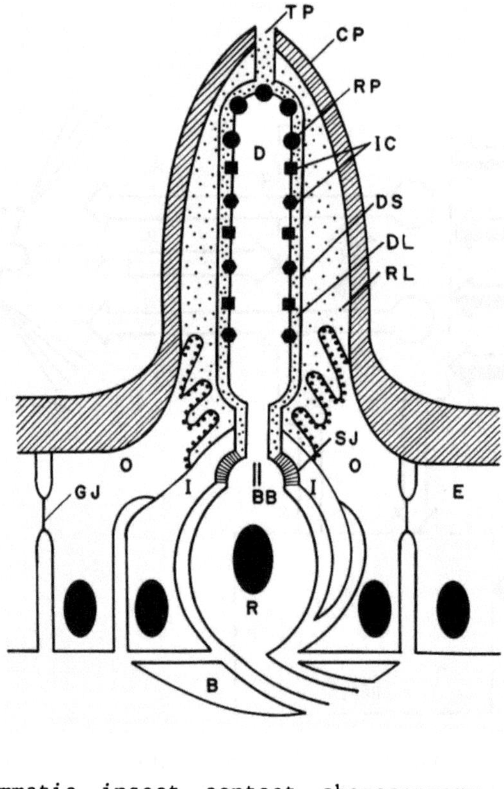

Fig. 6. A diagrammatic insect contact chemosensory sensillum showing major ultrastructural features. The cuticular process (CP) contains a tip pore (TP) with a single sensory cell (R) shown. The dendrite (D) contains receptor proteins (RP) and ion channels of at least two types (IC). A cuticular dendrite sheath (DS) separates the dendrite from the extracellular receptor lymph fluid (RL). An inner sheath cell (I) often makes septate junctions (SJ) with the receptor cell and the outer sheath cells (O). The outer sheath cells have gap junctions (GJ) with both the inner sheath cell and other epidermal cells (E). Microvilli of the outer sheath cells contain 10-nanometer particles (dots). The basal sheath cell (B) encloses the proximal axon of the receptor cell. The basal bodies (BB) are located near the constriction of the proximal dendrite.

cellular space. These microvilli contain 10-nanometer particles on their inner surface that may be part of a potassium-specific ATPase pump system (Kuppers and Thurm, 1982; Wieczorek and Gnatzy, 1985).

Stimulus molecules are believed to reach the dendrite membrane by diffusion and to reversibly combine with protein receptors in the membrane. Evidence that protein receptors exist in the dendrite membrane includes both enzyme and protein reagent disruption of stimulus effectiveness in the fly sugar cell (Shimada, et al., 1974; Shimada, 1975). Receptor occupation results in a depolarization of the dendrite membrane. The resulting receptor potential is propagated decrementally to the proximal dendrite region where the spike potential is produced (Morita and Shirashi, 1985). The spike potential propagates down the axon to the central nervous system (CNS), and antidromically back up the dendrite via ion channels independent from those for receptor potential generation. The sodium channel toxin, tetrodotoxin, abolished the negative half of the spikes, but did not reduce the frequency or positive half of spikes from the sugar-sensitive cell of the black blow fly (Wolbarsht and Hanson, 1965).

The accessory cells contribute to a metabolically supported transepithelial potential that can interact with the sensory cells and modify their activity (Kuppers and Thurm, 1982). Recent biochemical evidence suggests that these accessory cells are the site of a potassium-specific ATPase pump system that furnishes ions to the extracellular fluid surrounding the dendrites (Wieczorek and Gnatzy, 1985). These ions could serve as the current carrying ions for receptor potential and spike potential back-conduction. The sodium/potassium ratio of this extracellular fluid is distinctly different from the hemolymph in adult olfactory hairs (Kaissling and Thorson, 1980) and mechanosensory sensilla (Kuppers and Thurm, 1982), suggesting that potassium ions may have a special role in the sensillum. This interdependence between the receptor cells and the accessory cells indicates that the entire sensillum should be considered as a functional unit as proposed for mechanosensilla by Thurm and coworkers (Kuppers and Thurm, 1982).

The mechanisms by which stimulus molecules are removed or inactivated after they have combined with receptors are currently unknown. Pheromone-degrading esterases are known in some olfactory systems and glycosidases are known to occur in taste sensilla of several species (Zacharuk, 1985; Morita and Shirahashi, 1985; Vogt et al., 1985). The extracellular fluid surrounding the dendrites may participate in this stimulus inactivation. A few studies indicate that this fluid dynamically exchanges with the hemolymph through the sheath cells, but much remains to be learned about its contents and functional roles (Phillips and VandeBerg,1976; VandeBerg, 1981).

3.2. Function of Chemosensilla

3.2.1. Contact Chemosensory Systems. The unique properties of chemosensory cells are their sensitivity to the quantity of a chemical, and their response capacity for the quality or type of chemical. As evolved receptors, they should have increased sensitivities, i.e. lowered thresholds, and reduced numbers of effective stimulus compounds, i.e. limited response spectra. Insect chemosensory cells exhibit both of these specializations as indicated in the following examples.

The entire chemosensory complement of a given insect determines its sensitivity and selectivity for behavioral responses to stimuli. One of the goals of chemosensory research is to understand how chemosensory inputs regulate feeding behavior. It is obvious that many cells must be furnishing information during each aspect of the feeding sequence in Fig. 1. In some phytophagous insects like grasshoppers, the receptor complement is large in number and low in specificity (Blaney, 1975; Winstanley and Blaney, 1978). In caterpillars by contrast, the receptor complement is low in number and relatively high in specificity (Ma, 1972; Schoonhoven, 1972b). In both extremes there is redundancy among chemosensory cells both for specificity and in bilateral position, as well as overlap of the sensitivity ranges of individual receptor cells (Blom, 1978). In *Mamestra brassicae* L. (Lepidoptera: Noctuidae) caterpillars, there are three types of sugar-sensitive cells and inputs from all three are required to correlate with the amount of sugar-containing diet eaten over a selected concentration range (Blom, 1978). In *Pieris brassicae* caterpillars the maxillary palpal tip chemosensilla appear to produce inputs that inhibit feeding (Ma, 1972). The portion of the total sensory system that is required for normal behavioral responses is largely unknown in most phytophagous insects, yet such information is vital for determining the sensory basis for a plant allelochemical in reducing or inhibiting feeding. This link between single chemosensory cell input and behavioral output must be known before we are able to correlate the effects of allelochemicals on single cells in electrophysiological studies with their effects on the feeding behavior of the whole insect.

3.2.2. Sensitivity of contact chemosensory cells. Individual chemosensory cells are characterized for their sensitivity to selected stimuli by determining their dose response relationship to each compound. The result is the usual sigmoid semilog plot as shown in Fig. 7. The threshold of response is usually determined by extrapolation of the linear portion to the baseline, since this is more reliable than experimentally derived direct measurements. The response maximum (Rmax) and, more importantly,

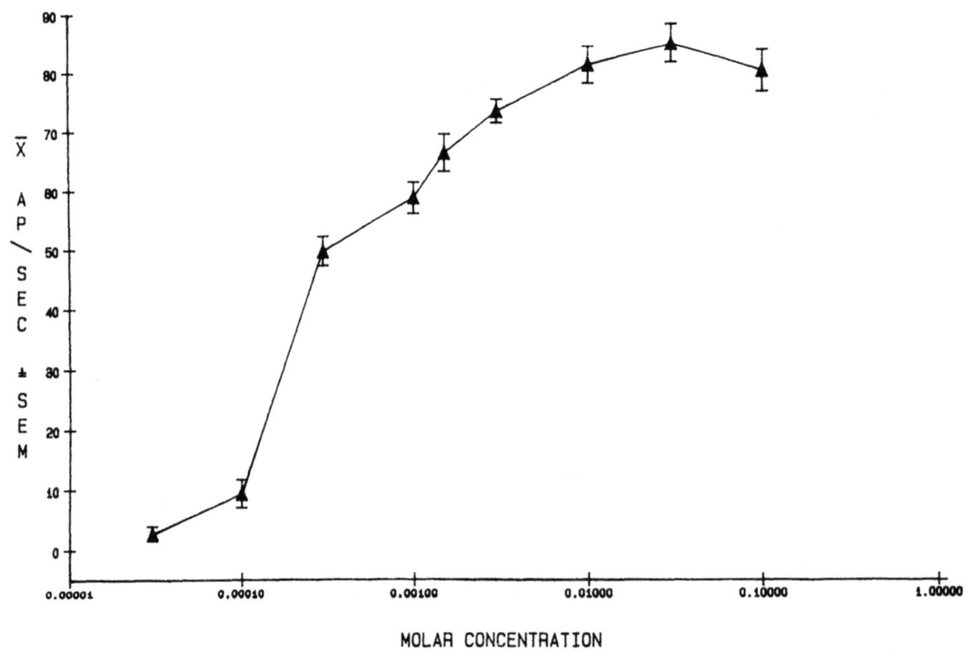

Fig. 7. The dose response curve of an inositol-sensitive cell in the medial styloconicum of the tobacco hornworm caterpillar.

the concentration at half maximum response (Kb) are used to compare stimulating effectiveness across different stimuli. This Kb represents the level at which half of the receptor sites are occupied (Hansen and Wieczorek, 1981). The slope and length of the linear portion are also characteristic of the response, but are much less often used. The slope of a specific log-log plot gives the Hill coefficient which indicates the number of sites that must be occupied to give a response; it is also indicative of positive or negative cooperativity (Wieczorek, 1976). A recent study indicates that response variability seriously affects the graphical evaluation of dose-response curves for insect chemosensory cells and suggests that curve fitting to the data may be a more reliable method than the linearizing plots used in the past (Maes, 1985a).

3.2.3. Specificity of contact chemosensory cells. The specificity of a chemosensory cell results from both the specificity of a given receptor site and from the composition of the population of receptor sites on the dendrite. Thus a cell may be responsive to many kinds of plant compounds as a result of either a uniform population of sites with wide specificities, or from a mixed population of different receptor sites each with narrow specificities. The situation is identical for taste or olfactory cells. The latter have been labeled as generalists or specialists according to their responses at the cellular level. The specificity of receptor sites of insect olfactory cells is known to a limited extent and represents specialization, even to the extent of optical isomers (Mustaparta, 1984).

Chemosensory cells are usually classified according to the magnitude of response to a fixed concentration of each compound in a series of structural analogs and expressed relative to some arbitrarily chosen standard (Maes, 1985b). The obvious limitation of this method is that the number of compounds chosen by the investigator limits the results. The cell is then labeled according to the best stimulus and the compounds to which the cell responds are known as the response spectrum or breadth of tuning (Dethier, 1977). Thus, sugar-sensitive or salt-sensitive cells have been most often described in the literature. The number of cells that have been investigated with a complete series of closely related stimuli is extremely small, and does not compare favorably with the number of vertebrate chemosensory cells receiving such attention (Smith and Travers, 1979; Boudreaux, et al., 1985; Jakinovitch, 1985). Thus the characterization of insect chemosensory cells in the true pharmacological sense must be considered grossly incomplete. Many types of plant allelochemicals may be detected by phytophagous insect taste cells, but the currently available list is far from complete. Our current knowledge is best summarized in the recent reviews by Stadler (1984) and Norris (1986).

3.2.4. Specificity of Chemoreceptor Sites. The most specific chemosensory receptor sites so far described in insects are the olfactory pheromone sites of several male moths and the gustatory sugar site of flies. Reviews of sugar-sensitive cells in insects by Hansen and Wieczorek (1981), Wieczorek and Koppl (1982), Mitchell and Gregory (1979), and Morita and Shirahashi (1985) account for most of what is known. The fly site appears to have both a pyranose and a furanose subsite, which can be differentiated on the basis of sensitivity to the protein reagents p-chloromercuribenzoate (PCMB) and 2,4,6-trinitrobenzene sulfonic acid (TNBS) (Shimada et al., 1974). The pyranose subsite is inhibited by PCMB but not TNBS, while the furanose subsite has the opposite sensitivities. The pyranose subsite requires equatorial hydroxyls at the C2, C3, and C4 positions with the ring oxygen contributing less to binding. The furanose subsite requires the α-D-fructofuranose structure with rigid stereospecific requirements as well (Shimada et al., 1985; Ohrui et al., 1985). The specificity of this sugar site appears greater in flies than in vertebrates (Jakinovitch, 1985).

Table 1 - Dose Response Characteristics of Sugar - Sensitive Chemosensory Cells of Phytophagous Insects

Insect	Stimulus	Location of cell	Threshold (mM)	Kb (mM)	Rmax (spikes/ sec.)	Slope (spikes/ sec.)	References
Pieris brassicae	sucrose	lateral galeal styloconica	1.0	50	120	60	Ma, 1972
Pieris brassicae	sucrose	epipharyngeal papilla	1.0	8.0	70	42	Ma, 1972
Mamestra brassicae	1-napthyl β-glycoside	lateral galeal styloconica	0.1	1.0	100	63	Wieczorek,1976
Entomoscelis americana	sucrose	maxillary palpi	0.01	1.0	52	21	Mitchell & Gregory,1979
Leptinotarsa decemlineata	sucrose	α-senilla	0.03	0.05	25	10	Mitchell & Harrison,1984
Rhagoletis pomonella	sucrose	tarsal hair	10.0	250	35	25	Bowdan,1984
Manduca sexta	glucose	medial galeal styloconica	0.5	8.0	43	11	Frazier (unpublished)

The specificity of sugar sites among phytophagous insects is much less well documented than for flies. Table 1 contains most of the comparative data on sugar sites in the species studied to date. It is apparent that the response properties of these cells vary widely. The actual specificities and reagent sensitivities of these sites have yet to be detailed. Clearly, we are far from a complete picture of the response capabilities of phytophagous insects for detecting plant sugars and sugar alcohols and our understanding of their role in regulating insect feeding behavior is even less understood.

Phytophagous insects may also have taste cells which are narrowly tuned for unique plant allelochemicals. Such indications have been reported for some time, but again the extent of the specificity of individual receptor sites has yet to be adequately documented. Table 2 lists some examples of insects with taste cells that may be specific for unique plant compounds. In some cases these compounds are known to stimulate feeding in the species listed, and to deter feeding in others.

4. INHIBITORY PLANT ALLELOCHEMICALS

4.1. Major Classes of Allelochemicals

Table 3 lists examples of the major classes of plant allelochemicals that inhibit the feeding of phytophagous insects. This table is based on compounds of known structure that have been shown to inhibit the feeding of one or more species of insects. The table is not intended to be comprehensive, but to compare relative potencies of different classes on different insect species. Plant fractions shown to exhibit inhibitory effects on feeding have not been included. It should also be noted that many of the compounds listed have activity on organisms other than insects. Other reviews on inhibitory plant allelochemicals may be consulted for further details (Adityachaudhury et al., 1985; Chapman, 1974; Dethier, 1982; Koul, 1982; Norris, 1986; Schoonhoven, 1981, 1982, 1986).

The term antifeedant has become widely used in the literature dealing with plant compounds that inhibit insect feeding. The earliest use of this term defined it as a feeding deterrent or rejectant and was proposed to be equivalent to the definition of Dethier et al., 1960 (Wright, 1967). Later it was recognized that some plant compounds produced irreversible feeding inhibition and would result in the death of the insect by starvation; the term absolute antifeedant was used to distinguish this special case (Wada and Munakata, 1968). Since then, the term has been used to cover all types of reduction in feeding regardless of the mode of action. The usual measure of feeding inhibition by candidate antifeedants is to determine the reduction in feeding in a choice test between control host plant leaves and those treated with the test compound over a 24 to 48 hour period (Bernays 1985). Such a measurement gives no information as to the mode of action of an antifeedant. The following sections will discuss the current information on the effects of plant compounds on insect chemosensory cells, and indicate the multiplicity of actions that can result in reduced feeding.

4.2. Modes of Action on Chemosensory Cells

4.2.1. Potential Types of Action. Feeding by phytophagous insects is dependent on currently unknown patterns of chemosensory input. Potentially any disruption in the integrity of these messages could result in the disruption of feeding. Schoonhoven (1982) indicated that plant chemicals

Table 2. Contact Chemosensory Cells of Phytophagous Insects specific for Plant Allelochemicals.

Insect	Location of sence cell	Plant compound	Reference
Pieris brassicae	maxillary styloconica	sinigrin	Ma, 1972
Delia brassicae	tarsal D	sinigrin	Stadler, 1984
Chrysolina brunsvicensis	tarsal	hypericin	Reese, 1969
Yponomeuta evonymellus	maxillary styloconica	prunasin	van Drongelen, 1978
		sorbitol	van Drongelen & van Loon, 1980
Leptinotarsa decemlineata	galeal α-sensilla	γ-amino-butyric acid L-alanine	Mitchell, 1985
Spodoptera exempta	maxillary styloconica	adenosine	Ma, 1976

could prevent feeding by altering chemosensory cell responses in any one of five ways. These are:

- stimulating a cell which is specifically tuned to plant compounds that deter feeding,
- stimulating a cell that is broadly tuned to plant compounds,
- stimulating the activity of some cells and inhibiting the activity of others, thereby changing complex and subtle codes,
- inhibiting the responses of cells that are stimulated by compounds that also stimulate feeding,
- or by evoking highly unnatural impulse patterns often at high frequencies and/or "bursting".

All five modes of alteration of sensory input have been confirmed by various studies of the effects of deterrents on chemosensory cells.

4.2.2. Deterrent Chemosensory Cells. The description of a bitter or deterrent cell in caterpillar maxillary styloconica was first given by Ishikawa (1966) and expanded to include many other species (Schoonhoven, 1972b, 1981, 1982; Schoonhoven and Jermy, 1977; Clark, 1981; Dethier, 1982). Since the 1960s there have been several examples of plant compounds that cause a cell to increase firing and also cause decreased consumption over the same concentration range, leaving little doubt that the concept of a deterrent cell is on fairly firm ground. The input from such a cell does not cause disrupted feeding in all cases, but is effective relative to other (stimulant cell) inputs. What is in question is exactly how do plant compounds cause these cells to fire spikes.

Many plant compounds appear to increase the rate of firing of insect

Table 3. Representatives of the Major Biosynthetic Classes of Plant Allelochemicals that Inhibit Insect Feeding

CHEMICAL CLASS	COMPOUND	STRUCTURE	INHIBITORY CONC. (PPM/ % FEEDING)	INSECTS AFFECTED	REFERENCES
1. Alkaloids					
aporphine alkaloid	isoboldine	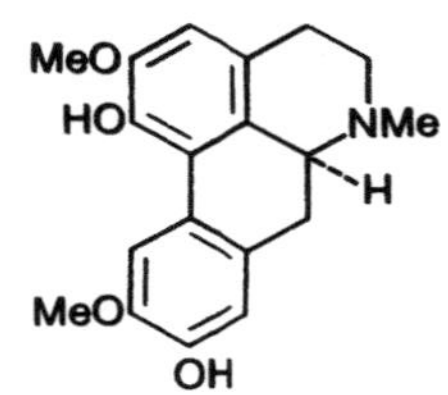	200/0	*Prodenia litura*	Wada & Munakata, 1968
benzphen- anthridine alkaloid	N-methyl flindersine	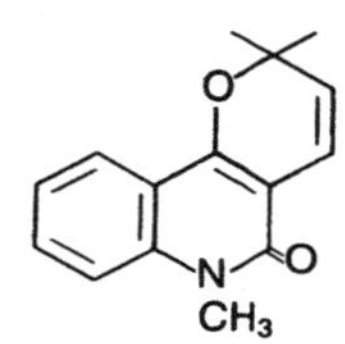	100/0	*Spodoptera exempta* *S. littoralis*	Chou et al., 1977

(continued)

Table 3 (cont.)

CHEMICAL CLASS	COMPOUND	STRUCTURE	INHIBITORY CONC. (PPM/ % FEEDING)	INSECTS AFFECTED	REFERENCES
steroidal alkaloid	tomatine	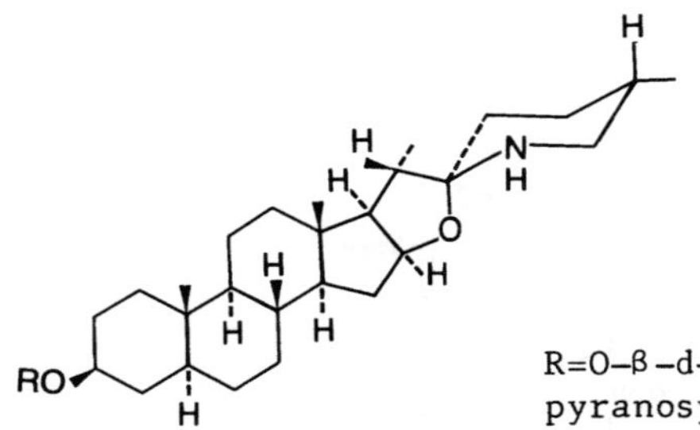 R=O-β-d-gluco-pyranosyl-(1→2glu)O-β-d-xylo-pyranosyl-(1→3glu)O-β-D-gluco-pyranosyl-(1→4gal)-β-D-galacto-pyranoside	40/0	*Pieris brassicae*	Ma, 1972
phenanthridine alkaloid	lycorine	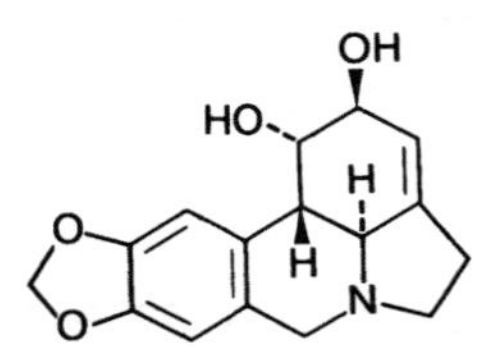	250/10	*Schistocerca gregaria*	Singh & Pant, 1980

pyrrolizidine alkaloid	senkirkine	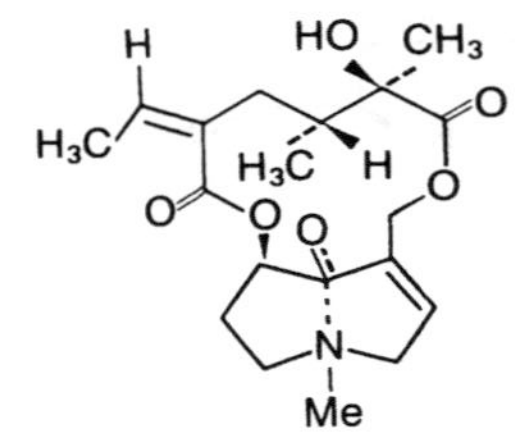	180/11	<u>Choristoneura fumiferana</u>	Bentley et al., 1984
macrocyclic maytansinoid alkaloid	trewiasine	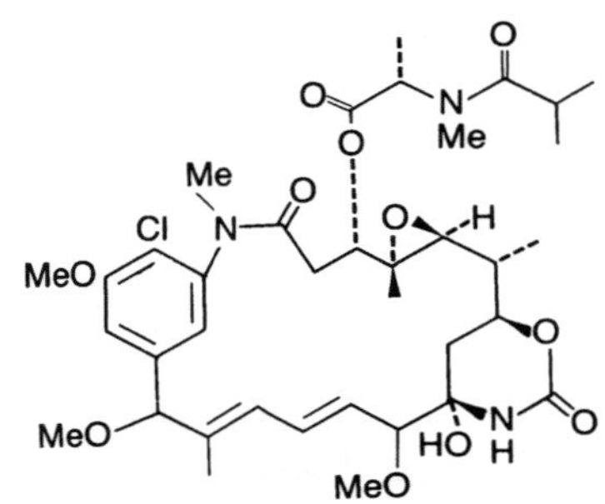	50/4	<u>Acalymma vittatum</u>	Reed et al., 1983

(continued)

Table 3 (cont.)

CHEMICAL CLASS	COMPOUND	STRUCTURE	INHIBITORY CONC. (PPM/ % FEEDING)	INSECTS AFFECTED	REFERENCES
alkaloid	quinine	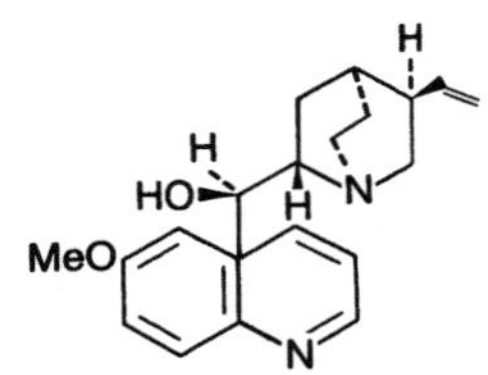	5/50	*Pieris brassicae*	Ma, 1972
alkaloid	strychnine	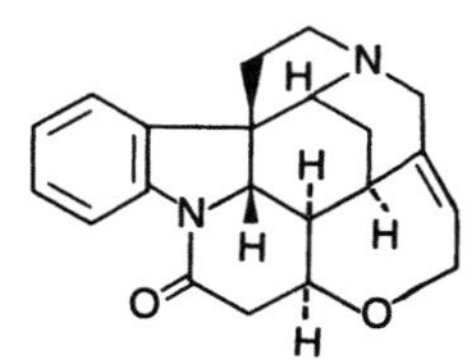	30/0	*Pieris brassicae*	Ma, 1972
alkaloid	2,5-dihydroxymethyl 3,4-dihydroxypyrrolidone	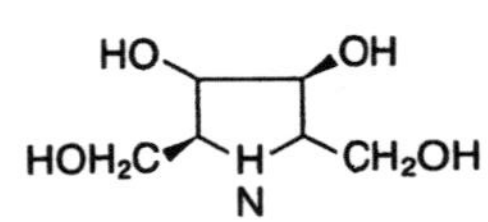	1000/30	*Schistocerca gregaria*	Blaney et al., 1984

2. Phenylpropanoids

cinnamic acid	cinnamonitrile	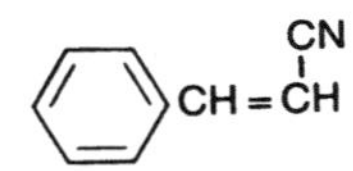	100/5	*Pieris brassicae*	Jones & Firn, 1979
furanocoumarin	isopimpinellin	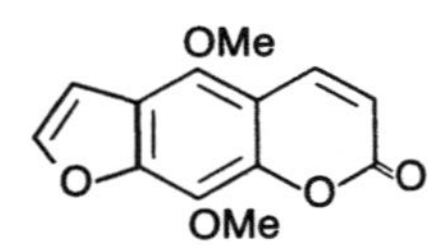	1/19	*Spodoptera litura* (other spp)	Yajima & Munakata, 1979
flavonoid, dihydro-chalcone	phlorizin		200/50	*Schizaphis graminum*	Dryer & Jones, 1981
	R= 2'-glucosyl	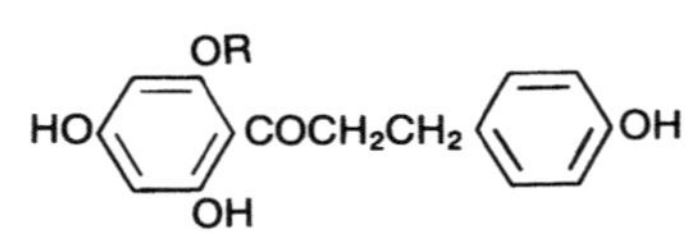			

(continued)

Table 3 (cont.)

CHEMICAL CLASS	COMPOUND	STRUCTURE	INHIBITORY CONC. (PPM/ % FEEDING)	INSECTS AFFECTED	REFERENCES
flavonoid	procyanidin	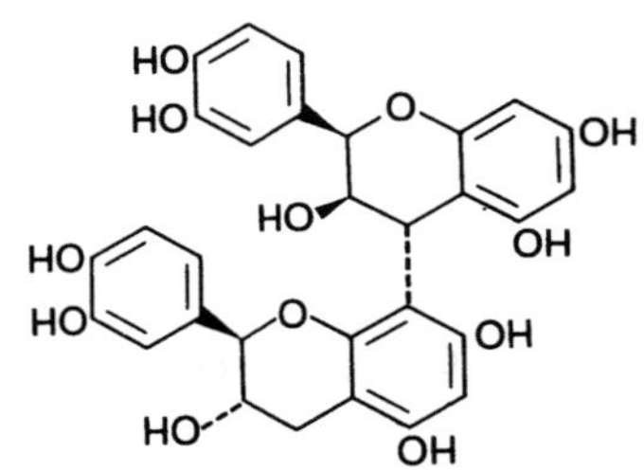	800/50	*Schizaphis graminum*	Dryer et al., 1981
isoflavonoid	(-)-vestitol	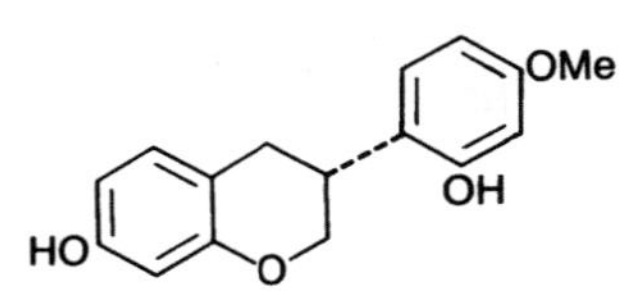	100/0	*Costelytra zealandica*	Russell et al., 1978

phenol	chlorogenic acid	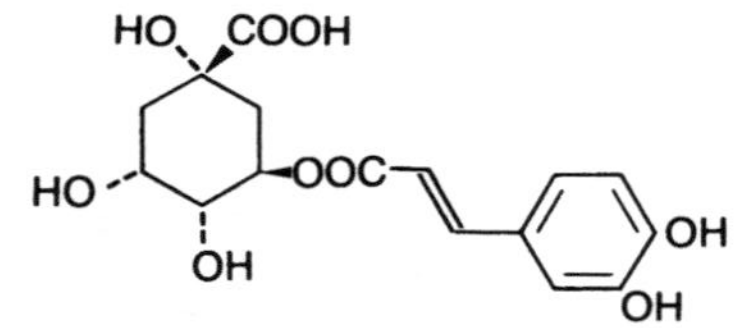	50/54	*Acalymma vittatum*	Reed et al., 1983
			570/60	*Pieris brassicae*	Jones & Firn, 1979
polyacetylene	phenylheptatriyne	$Ph(C{\equiv}C)_3CH_3$	10/40	*Euxoa messoria*	McLachlan et al., 1982
quinone	juglone	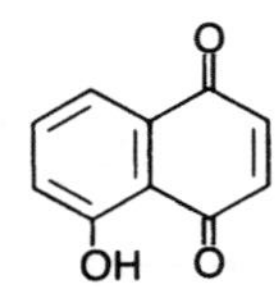	1600/0	*Scolytus multistriatus*	Gilbert et al., 1968
3. Monoterpenes					
iridoid	specionin	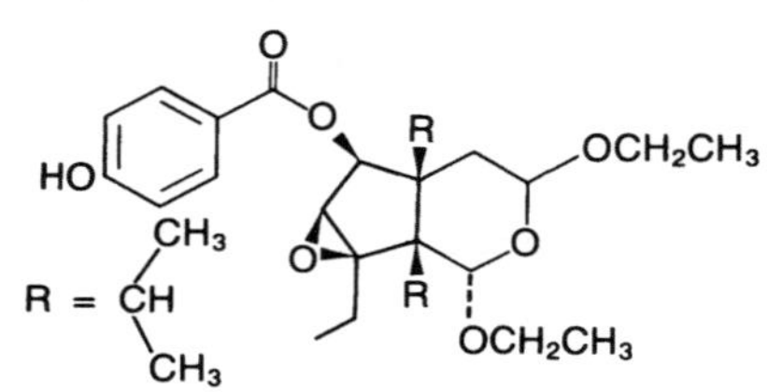	100/95	*Choristoneura fumiferana*	Chang & Nakanishi, 1983

(continued)

Table 3 (cont.)

CHEMICAL CLASS	COMPOUND	STRUCTURE	INHIBITORY CONC. (PPM/ % FEEDING)	INSECTS AFFECTED	REFERENCES
4. Sesquiterpenes					
sesquiterpene diol	shiromodiol diacetate	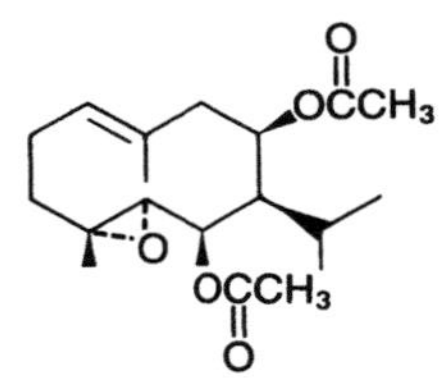	300/30	<u>Trimeresia miranda</u>	Wada et al., 1970
trans-4,cis-9 germacranolide	tulirinol	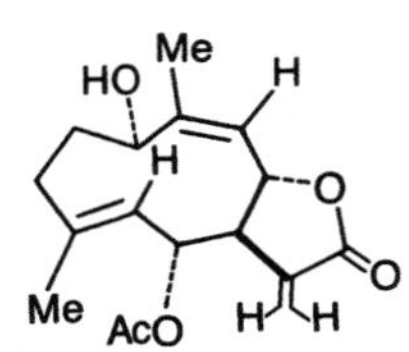	50/31	<u>Lymantria dispar</u>	Doskotch et al., 1986

sesquiterpene aldehyde	warburganal	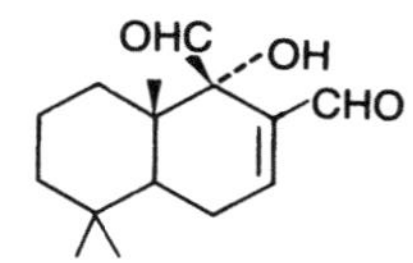	0.1/5	*Spodoptera exempta*	Kubo & Nakanishi, 1979
5. Diterpenes					
clerodane diterpene	clerodin	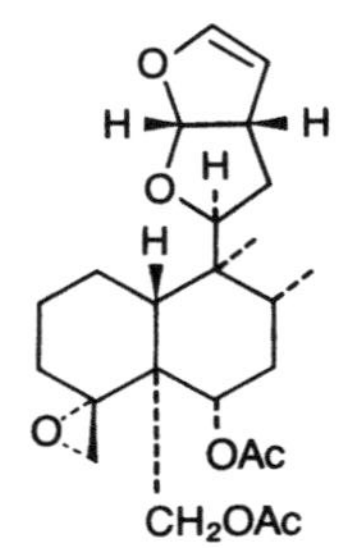	3.6/50	*Spodoptera litura*	Antonious & Saito, 1981
			50/0	*S. littoralis*	Hosozawa et al., 1974

(continued)

Table 3 (cont.)

CHEMICAL CLASS	COMPOUND	STRUCTURE	INHIBITORY CONC. (PPM/ % FEEDING)	INSECTS AFFECTED	REFERENCES
neoclerodane diterpene	ajugarin I	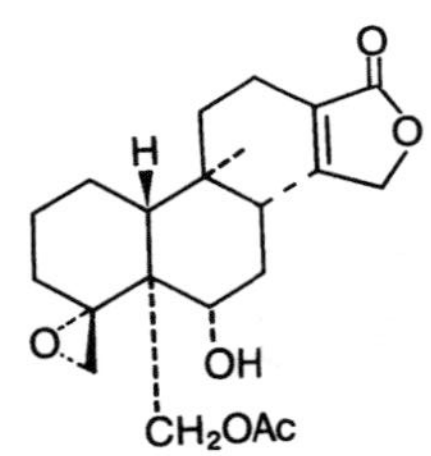	6/5	Schistocera gregaria Spodoptera exempta Pieris brassicae	Kubo & Nakaniski, 1979 Geuskens et al., 1983
grayanoid diterpene	kalmitoxin I	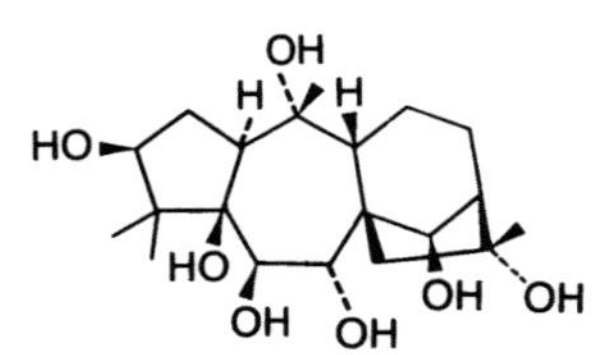	600/80	Lymantria dispar	Shaaban & Doskotch, 1980

limonoid triterpene	azadirachtin	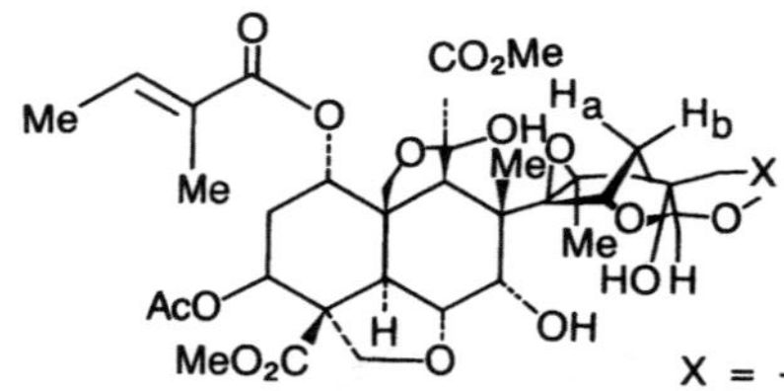 X = -CH=CH-	0.01/0	<u>Spodoptera frugiperda</u> (many others)	Warthen et al., 1978 Broughton et al., 1986 (structure)
limonoid triterpene	harrisonin	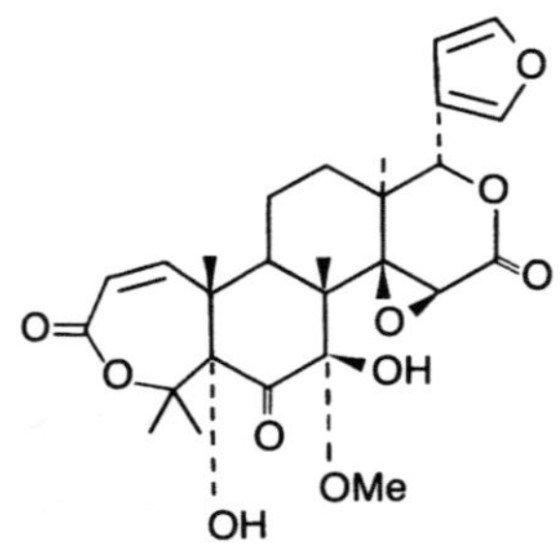	20/0	<u>Spodoptera exempta</u>	Kubo & Nakaniski, 1979

(continued)

Table 3 (cont.)

CHEMICAL CLASS	COMPOUND	STRUCTURE	INHIBITORY CONC. (PPM/ % FEEDING)	INSECTS AFFECTED	REFERENCES
steroidal triterpene	cucurbitacin E	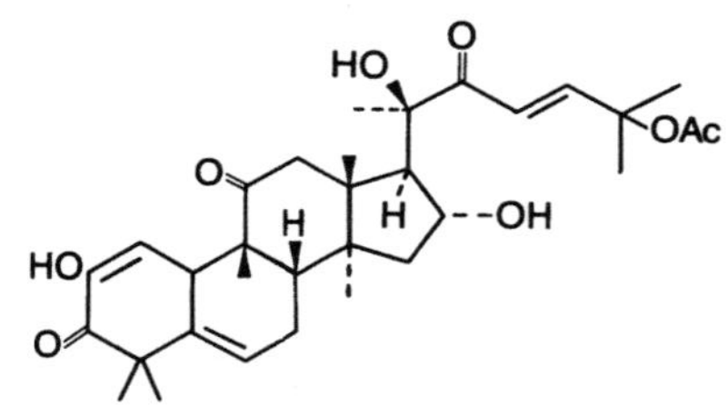	187/0	*Phyllotreta nemorum*	Nielsen et al., 1977
quassinoid triterpene	chaparrinone	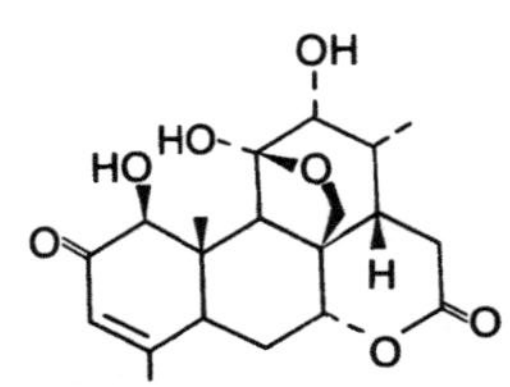	8.5/50	*Heliothis virescens*	Klocke et al., 1985

triterpene	withanolide E	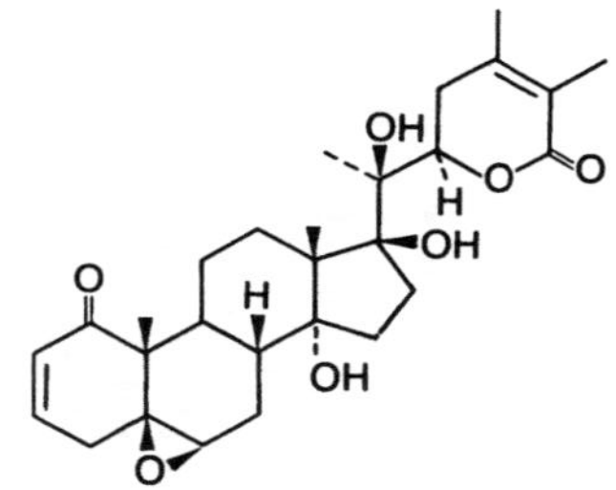	10/32	<u>Spodoptera littoralis</u>	Ascher et al., 1980

deterrent cells in a dose-dependent manner. Within a single insect species, the number of compounds that are effective are of extremely diverse structures. It is therefore difficult to imagine protein receptor sites that could accomodate such a wide variety of molecular shapes, yet, the occurrence and evolutionary specialization of such sites can be successfully argued (Dethier, 1980a). The data available in support of deterrent receptor sites are from applications of a single dose of many compounds to a given cell, and recordings of the number of spikes produced. This is adequate to determine that there is an effect on cell activity, but it does not establish that there is a membrane receptor protein analogous to those proposed for sugar-sensitive cells. In fact, there are no definitive data supporting the existence of a deterrent receptor site in any insect.

Many of the plant compounds that have been described as feeding deterrents are now known to have pharmacological action on nerve cells in the insect's central nervous system (Pelhate and Sattelle, 1982). Thus compounds like strychnine or quinine that modify potassium channel kinetics can be expected to modify chemosensory input in any of the ways mentioned above, by potentially acting not at specific chemoreceptor sites, but at one or more of the many macromolecular sites contained in the sensory or accessory cell membranes. The existence of intracellular targets is also quite possible. Since the normal functioning of the sensillum requires the functional integrity of its cellular constituents, disruption of any one of the components can result in altered chemosensory responses. The reduction in response of a glucose-sensitive cell to its normal stimulus, or the increase in firing of a cell that normally responds to plant compounds that deter feeding can both result from interactions of an allelochemical with sites other than the cell's chemoreceptor sites. Likely candidates are dendrite membrane ion channels or the ATPase pump in accessory cells.

4.2.3. Effects of Alkaloids. Since the alkaloids are such a large and diverse group of allelochemicals, it is not surprising that we have accumulated more data on their insect chemosensory actions than for any other group. An early review indicated their defensive roles in plants and summarized their known actions in deterrent cells (Levinson, 1976). Quinine inhibits impulse generation in sugar-sensitive cells of the gypsy moth, *Lymantria dispar* (L.) (Lepidoptera: Lymantriidae), and eastern tent caterpillar, *Malacosoma americanum* (Fabr.) (Lepidoptera: Lasiocampidae), (Dethier, 1982) and competitively blocks sucrose response in the the flesh fly, *Boettcheria peregrina* (Diptera: Sarcophagidae), (Morita et al., 1977). Quinine, nicotine, and sparteine were shown to inhibit a sugar-sensitive cell in the galeal sensilla of adults and larva of the red turnip beetle, *Entomoscelis americana* Brown (Coleoptera: Chrysomelidae) (Mitchell and Sutcliff, 1984). These same compounds also stimulated a glucosinolate-sensitive cell in these beetles, but the feeding deterrency of these compounds was proposed to be the result of reducing the firing of the sugar-sensitive cell. Both quinine and sparteine reduce the frequency of impulses produced by the sugar cell within one second when presented in combination with sugar. Since sparteine and quinine are also known to reduce potassium ion conductance in a number of other excitable cells, their sites of action on the sugar-sensitive cells could well be dendrite ion channels, rather than sugar receptor proteins (Mitchell and Sutcliff, 1984; Hille, 1984).

The steroidal glycoalkaloids tomatine, solanine, and chaconine elicited irregular firing from several cells in the galeal and tarsal sensilla of adult, and the larval galeal α-sensilla of the Colorado potato beetle, *Leptinotarsa decemlineata* Say (Coleoptera: Chrysomelidae), (Mitchell and Harrison, 1984, 1985). After a latency of one to 30 seconds, the response of the γ-aminobutyric acid (GABA)-sensitive cell was inhibited, and several cells fired irregularly. These effects were not dose dependent and were extremely variable in their onset time and in the number of preparations exhibiting effects. These same authors produced similar irregular firing of the sugar-sensitive cell in the black blow fly by treatment with the saponin digitonin. Their suggestion was that the Solanum alkaloids produce the observed effects by a rather nonspecific membrane disruption, similar to that of digitonin, and are not detected by a specific chemosensory cell receptor site. This raises the question of the exact mode of action of the wide variety of plant alkaloids that have been shown to cause the "deterrent cell" of caterpillars to fire. Are there actually specific receptor proteins for these substances, or do they cause the cell to fire by acting directly on membrane or intracellular components? Plant alkaloids also deter sucrose stimulated feeding in the black blow fly (Blades and Mitchell, 1986).

The plant alkaloid 2,5-dihydroxymethyl-3,4-dihydroxypyrrolidone (DMDP), a structural analogue of α-D-fructofuranose was shown to inhibit feeding and to be toxic to several species of insects by Blaney et al. (1984). This compound when combined with sucrose disrupted the response of the sugar-sensitive cell in the lateral styloconicum of Spodoptera exempta (Walker) (Lepidoptera: Noctuidae). It produced irregular firing of the cell, and was suggested to be acting as a sugar antagonist due to its close structural similarity to fructose.

4.2.4. Effects of Terpenes. Azadirachtin from the neem tree Azadirachta indica and several related trees is one of the most potent and universally active plant compounds that inhibit insect feeding. Recent structural studies indicate that a number of isomers of azadirachtin may account for the diversity of activity seen with crude neem extracts (Broughton et al., 1986). Neem extract induces firing of one cell in the labial palps and one in the A3 sensillum of the clypeo-labrum of Schistocerca gregaria L. (Orthoptera: Acrididae) and causes immediate cessation of feeding upon contact (Haskell and Schoonhoven, 1969). The cells firing in the presence of azadirachtin were not identified for their specificities to other stimuli. The medial styloconicum in Pieris brassicae and Lymantria dispar larvae, contain cells that fire in the presence of azadirachtin, (Schoonhoven and Jermy, 1977; Schoonhoven, 1982).

A more comprehensive study of the sensory effects of azadirachtin on caterpillars revealed that a 10 micromolar concentration of azadirachtin induces the firing of a large spike in the lateral and medial styloconica of S. exempta, and M. brassicae. This cell appears to fire independently of the sugar-sensitive cell (Simmonds and Blaney, 1983). Azadirachtin also stimulates a cell in the lateral styloconica of S. exempta, but there was an interaction between this cell and the response of the sucrose-sensitive cell. Thus the effects of azadirachtin are different on the two styloconica of the same insect. In two economically important species, the cabbage looper, Trichoplusia ni Hubner (Lepidoptera: Noctuidae), and the corn earworm, Heliothis zea (Boddie) (Ibid.), the responses of cells in the styloconica are unaffected by azadirachtin, while in the more polyphagous species, Spodoptera littoralis (Boisd.) (Ibid.), S. frugiperda (J. E. Smith) (Ibid.), and Heliothis armigera (Ibid.) cells in both styloconica are stimulated by azadirachtin together with interactions with sucrose-sensitive cells. The cells that fire to azadirachtin may be either the salt-sensitive ones or those previously labelled as responding to deterrent compounds. Irregular firing of other cells of unknown specificities was also observed. Clearly, azadirachtin is affecting more than one chemosensory cell type in more than one way, and the effects are different for caterpillars with different host plant ranges.

The sesquiterpene dialdehyde warburganal produces irregular firing of more than one cell, and then blocks stimulation in the sucrose- and inositol-sensitive styloconic cells of S. exempta (Ma, 1977). These

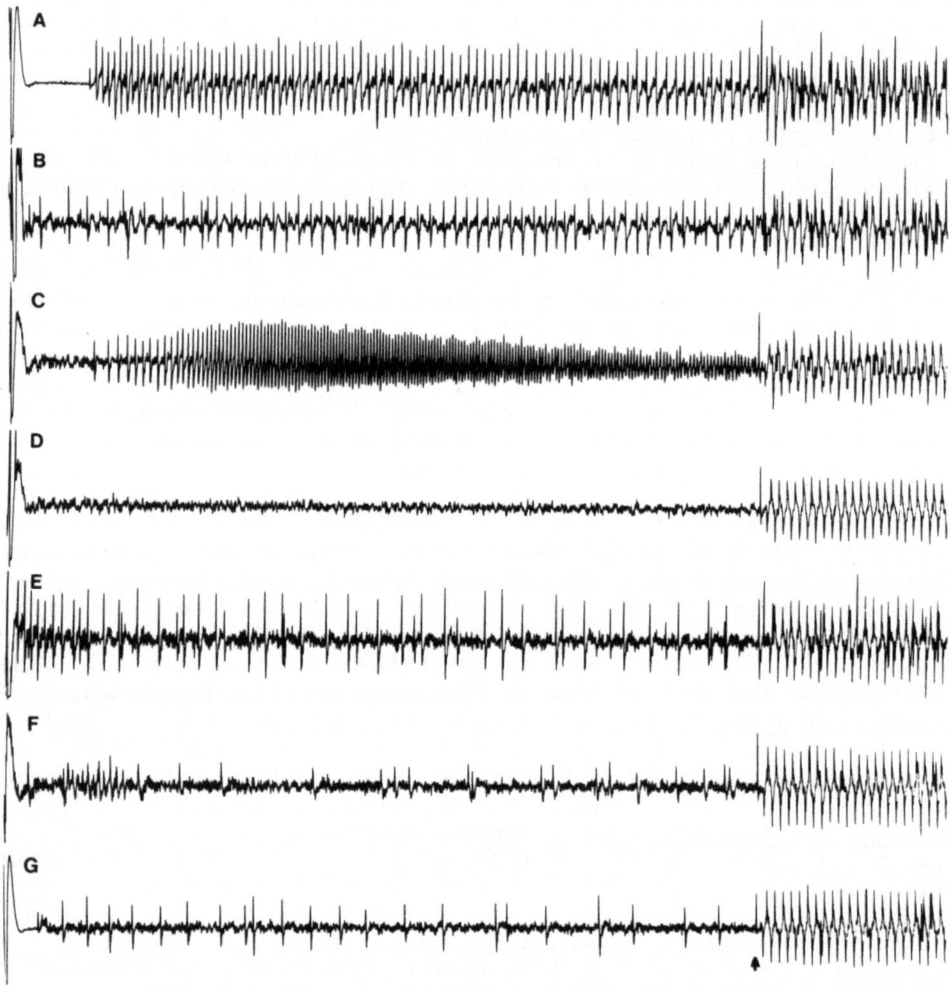

Fig. 8. Effects of some plant allelochemicals on the styloconic taste cells of the tobacco hornworm caterpillar. Results of single trials shown are representative of replicated observations.

A. Response of a cell in the lateral styloconica to 10 mM caffeine + 0.1 M NaCl.

B. Response of a cell in the lateral styloconica to 10 mM 1,2-naphthoquinone + 0.1 M NaCl.

C. Response of a cell in the lateral styloconica to 1000 ppm aristolochic acid + 0.1 M NaCl.

D. Lack of response to 0.3 M glucose stimulation following aristolochic acid (same cell as record C).

E. Response to 0.3 M glucose + 0.1 M NaCl in medial styloconica.

F. Response to 0.3 M glucose + 0.1 M NaCl following a three-minute treatment by 1000 ppm warburganal.

G. Response to 0.1 M NaCl in medial styloconia. Arrow denotes onset of 200 μV calibration pulses in all records.

effects are prevented by first mixing warburganal with sulfhydryl containing amino acids, but not with other amino acids. The inhibitory effects of warburganal are mimicked by the sulfhydryl-specific reagents PCMB and N-ethyl maleimide. It was suggested that the action of warburganal on these taste cells is mediated through interactions with membrane protein sulfhydryl groups (Ma, 1977).

Further elucidation of the mode of action of warburganal was provided for the styloconic cells of M. sexta (Frazier, unpublished). In this species as well, the glucose- and inositol-sensitive cells of both styloconica are blocked following warburganal treatment, Fig. 8, E and F. The receptor site for myoinositol could be specifically protected from the blocking effect of warburganal, by first saturating it with a high concentration of myoinositol. This protection of the receptor site for myoinositol indicates that warburganal either is binding at the receptor site for myoinositol or at an overlapping site. In addition to this action of warburganal at the receptor site, the shapes of spikes produced by both the glucose- and myoinositol-sensitive cells are greatly altered by warburganal treatment, indicating that a second site of action independent of the receptor sites is involved. The warburganal block of the myoinositol cell could not be reversed by treatment with either dithiothreitol (DTT), or with cysteine indicating that warburganal is not binding via sulfhydryl groups. A partial block of the myoinositol site with PCMB could be reversed with excess cysteine, but not with excess DTT. Although it is clear that warburganal can reduce the responsiveness of the glucose- and myoinositol-sensitive cells, and is shown to act at two separate sites, the requirement of protein sulfhydryl groups for binding by warburganal is not clearly substantiated.

A unique approach to understanding the chemosensory effects of clerodin, an antifeedant diterpene, was used by Antonious and Saito (1983). Clerodin was carefully applied to various sensilla of Spodoptera litura (Fabr.) (Lepidoptera: Noctuidae) with a micropipette, and the resulting changes in feeding behavior recorded within a 30-minute period. Clerodin produced greater antifeedant effects when applied to the maxillary palps than the styloconica, but no effects were observed when it was applied to the hypopharynx or antennae. Continuous feeding was resumed in 30 minutes after treatment with 1250 ppm. These effects were distinctly different from those produced by the formamidine insecticide, chlordimeform which acted only when applied to the hypopharynx producing delayed hyperactive locomotion followed by uncoordinated movements and tremors of the maxillae and mandibles.

4.2.5. Effects of other Allelochemicals. Both ziziphins and gymnemic acids are triterpene saponin glycosides that modify sweet taste perception in humans, and the response of sugar-sensitive cells in the black blow fly (Kennedy et al., 1975; Kennedy and Halpern, 1980a). The spike frequency of the sugar-sensitive cell is depressed following a three minute treatment as well as the behavioral response of labellar lowering. The spike frequency of the salt-sensitive cell is also partially depressed shortly after treatment, but both types of cells fire irregularly at longer intervals after treatment. These results have led to a proposed model of biphasic action involving a selective interaction with receptor proteins from the intramembrane side, followed by nonselective interactions with other membrane components (Kennedy and Halpern, 1980b). The observed effects of these compounds are quite similar to those for fatty acids and straight chain hydrocarbons on fly taste cells and for alkaloids on Colorado potato beetle taste cells (Steinhardt et al., 1967; Dethier and Hanson, 1968; Mitchell and Harrison, 1985).

From the foregoing examples of allelochemicals acting at several

Table 4. Plant Compounds that Induce Firing of Insect Deterrent Cells and Their Known Effects on Other Nerve Cells

COMPOUND	SPECIES WITH DETERRENT CELL	ACTION ON OTHER NERVE CELLS	REFERENCES
Quinine	*P. brassicae*, others	Block K+ channel conductance	Fishman & Spector, 1981
Strychnine	*P. brassicae*, others	Reduce K+ and Na+ channel conductance	Pelhate & Sattelle, 1982
Caffeine	*M. sexta*, others	Phosphodiesterase inhibitor	Nathanson, 1984
Atropine	*P. brassicae*	Reduce Na+, K+ channel conductance Reduce Na+, K+ channel inactivation	Pelhate and Satelle, 1982
Sparteine	*P. brassicae*	Reduce Na+, K+ channel conductance	Ohta et al., 1973
Nicotine	*P. brassicae*	Reduce K+ channel conductance	Frazier et al., 1973

sites other than receptor sites in several types of nerve cells, it seems highly likely that many of these compounds act in a similar manner on insect chemosensory cells (see also Kurihara, et al., 1986). If the resulting irregular patterns of spikes that have been recorded by several investigators represent "nonsense" to the CNS, or if it signals "deterrence", then a large number of chemosensory patterns of spikes may have the same behavioral meaning. Clearly much exciting research in chemosensory physiology awaits us.

4.3. Modes of Action on Other Excitable Cells

The thesis that plant allelochemicals can effectively prevent insect feeding damage by acting directly on chemoreceptors, or on other excitable cells within the insect has recently been put forth by Mitchell and Harrison (1985) and Murdock et al. (1985). In addition to effects such as those listed in Table 4, compounds like the pyrethroids, nicotine, reserpine, and physostigmine are well known for their toxic action on neurons and their connections. Exactly what neural pathways are responsible for controlling feeding behavior directly, and what additional levels of neuroendocrine modulation there are upon these primary neural circuits remain to be determined. However, there have been recent reports on octopaminergic compounds enhancing the behavioral response of labellar lowering to sugar cell inputs from tarsal receptors in the black blow fly, supporting the concept that octopaminergic receptors in the CNS positively modulate feeding and drinking behavior (Long and Murdock, 1983).

Behavior mediated by olfactory receptors in insects has also been shown to be modified by several types of neuroactive compounds, including those specific to octopaminergic systems (Mercer and Menzel, 1982; Linn and Roelofs, 1984, 1986). It is interesting that the understanding of the modes of action of olfactory repellents closely parallels our understanding of plant allelochemical effects on feeding behavior (Davis, 1985).

Clearly many possibilities exist for plant allelochemicals to modify insect feeding behavior by directly influencing the activity of chemosensory cells as well as their junctions with central neurons. The plethora of excitable cells and junctions in the CNS that may be directly involved in regulating the motor patterns that constitute feeding behavior may also be directly affected. Given the multiple levels of neural integration and modulation within the insect nervous system, the interactions with plant allelochemicals at the molecular level seem to offer almost limitless possibilities for challenging the adaptive mechanisms of insects. Insect feeding behavior is certainly a central focus in this arena, yet it is one about which we have much to learn.

5. SUMMARY

Insect feeding behavior is regulated by specific messages from the chemosensory cells associated with the antennae and mouthparts. Although the underlying neural circuits integrating this chemosensory input and the associated motor output elements have not been identified, the description of feeding behavior in specific behavioral acts, or fixed motor patterns is seen as a necessary step in this endeavor. Such a detailed approach to behavior is required to identify simple motor outputs that can be correlated with measurable chemosensory patterns. This approach allows progress in understanding sensory coding for both the messages that promote feeding as well as those that prevent it.

Plant allelochemicals that inhibit insect feeding may be detected by chemosensory cells with evolved receptor sites for these compounds, or they may block or evoke irregular patterns of firing from one or several cells. Additionally, other excitable cells in the insects neural and muscular systems are also possible targets. The potency of some allelochemicals in preventing insect feeding suggests that they are a highly evolved mechanism for plant protection. A further understanding of the mode of action of such plant compounds seems best approached with combined cellular and biochemical techniques.

6. ACKNOWLEDGEMENT

The helpful comments of E. A. Bernays, F. E. Hanson, and R. Y. Zacharuk on earlier drafts are greatly acknowledged. I would especially like to thank Sherry Catts for cheerfully typing the seemingly endless revisions, and L. B. Brattsten and S. Ahmad for involving me in this project.

7. REFERENCES

Adityachaudhury, N., A. Bhattacharyya, A. Chowdhury and S. Pal, 1985. Chemical constituents of plants exhibiting insecticidal antifeeding and insect growth regulating activities, J. Sci. Ind. Res., 44:85-101.

Altner, H., and L. Prillinger, 1980. Ultrastructure of invertebrate chemo-, thermo-, and hygroreceptors and its functional significance, Intern. Rev. Cytol., 67:69-139.

Ascher, K. R. S., N. E. Nemny, K. I. Eliyahu, A. Abraham and E. Glatter, 1980. Insect antifeedant properties of withanolides and related steriods from Solanaceae, Experientia, 36:998-999.

Antonious, A. G. and T. Saito, 1983. Mode of action of antifeeding compounds in the larvae of the tobacco cutworm, Spodoptera litura (F.) (Lepidoptera: Noctuidae) III. Sensory responses of the larval chemoreceptors to chlordimeform and clerodin, Appl. Entomol. Zool., 1:40-49.

Barnes, D. M., 1986. Lessons from snails and other models, Science, 231:1246-1249.

Bell, R. A., 1984. Role of frontal ganglion in lepidopterous insects, in: "Insect Neurochemistry and Neurophysiology", A. Borkovec and T. Kelly, eds., pp 321-324, Plenum Publ. Corp., New York.

Bentley, M. D., D. E. Leonard, W. F. Stoddard and L. H. Zalkow, 1984. Pyrrolizidine alkaloids as larval feeding deterrents for spruce budworm, Christoneura fumiferana (Lepidoptera: Tortricidae), Ann. Entomol. Soc. Amer., 77:393-397.

Berenbaum, M., 1985. Brementown revisited: interactions among allelochemicals in plants, Rec. Adv. Phytochem., 19:139-169.

Bernays, E. A., 1985. Regulation of feeding behavior in: "Comprehensive Insect Physiology, Biochemistry, and Pharmacology", G. A. Kerkut and L. I. Gilbert, eds., Vol. 6, pp. 1-32, Pergamon Press, New York.

Bernays, E. A. and S. J. Simpson, 1982. Control of food intake, Adv. Insect Physiol., 16:59-118.

Bernays, E. A. and R. R. Wrubel, 1985. Learning by grasshoppers: association of color/light intensity with food, Physiol. Entomol., 10:359-369.

Blades, D. and B. K. Mitchell, 1986. Effects of alkaloids on feeding by Phormia regina, Entmol. Exp. Appl., in press.

Blaney, W. M., 1975. Behavioural and electrophysiological studies of taste discrimination by the maxillary palps of larvae of Locusta migratoria (L.), J. Exp. Biol., 62:555-569.

Blaney, W. M. and A. M. Duckett, 1975. The significance of palpation by the maxillary palps of Locusta migratoria (L.): an electrophysiological and behavioral study, J. Exp. Biol., 63:701-712.

Blaney, W. M., M. S. J. Simmonds, S. V. Evans and L. E. Fellows, 1984. The role of the secondary plant compound 2,5-dihydroxmethyl-3,4-dihydroxy pyrrolidone as a feeding inhibitor for insects, Entomol. Exp. Appl., 36:209-216.

Blom, F., 1978. Sensory activity and food intake: a study of input-output relationships in two phytophagous insects, Netherl. J. Zool., 28:277-340.

Boudreau, J. C., L. Sivakumar, L. T. Do, T. D. White, J. Oravec and N. K. Hoang, 1985. Neurophysiology of geniculate ganglion (facial nerve) taste systems: species comparisons, Chem. Senses, 10:89-12.

Bowdan, E., 1984. Electrophysiological responses of tarsal contact chemoreceptors of the apple maggot fly Rhagoletis pomonella to salt, sucrose and oviposition deterrent, J. Comp. Physiol., 154:143-152.

Broughton, H. B., S. V. Ley, A. M. Z. Slawin, D. J. Williams and E. D. Morgan, 1986. X-ray crystallographic structure determination of detigloyldihydroazadirachtin and reassignment of the structure of the limonoid insect antifeedant azadirachtin, J. Chem. Soc. Chem. Commun., No. 1, pp. 46-47.

Chang, C. C. and K. Nakanishi, 1983. Specionin, an iridoid insect antifeedant from Catalpa speciosa, J. Chem. Soc. Commun., 956:605-606.

Chapman, R. F., 1974. The chemical inhibition of feeding by phytophagous insects: a review, Bull. Entomol. Res., 64:339-363.

Chapman, R. F., 1982. Chemoreception: the significance of receptor numbers, Adv. Insect Physiol., 16:247-35.

Chuo, F. Y., K. Hostettman, I. Kubo and K. Nakanishi, 1977. Isolation of an insect antifeedant N-methyl flindersine and several benzo[e] phenanthridine alkaloids from east African plants: a comment on chelerythrine, Heterocycles, 7:969-977.

Clark, J. V., 1981. Feeding deterrent receptors in the last instar African armyworm Spodoptera exempta: a study using salicin and caffein, Entomol. Exp. Appl., 29:189-197.

Davis, E. E., 1985. Insect repellents: concepts of their mode of action relative to potential sensory mechanisms in mosquitoes (Diptera: Culicidae), J. Med. Entomol., 22:237-243.

DeBoer, G., V. G. Dethier and L. M. Schoonhoven, 1977. Chemoreceptors in the preoral cavity of the tobacco hornworm, Manduca sexta, and their possible function in feeding behavior, Entomol. Exp. Appl., 22:287-298.

Dethier, V. G., 1976. "The Hungry Fly", Harvard University Press, Cambridge.

Dethier, V. G., 1977. Gustatory sensing of complex stimuli by insects, in: "Olfaction and Taste VI", I. LeMagnon and P. MacLeod, eds, pp. 323-332, Information Retrieval Press, London.

Dethier, V. G., 1980. Evolution of receptor sensitivity to secondary plant substances with special reference to deterrents, Amer. Nat., 115:45-66.

Dethier,V. G., 1982. Mechanism of host plant recognition, Entomol. Exp. Appl., 31:49-56.

Dethier, V. G., Crnjar R. M., 1982. Candidate codes in the gustatory system of caterpillars, J. Gen. Physiol., 79:549-569.

Dethier, V. G. and F. E. Hanson, 1968. Electrophysiological responses of the chemoreceptors of the blowfly to sodium salts of fatty acids, Proc. Nat. Acad. Sci., 60: 1296-1303.

Dethier, V. G., L. Barton Brown and C. N. Smith, 1960. The designation of chemicals in terms of the responses they elicit from insects, J. Econ. Entomol., 53:134-136.

Devitt, B. and J. J. B. Smith, 1985. Action of mouthparts during feeding in the dark-sided cutworm, Euxoa messoria, Can. Entomol., 117:343-349.

Doskotch, R. W., H. Y. Chen, T. M. O'Dell and L. Girard, 1980. Nerolidol: an antifeeding sesquiterpene alcohol for gypsy moth larvae from Melaleuca leucadendron, Chem. Ecol., 6:845-851.

Dryer, D. L. and K. C. Jones, 1981. Feeding deterrency of flavanoids and related phenolics towards Schizaphis graminum and Myzus persicae: aphid feeding deterrents in wheat, Phytochem., 20:2489-2493.

Fishman, M. C. and I. Spector, 1981. Potassium current suppression by quinidine reveals additional calcium currents in neuroblastoma cells, Proceed. Nat. Acad. Sci., USA, 78:245-249.

Fobes, J. F., J. B. Mudel and M. P. F. Marsden, 1985. Epicuticular lipid accumulation on the leaves of Lycopersicon penelli (Corr.) D'Arcy and Lycopersicon esculentum Mill, Plant Physiol., 77:567-570.

Frazier, J. L., 1986. Chemosensory regulation of mandibular motor output in the tobacco hornworm, Manduca sexta, submitted for publication.

Frazier, D. T., C. Sevik and T. Narahashi, 1973. Nicotine: effect on nerve membrane conductances, Europ. J. Pharmacol., 22:217-220.

Geuskens, R. B. M., J. M. Luteijin and L. M. Schoonhoven, 1983. Antifeedant activity of some ajugarin derivatives in three lepidopterous species, Experientia, 39:403-404.

Gibson, R. W. and J. A. Pickett, 1983. Wild potato repels aphids by release of aphid alarm pheromone, Nature, 302:600-609.

Gilbert, B. L., J. E. Baker and D. M. Norris, 1967. Juglone (5-hydroxy-1,4-napthoquinone) from Carya ovata, a deterrent to feeding by Scolytus multistriatus, J. Insect. Physiol., 13:1453-1459.

Hall, J. C. and R. J. Greenspan, 1979. Genetic analysis of Drosophila neurobiology, Annu. Rev. Genet., 13:127-195.

Hansen, K. and H. Wieczoreck, 1981. Biochemical aspects of sugar reception in insects, in: "Biochemistry of Taste and Olfaction", R. H. Cagan and M. E. Kare, eds., pp. 139-162, Academic Press, New York.

Hanson, F. E. 1983. The behavioral and neurophysiological basis of food plant selection by lepidopterous larvae, in: "Herbivorous Insects", S. Ahmad, ed., pp 3-23, Academic Press, New York.

Haskell, P. T. and L. M. Schoonhoven, 1969. The function of certain mouthpart receptors in relation to feeding in Schistocerca gregaria and Locusta migratoria migratorioides, Entomol. Exp. Appl., 12:423-440.

Hatfield, L. D., J. Ferrera, and J. L. Frazier, 1983. Host selection and feeding behavior by the tarnished plant bug, Lygus lineolaris (Hemiptera: Miridae), Ann. Entomol. Soc. Am., 76:688-691.

Hille, B., 1984. "Ionic Channels of Excitable Membranes", Sinauer Assoc., Sunderland.

Hirao, T., K. Yamaoka and N. Arai, 1976. Studies on mechanism of feeding in the silkworm, Bombyx mori L. II. Control of mandibular biting movement by olfactory information through maxillary sensilla basiconica, Bull. Sericul. Exp. Sta., 26:385-410.

Hodgson, E. S., J. Y. Lettvin and K. D. Roeder, 1955. Physiology of a primary chemoreceptor unit, Science, 122:417-418.

Hosozawa, S., N. Kato, K. Munakata and Y. L. Chen, 1974. Antifeeding active substance for insect, Agr. Biol. Chem., 38:1045-1048.

Huber, F., 1984. Insect neuroethology: approaches to the study of behavior and the underlying neural mechanisms, in: "Animal behavior", K. Aoki, S. Ishii and M. Morita eds., pp. 3-185, Japan Scientific Societies Press, Tokyo.

Ishikawa, S., 1966. Electrical response and function of a bitter substance receptor associated with the maxillary sensilla of the larva of the silkworm, Bombyx mori, J. Cell. Physiol., 67:1-12.

Jakinovitch, W., Jr., 1985. Sugar taste reception in the gerbil, in: "Taste, Olfaction, and the Central Nervous System", D. W. Pfall, ed., pp. 65-91, Rockefeller Press, New York.

Jones, C. G. and R. D. Firn, 1979. Some allelochemicals of Pteridium aquilinum and their involvement in resistance to Pieris brassicae, Biochem. Syst. Ecol., 7:187-192.

Kandel, E., 1979. "Behavioral Biology of Aplysia", W. H. Freeman and Co., San Francisco.

Kaissling, K. and J. Thorson, 1980. Olfactory sensilla: structural, chemical and electrical aspects of the functional organization, in: "Receptors for Neurotransmitters, Hormones and Pheromones in Insects", D. B. Sattelle, L. M. Hall, J. G. Hildebrand eds., pp. 261-282, Elsevier/North-Holland, New York.

Kennedy, L. M. and B. Halpern, 1980a. Fly chemoreceptors: a model system for the taste modifier ziziphin, Physiol. Behavior, 24:1-9.

Kennedy, L. M. and B. Halpern, 1980b. A biphasic model for action of the gymnemic acids and ziziphins on taste receptor cell membranes. Chem. Senses, 5:149-158.

Kennedy, L. M., B. Sturkchow and F. J. Waller, 1975. Effect of gymnemic acid on single taste hairs of the housefly Musca domestica, Physiol. Behavior, 14:755-765.

Klocke, J. A., M. Arisawa, S. S. Handa, A. D. Kinghorn, G. A. Cordell and N. R. Farnsworth, 1985. Growth inhibitory, insecticidal and antifeedant effects of some antileukemic and cytotoxic grass inoids on two species of agricultural pests, Experientia, 41:379-382.

Koul, O., 1982. Insect feeding deterrents in plants, Indian Rev. Life Sci., 2:97-125.

Kubo, I. and K. Nakanishi, 1979. Some terpenoid insect antifeedants, in: "Advances in Pesticide Science", H. Geissbuhler, ed., Part 2, pp. 284-294, Pergamon Press, Oxford.

Kuppers, J. and U. Thurm, 1982. On the functional significance of ion circulation induced by electrogenic tissue, in: "Exogenous and Endogenous Influcences on Metabolic and Neural Control", A. D. F. Addink and N. Spronk, eds, pp. 313-327, Pergamon Press, Oxford.

Kurihara, K., K. Yoshii and M. Kashiwayanagi, 1986. Transduction mechanisms in chemoreception, J. Comp Biochem. Physiol., in press.

Levinson, H.Z., 1976. The defensive role of alkaloids in insects and plants, Experientia, 32: 408-411.

Linn, E. C. and W. L. Roelofs, 1984. Sublethal effects of neuroactive compounds on pheromone response thresholds of male oriental fruit moths, Arch. Insect Biochem. Physiol., 1:331-344.

Linn, E. C. and W. L. Roelofs, 1986. Modulatory effects of octopamine and serotonin on male sensitivity and periodicity of response to sex pheromone in the cabbage looper moth, Trichoplusia ni, Arch. Insect Biochem. Physiol., 3:161-171.

Long, T. F. and L. L. Murdock, 1983. Stimulation of blow fly feeding behavior by octopaminergic drugs, Proc. Nat. Acal. Sci., USA, 80:4159-4163.

Ma, W. C., 1972. Dynamics of feeding responses in Pieris brassicae L. as a function of chemosensory input: a behavioral, ultrastructural and electrophysiological study, Meded. Landbouwhohgesch, 001 Wageningen, 7211:1-162.

Ma, W. C., 1976. Electrophysiological evidence for chemosensitivity to adenosine, adenine and sugars in Spodoptera exempta and related species, Experientia, 33:356-358.

Ma, W.C., 1977. Alterations of chemoreceptor function in armyworm larvae (Spodoptera exempta) by a plant-derived sesquiterpene and by sulfhydryl reagents, Physiol. Entomol., 2:199-207.

Maes, F. W., 1985a. Response noise affects the graphical evaluation of response versus concentration curves, Chem. Senses, 10:23-34.

Maes, F. W., 1985b. Improved best stimulus classifications of taste neurons, Chem. Senses, 10:35-44.

McLachlan, D., J. T. Arnason, B. J. R. Philogene and D. Champagne, 1982. Antifeedant activity of polyacetylene, phenylheptatriyne (DHT), from the Asteraceae to Euxoa messoria (Lepidoptera: Noctuidae), Experientia, 38:1061-1062.

Mercer, A. R. and R. Menzel, 1982. The effects of biogenic amines on conditioned and unconditioned responses to olfactory stimuli in the honey bee Apis mellifera, J. Comp. Physiol., 145:363-368.

Miller, J. R. and K. L. Strickler, 1984. Finding and accepting host plants, in: "Chemical Ecology of Insects", W. Bell and R. Carde, eds, pp. 127-158, Sinauer Assoc., Sunderland.

Mitchell, B. K., 1985. Specificity of an amino acid-sensitive cell in the adult Colorado beetle, Leptinotarsa decemlineata, Physiol. Entomol., 10:421-429.

Mitchell, B. K. and P. Gregory, 1979. Physiology of maxillary sugar sensitive cell in the red turnip beetle, Entomoscelis americana, J. Comp. Physiol., 132:167-178.

Mitchell, B. K. and G. D Harrison, 1984. Characterization of galeal chemosensilla in the adult Colorado potato beetle, Leptinotarsa decemlineata, Physiol. Entomol., 9:49-56.

Mitchell, B. K. and G. D. Harrison, 1985. Effects of Solanum glycoalkaloids on chemosensilla in the Colorado potato beetle; a mechanism of feeding deterrence, J. Chem. Ecol., 11:73-83.

Mitchell, B. K. and J. F. Sutcliffe, 1984. Sensory inhibition as a mechanism of feeding deterrence: effects of three alkaloids on leaf beetle feeding, Physiol. Entomol., 9:57-64.

Morita, H. and A. Shirahashi, 1985. Chemoreception physiology, in: "Comprehensive Insect Physiology Biochemistry and Pharmacology", G. A. Kerkut and L. I. Gilbert, eds, vol. 6, pp. 133-170, Pergamon Press, New York.

Morita, H., K. Enomoto, M. N. Kakashima, I. Shimada and H. Kijima, 1977. The receptor site for sugars in chemoreception of the flesh fly and the blow fly, in: "Proceed. of Sixth Intern. Symp. Olfaction and Taste", J. LeMagnen and P. MacLeod, eds., pp. 39-46, Information Retrieval, London.

Murdock, L. L., G. Brookhart, R. S. Edgecomb, T. F. Long and L. Sudlow, 1985. Do plants "psychomanipulate" insects, in: "Bioregulators for Pest Control", P. A. Hedin, ed., Symp. Ser. No.217, Amer. Chem. Soc., Press, Washington.

Mustaparta, H., 1984. Olfaction, in: "Chemical Ecology of Insects", W. Bell and R. Carde, eds, pp. 37-72, Sinauer Assoc. Sunderland.

Nathanson, J. A., 1984. Caffeine and related methylxanthines: possible naturally occurring pesticides, Science, 226:184-187.

Nielsen, J. K., L. M. Larsen and H. J. Sorenson, 1977. Curcubitacins E and I in Iberis amara, feeding inhibitors for Phyllotreta nemorum, Phytochem., 16:1519-1522.

Nolen, T. G. and R. R. Hoy, 1984. Initiation of behavior by single neurons: the role of behavioral context, Science, 226:992-994.

Norris, D. M., 1986. Antifeeding, in: "Chemistry of Plant Protection", G. Hang and H. Hoffman, eds, pp. 99-146, Springer-Verlag, New York.

Ohrui, H., H. Hiroyuki, I. Shimada and H. Meguro, 1985. Sweet taste response of fly to furanoses and their analogues, Agric. Biol. Chem., 49:3319-3321.

Ohta, M., T. Narahashi and R. F. Keeler, 1973. Effects of veratrum alkaloids on membrane potential and conductance of squid and crayfish giant axons, J. Pharmac. Exp. Therap., 184:143-154.

Pelhate, M. and D. Sattelle, 1982. Pharmacological properties of insect axons, J. Insect Physiol., 11:889-903.

Phillips, C. and J. Vandeberg, 1976. Mechanism of fluid flow in trichogen and tormogen cells of Phormia regina (Meigen), Int. J. Insect Morphol. Embryol., 5:423-431.

Prokopy, R. J., R. H. Collier and S. Finch, 1983. Leaf color used by cabbage root flies to distinguish among host plants, Science, 221:190-192.

Reed, D. K., W. F. Kwolek and C. R. Smith, Jr., 1983. Investigation of antifeedant and other insecticidal activities of trewiasine towards the striped cucumber beetle and codling moth. J. Econ. Entomol., 76:641-645.

Rees, C. J. C., 1969. Chemoreceptor specificity associated with choice of feeding site by the beetle, Chrysolina brunsvicensis on its food plant, Hypericum hirsutum, Entomol. Exp. Appl., 12:565-583.

Russell, G. B., D. R. W. Sutherland, R. F. N. Hutchins and P. E. Christmas, 1978. Vestitol: a phytoalexin with insect feeding-deterrent activity, J. Chem. Ecol., 4:571-579.

Seath I., 1977. Sensory feedback in the control of mouthpart movements in the desert locust Schistocerca gregaria, Physiol. Entomol., 2:147-156.

Schoonhoven, L. M., 1972a. Secondary plant substances and insects, Rec. Adv. Phytochem., 5:197-224.

Schoonhoven, L. M., 1972b. Plant recognition by lepidopterous larvae, in: "Symposia of the Royal Entomological Society", H. F. Van Emden, ed., pp. 87-99, Blackwell Scientific, Oxford.

Schoonhoven, L. M., 1981. Chemical mediators between plants and phytophagous insects, in: "Semiochemicals: their Role in Pest Control", D. A. Nordlund, R. L. Jones and W. J. Lewis, eds, pp. 33-50, John Wiley and Sons, Inc., New York.

Schoonhooven, L. M., 1982. Biological aspects of antifeedants, Entomol. Exp. Appl., 31:57-69.

Schoonhoven, L. M. 1986. What makes a caterpillar eat? The sensory code underlying feeding behavior, in: "Perspectives in Chemoreception and Behavior", E. A. Bernays and R. F. Chapman, eds., Springer-Verlag, in press.

Schoonhoven, L. M. and T. Jermy, 1977. A behavioral and electrophysiological analysis of insect feeding deterrents, in: "Crop protection agents", McFarlane ed., pp. 133-146. Academic Press, New York.

Shaaban, F. El-D. and W. R. Doskotch, 1980. Antifeedant diterpenes for the gypsy moth larva from Kalmia latifolia: isolation and characterization of ten grayanoids, J. Nat. Prod., 43:617-631.

Shimada, I., 1975. Chemical treatments of the labellar sugar receptor of the fleshfly, J. Insect. Physiol., 21:1565-1574.

Shimada, I., H. Horiki, O. Hiroshi and H. Meguro, 1985. Taste responses to 2,5-anhydro-D-hexitols; rigid stereospecificity of the furanose site in the sugar receptor of the flesh fly, J. Comp. Physiol., 157:477-482.

Shimada, I., A. Shirahashi, H. Kijima and H. Morita, 1974. Separation of two receptor sites in a single labellar sugar receptor of the flesh fly by treatment with p-chloromercuribenzoate, J. Insect Physiol., 20:605-621.

Shirahashi, A. and T. Yano, 1984. Neuronal control of the feeding behavior in the blow fly, in: "Animal behavior", K. Aoki, S. Ishi and H. Morita, eds., pp. 83-93. Japan Scientific Soc. Press, Tokyo.

Simmonds, M. J. J. and W. M. Blaney, 1983. Some neurophysiological effects of azadirachtin on lepidopterous larvae and their feeding response, Proceed. 2nd Int. Neem Conference, Rauischdzhausen, pp. 163-180.

Singh, R. P. and N. C. Pant, 1980. Lycorine - a resistance factor in plants of subfamily amaryllidoideae (Amaryllidaceae) against desert locust Schistocerca gregaria, Experientia, 36:525-553.

Smith, D. V. and J. B. Travers, 1979. A metric for breadth of tuning of gustatory neurons, Chemical Senses and Flavour, 4:215-229.

Stadler, E., 1984. Contact chemoreceptors, in: "Chemical Ecology of Insects", W. Bell and R. Carde eds., pp. 3-36, Sinauer Associates, Sunderland,.

Steinhardt, R. A., H. Morita and E. S. Hodgson, 1967. Mode of action of straight chain hydrocarbons on primary chemoreceptors of the blow fly, Phormia regina, J. Cell Physiol., 67:53-62.

VandeBerg, J. S., 1981. Ultrastructural and cytochemical parameters of chemical perception, in: "Perception of behavioral chemicals", D. M. Norris ed., pp. 105-131., Elsevier/North-Holland, New York.

van Drongelen, W., 1978. The significance of contact chemoreceptor sensitivity in the larval stage of different Yponomeuta species, Entomol. Exp. Appl., 24:143-147.

van Drongelen, W. and J. J. van Loon, 1980. Inheritance of gustatory sensitivity in F1 progeny of crosses between Yponomeuta cagnagellus and Y. malinellus, Entomol. Expt. Appl., 28:199-203.

van der Wolk, F. M., B. Menco, and H. van der Starre, 1984. Freeze-fracture characteristics of insect gustatory and olfactory sensilla. II. Cuticular features, J. Morph., 179:305-321.

Vogt, L. M., L. M. Riddiford and G. D. Prestwich, 1985. Kinetic properties of a sex pheromone-degrading enzyme: the sensillar esterase of Antheraea polyphemus, Proc. Nat. Acad. Sci., U.S.A., 82:8827-8831.

Wada, K. and K. Munakata, 1968. Naturally occurring insect control chemicals. Isoboldine, a feeding inhibitor and cocculodine, an insecticide in the leaves of Cocculus trilobus, J. Agr. Food Chem., 16:471-474.

Wada, K., Y. Enomoto and K. Munakata, 1970. Insect feeding inhibitors in plants II. Structures of shiromodiol diacetate, shiromool and shiromodiol monoacetate, Agric. Biol. Chem., 34:946-953.

Warthen, J. D. Jr., R. E. Redfern, E. C. Vebel and G. D. Mills, Jr., 1978. An antifeedant for fall armyworm larvae from neem seeds, Agric. Res. Results, 1:1-9.

Wieczorek, H., 1976. The glycoside receptor of the larvae of Mamestra brassicae (Lepidoptera: Noctuidae), J. Comp. Physiol., 106:153-176.

Wieczorek, H. and L. R. Koppl, 1982. Reaction spectra of sugar receptors in different taste hairs of the fly, J. Comp. Physiol., 149:207-213.

Wieczorek, H. and W. Gnatzy, 1985. The electrogenic pump of insect cuticular sensilla, Insect Biochem., 15:225-232.

Winstanley, C. and W. M. Blaney, 1978. Chemosensory mechanisms of locusts in relation to feeding, Entomol. Exp. Appl., 24:550-558.

Wolbarsht, M. L. and F. E. Hanson, 1965. Electrical activity in the chemoreceptors of the blowfly. III. Dendritic action potentials, J. Gen. Physiol., 48:673-683.

Woodhead, S., 1983. Surface chemistry of Sorghum bicolor and its importance in feeding by Locusta migratoria, Physiol. Entomol., 8:345-352.

Wright, D. P., Jr., 1967. Antifeedants, in: "Pest Control", W. Kilgore and R. Doutt, eds., pp. 287-293, Academic Press, New York.

Yajima, T. and K. Munakata, 1979. Phloroglucinol-type furanocoumarins: a group of potent naturally-occurring insect antifeedants, Agric. Biol. Chem., 43:1701-1706.

Zacharuk, R. Y., 1985. Antennae and sensilla, in: "Comprehensive Insect Physiology Biochemistry and Pharmacology", G. A. Kerkut and L. I. Gilbert, eds, Vol. 6, pp. 1-69, Pergamon Press, New York.

ALLELOCHEMICALS AND ALIMENTARY ECOLOGY: HETEROSIS IN A HYBRID ZONE?

J. Mark Scriber

Department of Entomology
University of Wisconsin
Madison, WI 53706

1. INTRODUCTION

Proper interpretation of differential survival, growth, and reproduction of phytophagous insects on various host plants depends on our ability to discriminate between a large number of plant characteristics, insect characteristics, and environmental factors that influence the pre-ingestive acceptability (Ahmad, 1983; Miller and Strickler, 1984) and post-ingestive suitability of food plants (Scriber and Slansky, 1981; Berenbaum, 1985, 1986). Research over the last several years has addressed these various concerns in considerable detail for leaf-chewing Lepidoptera and has made it more feasible to differentiate between environmental and/or food plant effects and heritable physiological adaptations (Scriber, 1983, 1984a; Rausher, 1984; Whitham et al., 1984; Slansky and Scriber, 1985; Mattson and Scriber, 1985).

This chapter will address the genetic and physiological bases of differential food plant use by North American tree-feeding Papilionidae species. Particular attention will be given to the potential influence of heterosis (hybrid vigor) between these taxa with regard to differential food utilization abilities, growth versus consumption rates and digestive efficiencies. While these results have implications in biogeographic distribution and in population ecology, the focus of this chapter is on the genetic control of insect physiological responses to different phytochemistry. The general lack of, but need for, studies of the importance of bioenergetics in evolutionary genetics has recently been reviewed by Watt (1985).

1.1. A case study of insect-plant interactions. The insects.

The North American tree-feeding swallowtail butterflies (*Papilio spp.* Lepidoptera: Papilionidae) offer a unique combination of research opportunities because of their geographic distribution, feeding specializations and restrictions, and mating compatibilities which permit hybridizations by hand-pairing. It is feasible that a simple genetic mechanism for detoxification may be responsible for the basic pattern of parapatric distribution of the North American *Papilio glaucus* L. taxa.

Our research on the interactions of the North American *Papilio glaucus* group and their food plants began in 1971, with the specific objectives of determining the physiological and/or ecological advantages of feeding specialization. An early study (Scriber and Feeny, 1979) of 20 species of Lepidoptera on a variety of host plant species revealed no consistent trend in the efficiency of biomass or nitrogen use for those species feeding on one plant family (arbitrarily called "specialists") versus those feeding on 2-10 families ("intermediates") or more than 10 families ("generalists"). The primary conclusion presented by Scriber and Feeny was that phytochemical factors such as nitrogen and leaf water content (Scriber, 1977) which are correlated with plant growth form (e.g. herb, shrub, tree) account for a major portion of the variation in larval efficiency and rate of growth (see also Mattson and Scriber, 1985).

Subsequently, with a better understanding of the plant growth form as a source of variability in larval growth responses (Scriber, 1984a), we have been assessing the genetic basis of geographic variability in food plant utilization abilities for a number of Lepidopteran groups. Many genetic growth response characteristics are important in the interpretation of biogeographic distribution patterns (Scriber, 1982a, 1983).

Within Section III of the Papilionidae (the glaucus and troilus groups; Munroe, 1960; cf. Hancock, 1983) it is tempting to assume that evolutionary resource partitioning of food plants via specialization at the plant family level has occurred (Brower, 1958; Scriber, 1984b). For example, we generally observe P. troilus L. on spicebush, Lindera benzoin, and P. palamedes Drury on red bay, Persea borborna (both in the Lauraceae); P. eurymedon Lucas on Rhamnaceae; P. multicaudatus Kirby on Rutaceae; the California P. rutulus Lucas on Plantanaceae; P. g. canadensis R & J and Rocky Mountain P. rutulus on Betulaceae and Salicaceae; and P. g. australis Maynard only on sweet bay, Magnolia virginiana, of the Magnoliaceae. This pattern of feeding specialization with a variety of mutually exclusive food plant utilization preferences and/or abilities is depicted in Table 1 and Fig. 1. Various foods are also used in particular locations, and some regional (local) specialization occurs even in the most polyphagous of the 563 species of Papilionidae, P. glaucus L. (Scriber and Evans, in preparation).

We observe excellent neonate larval survival of P. g. australis and P. g. glaucus on all of the Magnoliaceae tested, but poor survival on all of the Salicaceae tested, Table 1. Survival of P. g. canadensis, P. rutulus (from California), and P. eurymedon is greatest on their "favorite" families (i.e. Salicaceae, Plantanaceae, and Rhamnaceae, 80%, 93%, and 43% respectively) but very poor on the Magnoliaceae, Table 1. The Betulaceae, Rosaceae, and Rutaceae appear to be generally acceptable to all of the species in the P. glaucus complex. In addition to inter-family differences, significant interspecific differences in key phytochemicals (plant allelochemicals) within the Magnoliaceae family and within the Salicaceae family are suggested by differential performances of P. g. canadensis and P. rutulus across the plant species. For example P. g. canadensis does as well as P. g. glaucus and P. g. australis on mountain magnolia, Magnolia acuminata, Table 1; and P. rutulus appears unadapted to big-toothed aspen, Populus grandidentata (i.e. 0% survival; Table 1), a Great Lakes region plant species which is used as a host by P. g. canadensis (Scriber et al., 1982). Mortality of P. rutulus, P. eurymedon, and P. g. canadensis on the Magnoliaceae and mortality of P. g. glaucus and P. g. australis on the Salicaceae is generally characterized by larval feeding with some feces produced, regurgitation, and death within three days. These symptoms strongly suggest a toxic mechanism (not starvation). Essentially all of the six to seven percent of P. g. glaucus and P. g. australis surviving the first instar on quaking aspen, Populus tremuloides, died shortly into the second instar; very few survived to pupation.

Table 1. Percent first instar (neonate) survival in no-choice laboratory feeding tests (1979-1985).

Plant	No. of larvae	*Papilio* taxon: eurymedon (1)	rutulus (6)	g. canadensis (121)	g. glaucus* (207)	g. australis (20)	troilus (5)
Lauraceae:							
spicebush	686	0	0	1	19	19	70
Magnoliaceae:							
sweet bay	1677	?	0	0	64	79	0
tulip tree	2710	0	0	1	80	75	0
mt. magnolia	310	0	9	60	67	62	0
Salicaceae:							
quaking aspen	2648	72	74	80	7	6	0
cottonwood	308	?	70	33	2	0	0
bigtoothed aspen	268	?	0	48	8	0	0
balsam poplar	186	?	?	68	12	0	?
Plantanaceae:							
sycamore	282	?	93	37	34	18	?
Rhamnaceae:							
Rhamnus and Ceanothis spp.	140	45	0	9	0	0	0
Betulaceae:							
paper birch	199	25	67	73	35	27	0
Rutaceae:							
hop tree	174	?	67	70	67	36	0
Rosaceae:							
black cherry	7050	75	87	74	81	82	0

Relations indicated by a question mark have yet to be assessed. (n)indicates number of mothers. *Data from 1979-1984.

The general abilities of neonate larvae of P. g. glaucus and P. g. australis to survive on the Magnoliaceae, but not on the Salicaceae, and of P. g. canadensis and P. rutulus to survive on the Salicaceae, but not on the Magnoliaceae, appear to be genetically based. The reciprocal inabilities of these taxa to handle each other's favorite food plant is closely paralleled by other biological differences such as color polymorphisms (Scriber and Evans, unpublished) and voltinism/diapause differences (Hainze and Scriber, 1985; Hagen and Lederhouse, 1985). Furthermore, these abilities segregate relatively sharply along a phytogeographic zone of parapatry, e.g. as in the case of P. g. glaucus and P. g. canadensis, from central Minnesota and Wisconsin, across the Great Lakes region, down into the Appalachian Mountains, and north into central New England, Fig. 1 (Scriber, 1983 and current work in our lab). The degree of gene flow between these parapatric taxa has not been completely assessed, but natural hybrids are strongly suspected in zones of overlap (Scriber and Evans, unpublished; Scriber and Hainze, 1986; Hagen, pers. comm.; Luebke, pers. comm.).

Occasionally in this zone of parapatry (Scriber, 1975), and in essentially all (more than 90 percent) of our laboratory hybrids (Scriber, 1982), we observe excellent survival of hybrid individuals (from a single brood/mother) on both tulip tree, Liriodendron tulipfera, and quaking aspen, as well as on black cherry, Prunus serotina, Table 2. Black cherry serves as a general common denominator and therefore is useful as a control for assessment of F2 breakdown ("hybrid dysgenesis") and interpretation of backcross studies. P. eurymedon survival parallels that of P. rutulus on these three food plants (see also Table 1) and the various crosses also reflect a virtually identical parallel pattern (e.g. P. g. glaucus x P. eurymedon hybrids exhibit 64 percent survival on tulip tree). Also, P. eurymedon hybrids with P. rutulus and with P. g. canadensis

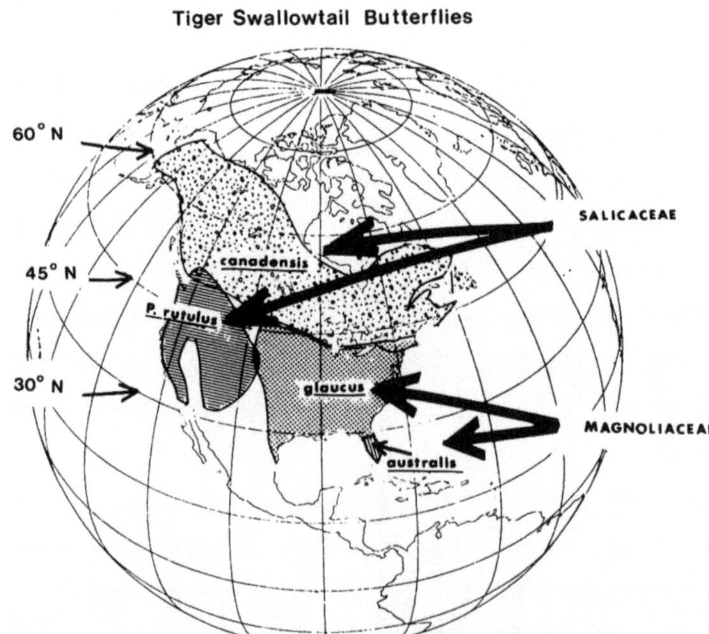

Fig. 1. The geographic distribution and favorite food plant families of the three putative subspecies of the eastern tiger swallowtail butterfly, Papilio glaucus and the western P. rutulus.

Table 2. Percent first instar survival of *Papilio* spp. and their primary hybrids.

Parents and hybrids	No. of mothers	Hypothetical genotype	Food plant: Quaking aspen	Food plant: Black cherry	Food plant: Tulip tree	Total no. of larvae
P. rutulus (r)	6	QQtt	68 (38)	87 (30)	0 (10)	(78)
P. g. canadensis (c)	206	QQtt	79 (358)	74 (2872)	1 (420)	(3650)
cxr	11	QQtt	69 (52)	66 (100)	0 (46)	(198)
gxr	16	QqTt	66 (190)	79 (434)	65 (234)	(858)
axc	4	QqTt	94 (46)	76 (109)	66 (92)	(247)
cxa	1	QqTt	-	-	100 (2)	(2)
cxg	10	QqTt	78 (88)	84 (117)	76 (129)	(362)
gxc	91	QqTt	61 (690)	80 (3020)	75 (885)	(4595)
axr	9	QqTt	50 (96)	69 (129)	84 (147)	(372)
gxa	2	qqTT	11 (9)	47 (15)	71 (7)	(31)
axg	2	qqTT	0 (8)	64 (11)	90 (10)	(29)
P. g. australis (a)	31	qqTT	6 (124)	83 (304)	75 (276)	(704)
P. g. glaucus (g)	207	qqTT	7 (2101)	80 (4564)	80 (2085)	(8750)

Of the possible pairings, only *P. rutulus* (females) x *canadensis*, *rutulus* x *glaucus*, and *rutulus* x *australis* have not produced viable offspring. Figures in parentheses are total number of larvae of each species of hybrid.

exhibit absolutely no ability to use tulip tree leaves (survival is zero percent in all cases as it is with pure P. eurymedon stock). Conversely, P. eurymedon can survive well on quaking aspen and this ability is transferred to hybrid larvae (e.g. glaucus x eurymedon).

The following studies were conducted in order to determine whether this remarkable ability to feed on both plant species, one from the Magnolicaceae, the other from the Salicaceae, in our hybrids was due to "heterosis" or simply to "biochemical intermediacy" in which they perform somewhere between the two putatively homozygous parental types (Berger, 1976; Plapp and Wang, 1983; Mitton and Grant, 1984). We were also particularly interested in elucidating the specific physiological mechanisms involved in the successful survival and subsequent growth of such F1 hybrids, F2 hybrids, and backcrossed individuals and in evaluating the genetic influence of these responses.

Very little is known about the molecular and physiological mechanisms of heterosis in plants or animals. However, it has been estimated that 70-80 percent of the effects on growth and developmental stability can be attributed to heterozygosity (Mitton and Grant, 1984). These authors review the literature and describe a number of positive correlations with protein heterozygosity and growth rates for marine invertebrates, salamanders, mammals, quaking aspen, and several conifer species. They emphasize, however, that the mechanisms underlying the phenomenon remain generally unknown. The enzymes used in such studies are usually convenient chromosomal markers rather than important proteins directly affecting metabolic rates. In a rare and valuable ecological observation, Collins (1984) dramatically illustrates a natural case in which faster growth of hybrid larvae in a Hyalophora euryalis Boisduval (Lepidoptera: Saturniidae) - H. gloveri Strecker (Ibid.) hybrid zone conveys a direct selective advantage during years of early winters. The direct physiological effects of allozyme polymorphisms on survival, growth or reproduction are very rarely known (but see Burton and Feldman, 1983; Nei and Koehn, 1983). Therefore the molecular and biochemical mechanisms of heterosis are difficult to identify.

Heterosis may in part arise from combinations of alleles for active but relatively unstable enzyme forms with alleles that specify stable but inactive enzymes (Schwartz and Laugner, 1969; Watt, 1977). This or other unspecified mechanisms could result in a greater catalytic efficiency of enzymes in the heterozygotes (Berger, 1976). As a result, a reduced amount of metabolic energy would be required for sustained activity, leaving more for growth and reproduction. This theory holds that the selective value of heterozygosity at any particular locus might be generally quantified in terms of the energy rechanneling potential or "efficiency" of food processing.

1.2. Estimation of growth characteristics.

The following study addresses the genetic basis of the physiological phenomena involved in both the survival and growth of hybrids and backcrossed larvae on food plants with known and different antibiosis effects. Growth rates and efficiencies can be measured gravimetrically to provide insight into the physiological mechanisms responsible for observed growth rates (Waldbauer, 1968; Scriber and Slansky, 1981). A larval relative growth rate (R.G.R. = mg of dry weight gain per mg of larval tissue and day) is the product of the relative consumption rate (R.C.R. = mg dry weight of ingested leaf tissue per mg larval tissue and day), approximate digestibility (A.D. = mg food assimilated per mg food ingested), and the efficiency of conversion of digested food (E.C.D. = mg larval biomass gained per mg food digested). In formula form this is expressed as:

$$RGR = RCR \times AD \times ECD$$

or in terms of actual measurements:

$$\frac{B}{\bar{B}T} = \frac{I}{\bar{B}T} \times \frac{I-F}{I} \times \frac{B}{I-F}$$

where

B = larval weight gain,
$\bar{B}$ = mean larval weight, measured as initial weight plus final weight divided by two,
I = ingested biomass,
F = egested biomass, and
T = days

It should be pointed out that these studies are preliminary. While some of the "physiological phenotypes" can be discerned, the identification of their underlying "biochemical phenotypes" and actual genotypes (see Koehn et al., 1983; Hilbish and Koehn, 1985) must await additional studies. Cytochrome P-450 and other detoxifying enzyme activities need to be examined in these and other studies of feeding abilities to evaluate their ecological significance (Brattsten, 1979; Gould, 1984).

1.3. The plant allelochemicals.

Our knowledge of the phytochemical mechanisms involved in this differential antixenosis/antibiosis to *Papilio* is also only in its infancy. In the last two years we have investigated the phytochemical basis of these differential utilization abilities for larvae on the Salicaceae and Magnoliaceae. Several compounds have been isolated from leaves of both quaking aspen and tulip tree that exhibit biological activity against the highly polyphagous southern armyworm, *Spodoptera eridania* (Cramer) (Lepidoptera: Noctuidae) larvae (Sunarjo and Hsia, 1984; Manuwoto et al., 1985). While salicin was extracted in high concentrations from quaking aspen, the most active antifeedant extracted, isolated, and identified in our studies to date was pyrocatechol (1,2-benzenediol). Similar feeding preference tests have been conducted with penultimate and final instars of *P. g. glaucus* and suggest that pyrocatechol is indeed at least one of the important compounds involved in the non-usage of Salicaceae by larvae of this subspecies. For example, several different leaf surface concentrations from 24 $\mu g/cm^2$ to 196 $\mu g/cm^2$ all significantly altered preference and feeding rates, Table 3. On the basis of related research, we expect that additional plant phenolics may be of particular significance to *Papilio* larvae fed quaking aspen (Zucker, 1982; Palo, 1984).

Phenolics are the only major class of secondary plant compounds reported to occur in the Salicaceae to date (Pearl and Darling, 1968; Thieme and Benecke, 1971; Bryant, 1981; Lechowicz, 1983; Palo, 1984, and pers. comm.). Examples of potentially significant phenolics present in the Salicaceae, including quaking aspen, are listed in Palo (1984). The composition of secondary compounds in tulip trees and other Magnolicaceae is highly diverse, including sesquiterpene lactones (Doskotch et al., 1972, 1975, 1977b, 1980), benzylisoquinoline alkaloids, cyanogenic glycosides, and various essential oils (Taylor, 1962; Santamour and Treese, 1971; Miller and Feeny, 1983; D.S. Seigler and R.L. Lindroth, 1984, pers. comm.).

2. HYBRIDIZATION STUDIES AND EXPERIMENTAL APPROACH

We have developed methods for hand-pairing of adults and mass-rearing

Table 3. Preliminary leaf disc (black cherry) feeding preference[1] studies using fifth instar larvae of *Papilio g. glaucus* and various surface concentrations of 1,2-benzenediol (pyrocatechol) extracted from quaking aspen leaves. Data are mean ± s.d. based on experiments with six larvae per concentration (Madison, WI, 1984).

Pyrocatechol concentration on leaf surface ($\mu g/cm^2$)	Amount of leaf disc consumed (cm^2) treated (with extract)	control (solvent only)
24	0.32 ± 0.39	1.21 ± 0.75*
48	0.26 ± 0.25	1.51 ± 1.07*
96	0.60 ± 1.08	2.19 ± 1.47*
192	0.36 ± 0.43	1.62 ± 0.92*

*Indicates significant differences at the $P < 0.10$ level (paired t-test).

1 Four discs, two treated and two control were arranged in a petri dish with controls opposite each other and treated opposite each other such that larvae would encounter a different treatment to either side of its current feeding site (C-T-C-T).

2 An even distribution across the leaf disc surface was accomplished by using a calibrated microliter syringe to apply 3 µl (microliters) of acetone (control) or pyrocatechol in acetone dilutions on both the upper and lower leaf surfaces (Hainze and Sunarjo, unpublished).

of larvae in the laboratory that are dependable and repeatable. This has allowed us in the last four years to conduct inter-specific and inter-subspecific hybridization studies as well as inter-population analyses of differential food plant utilization abilities. While we have observed interesting variations in response to dozens of different food plant species (Scriber, work in progress), three key food plants provide an excellent system with which to investigate the genetic and physiological basis of differential survival and growth: tulip tree, quaking aspen, and black cherry.

Acute responses, i.e. death of the neonate larvae within the first few days, preclude assessment of the chronic physiological effects that might otherwise manifest themselves throughout subsequent larval stages. Two different experimental approaches help resolve this apparent dilemma. First, the widely acceptable and highly suitable food plant black cherry was used concurrently in all survival bioassays and feeding experiments as a control, i.e. as a "barometer" of hybrid breakdown, Fig. 2. For example, survival of F2 and backcross larvae fed either quaking aspen or tulip tree is significantly poorer than for those fed black cherry, Table 4. This suggests that it may, in fact, be differential adaptation to the phytochemical composition of aspen which is the source of the variation, rather than general "hybrid breakdown" in this particular case. The second approach, was to rear backcross larvae of all hypothetical genotypes, Table 5, on the acceptable black cherry leaves through the first three instars and then switch subsets of these larvae onto quaking aspen or tulip tree. In this manner it was possible to bypass the extremely susceptible neonate larval stage for the putative homozygous recessives. We could thus assess the differential physiological and growth abilities of these genotypes as

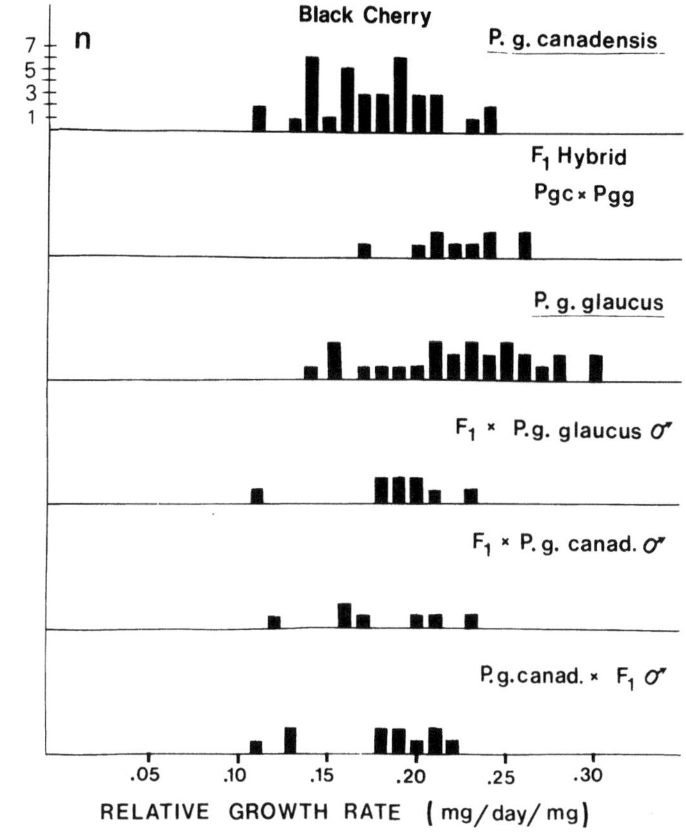

Fig. 2. The relative growth rates of penultimate instar individual larvae of *P. g. canadensis* (n = 31), F1 hybrids (10), *P. g. glaucus* (n = 29), and three backcrosses (n = 9, n = 7, n = 11) when fed black cherry in no-choice feeding experiments. The Y-axis is number of individuals (1 square = 1 individual). Siblings of these six phenotypes were also used in experiments on tulip tree, Fig. 3, and quaking aspen, Fig. 4.

well as those of the putative heterozygotes (hybrids and some backcross larvae) and homozygous dominants on each test plant.

We reasoned that this method would not only allow determination of whether the differential utilization abilities are restricted to the neonate stages, but would also resolve whether or not the heterozygous individuals (e.g. F1 hybrids and some backcrosses) grow more (or less) efficiently than the homozygous dominants. A complication here may be involvement of detoxifying enzymes which could differ, e.g. in terms of inducibility, in neonate and mature larvae (Ahmad et al., Chapter 3 in this text). It would be interesting to determine whether there is a genic dose-effect or gene amplification effect with a heterozygote less vigorous than a homozygous dominant, or whether there is a heterosis effect, with the dual-enzyme heterozygote more vigorous or efficient in its consumption and assimilation of plant tissue for larval growth than the homozygote (for additional discussion see Schwartz and Laughner, 1969; Berger, 1976; or Mitton and Grant, 1984).

The study was conducted with a subset of the various backcross larvae which survived the first three instars on tulip tree (thus eliminating homozygous recessives). This approach allowed a valid reference group, i.e., their siblings were switched from black cherry onto tulip tree at

Table 4. Preliminary results of neonate F2 hybrid larval survival on three different food plants.

Phenotype[1] (F1 x F1 = F2)	Number of fertile females	Percent eclosion (# larvae eclosed/ # ova produced)	Percent survival (# through 2nd instar/initial #) Quaking aspen	Tulip tree	Black cherry
(Pgc x Pgg)[2]	5	104/209 = 49.8	18/44 = 40.9	27/56 = 48.2	46/70 = 65.7
(Pgg x Pgc)[2]	7	479/665 = 72.0	19/131 = 14.5	77/98 = 78.6	198/261 = 75.9
(Pga x Pgc)[2]	2	123/280 = 43.9	2/28 = 7.1	5/27 = 18.5	18/38 = 47.4

[1] Pgg = *P. g. glaucus*; Pgc = *P. g. canadensis*; Pga = *P. g. australis*. The first listed is the female of the F1 cross.

Table 5. Relative fitness of nine possible P. glaucus genotypes on quaking aspen and tulip tree.

Putative genotype*	Fitness on Quaking aspen	Fitness on Tulip-tree
QQ TT	High	High
QQ Tt	High	High (?)
QQ tt	High	Low
Qq TT	High (?)	High
Qq Tt	High (?)	High (?)
Qq tt	High (?)	Low
qq TT	Low	High
qq Tt	Low	High (?)
qq tt	Low	Low

* Assumptions: Loci Q and T act additively; alleles Q and T are dominant to q and t respectively. Homozygous recessives qq and tt are unable to use quaking aspen or tulip tree, respectively (see Table 1).

(?) Question marks indicate that if alleles are not dominant at either locus, we may observe moderate survival percentages or "intermediate" growth performance, perhaps a heterosis effect.

the fourth (penultimate) instar. Subsequent extensive fourth and fifth instar gravimetric feeding experiments were done concurrently with both groups. Since the homozygous recessives were selected out before the fourth instar on their nonhost plant, in this case the tulip tree, it should be possible to determine if larval growth performance of the survivors follows either hypothesized genotype (see above and Table 4). In contrast, the cherry-reared "switched" cohorts might have three separate classes of growth performance corresponding to genotypes in either a non-dominant or dominant scheme, Table 4.

3. RESULTS

Growth performances (R.G.R.s) of individual larvae from three different backcross families on each of the three different food plants are represented with respect to F1 hybrids of P. g. glaucus and P. g. canadensis in the histograms in Figures 2-4. While sample sizes are much too small to be of use in quantitative genetic analyses, it is apparent that backcross larvae maintained on black cherry and assayed during the fourth instar grow at rates (0.11-0.23 mg/mg,day) intermediate between the parental-type larvae, Fig. 2. A significant portion of those P. g. canadensis backcross larvae switched to the tulip tree test plant, Fig. 3, were severely suppressed in growth rates (less than 0.09 mg/mg,day). These results, and the fact that the P. g. glaucus backcross larvae did not suffer severe growth suppression when switched from cherry to the test plant, are consistent with the working hypothesis of a simple genetic mechanism such as homozygous recessive lethals. However, the small sample sizes do not enable us to distinguish between this and the possibility of multiple factor (polygenic) control. The complete data for Fig. 3 are in Appendix 1.

In contrast, genetic control for growth on quaking aspen appears to be more complex, and hybrid vigor might be invoked as an explanation of the F1 hybrid larval growth rates, Fig. 4. Surprisingly, the poorest larval growth (less than 0.09 mg/mg,day) was observed for larvae of a P. g. canadensis backcross as well as the predicted P. g. glaucus backcross. The

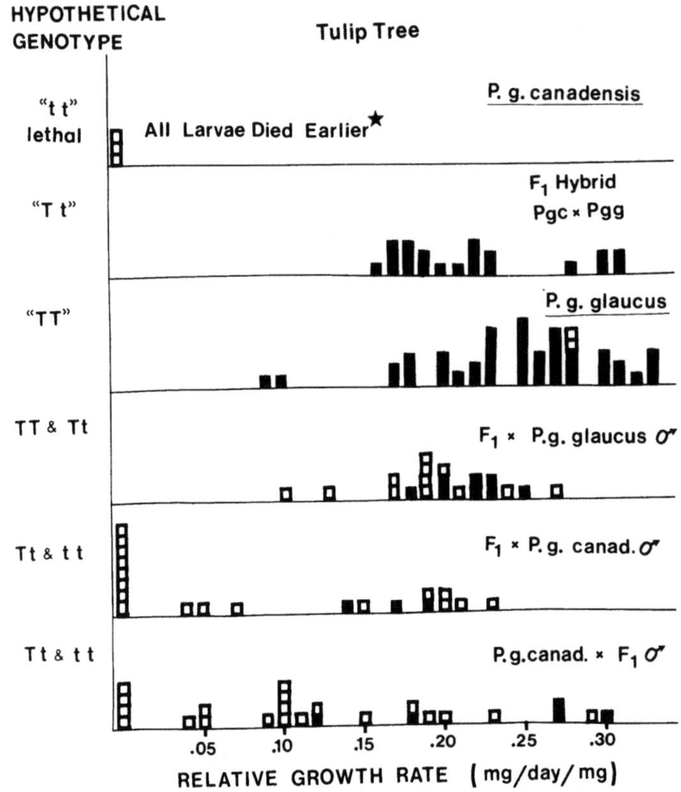

Fig. 3. The relative growth rates of penultimate instar individual larvae of P. g. canadensis, P. g. glaucus (n = 46), F1 hybrids (n = 21), and backcrosses on tulip tree. Most neonate larvae (*) of P. g. canadensis fail to survive the first instar on tulip tree, Table 2. Open squares signify penultimate instar larvae reared on black cherry and switched to tulip tree at the fourth instar. Solid squares designate larvae reared on tulip tree throughout, see also Table 6.

explanation for this is uncertain. The effect of switching all putative genotypes from cherry to aspen (open squares in Fig. 4) was generally more "stressful" compared to rearing genotypes through entirely on aspen (solid squares in Fig. 4). A similar but milder effect was observed for the switch from cherry to tulip tree (solid squares in Fig. 3). The complete data base for Fig. 4 is in Appendix 2. Such differences are within the range of performance responses known for other Lepidoptera (Scriber 1982b; Grabstein and Scriber, 1982b), but it remains unknown if these effects are due to induction of digestive and/or detoxifying enzymes, or whether they are due to homeostatic responses to behavioral antixenosis (Grabstein and Scriber, 1982a).

On the basis of the data for penultimate instar performances in Figures 2-4 and the data in Tables 1 and 2, a working hypothesis is constructed, Fig. 5. Backcross studies were used to evaluate the genetic mechanisms, Table 6, based on these hypothesized genotypes. Suppressed survival is observed as expected on tulip tree for the P. g. canadensis backcross and on quaking aspen for the P. g. glaucus backcrosses (with

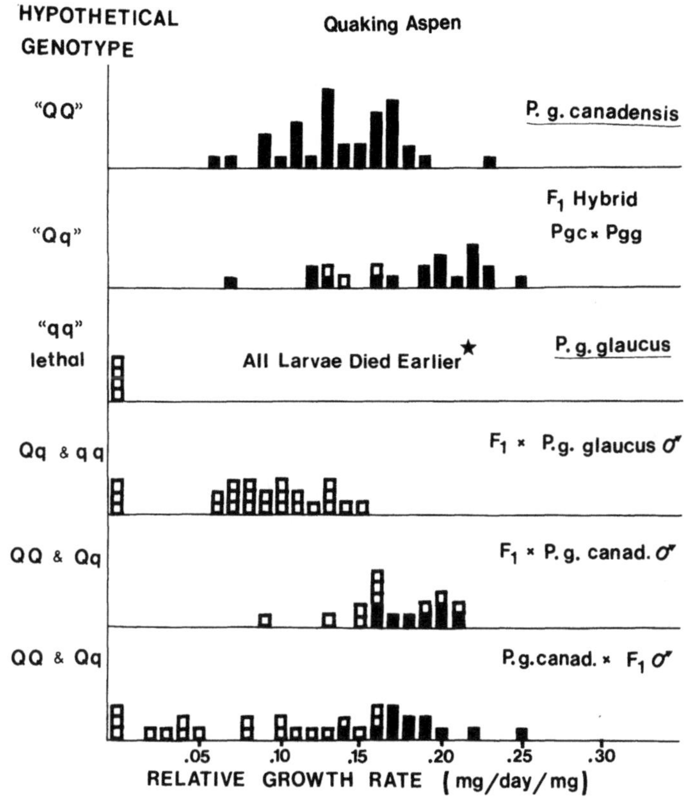

Fig. 4. The relative growth rates of penultimate instar individual larvae of P. g. canadensis (n = 37), P. g. glaucus, F hybrids (n = 22), and backcrosses on quaking aspen. Most neonate larvae (*) of P. g. glaucus fail to survive the first instar on quaking aspen, Table 2. Open squares signify larvae reared on black cherry and switched to quaking aspen at the fourth instar. Solid squares designate larvae reared on quaking aspen throughout, see also Table 6.

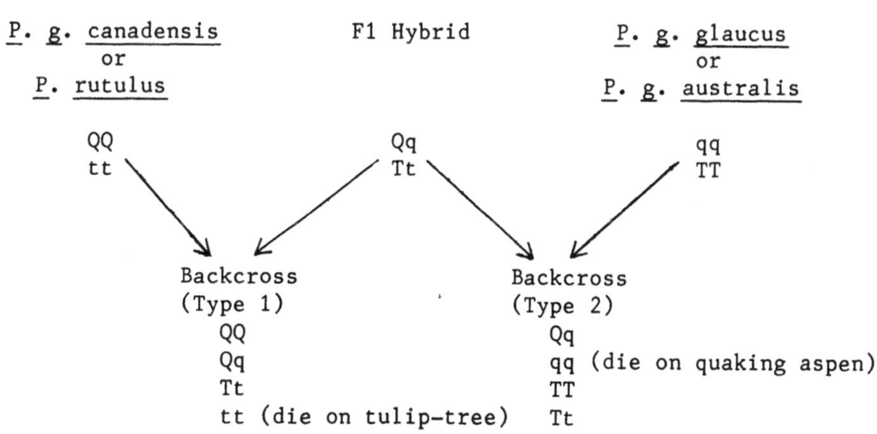

Fig. 5. Hypothesized genotypes of reciprocal backcross progeny and their first instar survival on tulip tree and quaking aspen (under the assumption of dominance). All genotypes are compared to larval performance on black cherry as a mutually acceptable control, Table 1.

Table 6. Percent survival of first instar hybrid backcross progeny (preliminary results, see Fig. 5).*

Reciprocal backcross	Fertile females	First Instar Survival (percent) Quaking aspen	Tulip tree	Black cherry
Type 1. (F1 with P. g. glaucus)				
F1 x canadensis	20	209/330 = 63.3	94/521 = 18.0	621/748 = 83.0
canadensis x F1	1	16/16 = 100	16/42 = 38.1	59/65 = 90.8
Total larvae		346	563	813
Type 2. (F1 with P. g. canadensis)				
F1 x glaucus	24	163/281 = 58.0	260/300 = 86.7	1206/1705 = 70.7
glaucus x F1	21	89/341 = 26.1	312/358 = 87.2	610/815 = 74.8
Total larvae		622	658	2520
Type 2. (F1 with P. rutulus)				
P. g. glaucus x F1	7	61/142 = 43.0	30/109 = 27.5	392/462 = 84.8

* Hybrid females between rutulus and glaucus or rutulus and australis, have never emerged in all of our mass-rearings to date; the female of the cross is listed first.

hybrid canadensis and rutulus). These results further suggest a definite genetic control as indicated with F2 crosses, Table 4, however the elucidation of simple versus polygenic aspects await additional studies.

4. DISCUSSION

We have observed a sharp differential in the ability of individual larvae of the North American tiger swallowtail butterfly group to survive and grow on plants in the families Magnoliaceae and Salicaceae. These differences are most pronounced with two plant species, namely, tulip tree and quaking aspen, Table 1. The western and northern taxa, P. rutulus and P. canadensis, respectively, lack the genetic or physiological ability to survive and grow on tulip tree, while the eastern taxa, P. g. glaucus and P. g. australis, lack the ability to survive and grow on quaking aspen. Laboratory-produced hybrids between these taxa, i.e. P. g. c. or P. r. x P. g. g. or P. g. a., are generally able to survive and grow very well on both of the plant species, Table 2.

It is unlikely that these broadened host utilization abilities in hybrid neonate larvae is simply due to heterosis because intrapopulation pairings and pairings of various populations of P. g. glaucus with Florida populations of the putative P. g. australis convey no improvement in survival on quaking aspen. On the other hand, either glaucus or australis crossed with either rutulus or canadensis result in improved survival on this plant, Table 2. Similarly, crosses between P. rutulus and P. g. canadensis do not improve survival on tulip tree or sweet bay, whereas crosses of either of these to either P. g. glaucus or P. g. australis do, Table 2.

Biochemical intermediacy is observed in the growth rates of penultimate instar hybrids on black cherry, Fig. 2 (see also Ritland and Scriber, 1985) and tulip tree, Fig. 3. However, heterosis cannot be ruled out for F1 hybrids on quaking aspen, Fig. 4, where the mean growth rate of hybrids is 0.186 ± 0.014 mg/day,mg compared to 0.133 ± 0.007 for P. g. canadensis. The physiological mechanism responsible for the heterotic growth is not increased consumption rates but is due to increased food utilization efficiencies.

Of the three possible contributing indices responsible for the suppressed growth, the most significant deleterious effects were evident in the drastically low E.C.D.s (compare Slansky and Scriber, 1985) which reflect tremendous metabolic (respiratory) costs for those larvae, Appendix 1 and 2. Consumption rates and digestibilities remain generally high, even for the drastically suppressed growth rates. Interpretations of whether these less than subtle "switching effects" on larval growth performance are behavioral or physiological in nature must be made with caution (see Schoonhoven and Meerman, 1978; Scriber, 1979; 1981, 1982b; Grabstein and Scriber, 1982a and 1982b). We tentatively conclude that significant differences do occur in metabolic (respiratory and/or detoxification) costs of the backcross genotypes, and this is the apparent cause of the slow growth and/or death. Much larger sample sizes are required to determine the number of alleles or loci controlling these differential detoxification or processing abilities (see Lande, 1982; Leslie and Dingle, 1983). Unfortunately, the mode of inheritance of behavioral and physiological adaptations of "biotypes" is poorly known for insects in general due to a lack of backcross and F2 studies (Futuyma and Peterson, 1985).

Natural hybridization is believed to occur across the area of parapatry for P. g. glaucus and P. g. canadensis based on evidence from diapause studies (in progress in our laboratory), electrophoresis (Hagen and

Lederhouse, 1985; Scriber and Collins, unpublished), multivariate morphometric discriminant analyses with known hybrid reference groups (Luebke, pers. comm.; Scriber unpublished), and studies of the melanic (mimetic) polymorphism in females (Scriber and Evans, in preparation). Some of the natural food plant utilization abilities reported here and elsewhere (Scriber, 1975 and unpublished; Hagen, pers. comm.) also suggest that introgression between P. g. canadensis and P. g. glaucus may explain the ability of some females to produce larvae able to survive and grow on both tulip tree and quaking aspen. The results presented here cannot justify the speculation that heterosis in the hybrid zone may be a selective advantage to Papilio as was the case for Hyalophora (Collins, 1984) and Drosophila (Nagle and Mettler, 1969), although we do know that it would broaden the range of acceptable/suitable host plants and could facilitate host shifts (Gould et al., 1982; Tabashnik, 1983; Diehl and Bush, 1984) and geographic range expansions (Scriber and Hainze, 1986).

At the biochemical level, we are uncertain of the phytochemical mechanisms involved in the observed differential antixenotic/antibiotic responses of Papilio taxa. We have observed evidence of different phytochemistry between plant species within each of the two key families (Salicaceae and Magnoliaceae) and have not ruled out different phytochemistry among individuals within a plant species (see Whitham et al., 1984) or seasonal/diurnal fluctuations (Scriber and Slansky, 1981; Schultz et al., 1982; Mattson and Scriber, 1985). Purified bioactive plant extracts have been identified and used with our system (e.g. pyrocatechol, Table 3). However a mixture of active ingredients is likely to be involved in the toxic component which kills the neonate larvae (Manuwoto et al., 1985; Sunarjo, 1985; R. Lindroth, unpublished).

In general, resistance to insecticides in insects may be due to a combination of changes in structural genes, i.e. those specifying the protein structure of the enzymes, and changes in regulatory genes or loci associated with the structural genes, i.e. changes affecting their expression (Plapp and Wang, 1983). If the change in F1 hybrids was due only to a structural change, we might expect the enzyme activity to be different and show no change in inducibility. On the other hand, if the change in F1 hybrids was associated with a separate regulatory gene, we might expect synthesis of increased or decreased amounts of detoxifying enzymes. Furthermore, it is frequently the case that mutations in regulatory genes in the F1 are either dominant or recessive instead of intermediate, as in structural genes, resulting in differences in inducibility (Plapp and Wang, 1983). While our F1 neonate utilization abilities may favor the regulatory gene hypothesis, our larval growth rate observations seem likely to involve both structural and regulatory gene control. The activities of detoxifying enzymes in hybrids compared to parent species may in part explain the different food plant utilization abilities observed. This deserves additional attention.

We observed little reduction of viability on the black cherry controls in our F2 crosses, Table 2, which suggests that selectively advantageous intermediate or heterotic effects are feasible in F1 hybrids. We are likely not observing any severe breaking apart of coadapted gene combinations as reported by Oliver (1979) and Dingle et al. (1982). The short term advantages of heterosis or heterozygosity must be weighed against the other ecological advantages and selective pressures that depend upon the entire suite of life history characteristics and co-adapted gene complexes (Barton and Hewitt, 1981). The relative importance of individual heterozygosity and population heterogeneity in hybrid blend zones such as in Wisconsin warrants additional research.

Determination of the specific chemical basis and genetic control of these observed differences in food plant utilization abilities will contribute to our understanding of the ecological processes involved in host race formation, coadaptation, and coevolution between insect herbivores and plants. Moreover, such studies will enable us to better understand the physiological mechanisms underlying the processes involved in differential utilization abilities or detoxification abilities at both the interpopulation and interspecific level. It is only with such understanding of variation at several levels of biological organization (from the biochemical to the ecological) that we will be able to appropriately distinguish between the various categories of "biotype", i.e. environmentally induced polyphenisms, polymorphic/polygenic variation, geographic races, host races, or species (Futuyma, 1983; Diehl and Bush, 1984). Until we can make these distinctions, advances in our understanding of the importance of molecular mechanisms underlying evolutionary processes will also be hindered.

5. SUMMARY

More than 15,000 neonate larvae of the P. glaucus species complex of butterflies have been bioassayed on various food plants from across North America. A general reciprocal inability to survive on each other's favorite food plant family was observed for several of the taxa. For example, the northern P. g. canadensis and the western P. rutulus grow well on quaking aspen but not on tulip tree. The reverse is true for the two putative subspecies in the eastern United States, P. g. glaucus and P. g. australis. Our hybridization studies indicate that these differences at least to some extent rely on a genetic component since hybrids between either P. rutulus or P. g. canadensis and either P. g. glaucus or P. g. australis are able to survive and grow very well on both the Salicaceae and Magnoliaceae.

While heterosis may be suggested by some hybrid growth rates, it is not the mechanism responsible for the high survival of all neonate F1 hybrid larvae. This is suggested by the observations that survival on tulip tree is not enhanced by intra-population or inter-population crosses of either P. rutulus or P. g. canadensis and by similar observations that survival on quaking aspen is not enhanced by intra- or inter-populaton crosses of either P. g. glaucus or P. g. australis. In the F2 crosses, it appears that coadapted gene complexes are not being severely disrupted since neonate survival is basically comparable to the primary (F1) hybrids and backcrosses in the black cherry controls.

The larval growth rates of the various putative genotypes of F1 and backcross hybrids on tulip tree, quaking aspen, and black cherry (the control) were studied gravimetrically in an experimental "switching" design in order to determine the particular physiological mechanism involved in the antibiosis. The individuals segregating with greatly reduced penultimate instar growth rates generally suffered significantly increased metabolic rates reflected in very low, i.e. decreased, efficiencies of conversion of digested food (E.C.D.s). The relative consumption rates (R.C.R.s) and assimilation efficiencies (A.D.s) were generally not suppressed in these cases. These extremely large respiratory (energy) expenditures may reflect many metabolic disruptions, perhaps including induction of detoxifying enzymes.

The enhanced growth rates and efficiencies of the F1 hybrids on quaking aspen relative to larvae of either parental population may reflect the results of molecular heterosis. Physiological systems involved in digestion, assimilation, and metabolism for larval growth may benefit from a unique combination of alleles specifying both active and stable enzymes, or from a greater catalytic efficiency requiring a reduced amount of metabolic energy for a given activity level. On the other hand, biochemical intermediacy is suggested in larval growth and efficiencies of hybrids on tulip tree and black cherry.

6. ACKNOWLEDGEMENT

This research was supported in part by grants from the National Science Foundation (DEB7921749; BSRS306060; BSR8503464), the USDA-SEA (85-CRCR-11598), the Graduate School and College of Agriculture and Life Sciences at the University of Wisconsin, Madison (Hatch 5134). I thank Mark Evans for his enthusiastic and valuable assistance throughout these studies. May Berenbaum, Lena Brattsten, Michael Collins, Hugh Dingle, Paul Feeny, John Hainze, Stephen Hsia, Syafrida Manuwoto, Richard Lindroth, Bill Plapp, and Puis Sunarjo have had a significant impact upon the direction of research with either their ideas or assistance.

7. REFERENCES

Ahmad, S. ed., 1983. "Herbivorous Insects: Host-Seeking Behavior and Mechanisms", Academic Press, New York. 257 pp.

Barton, N. H., and G. M. Hewitt, 1981. Hybrid zones and speciation, in: "Evolution and Speciation", W. R. Atchley and D. S. Woodruff, eds., pp. 109-145, Cambridge Univ. Press, Oxford.

Berenbaum, M. R., 1985. Brementown revisited: Interactions among allelochemicals in plants, Rec. Adv. Phytochem., 19:139-169.

Berenbaum, M. R., 1986. Post-ingestive effects of phytochemicals on insects: On Paracelsus and plant products, in: "Insect-Plant Interactions", T. A. Miller and J. R. Miller eds., Chapter 5, Springer Verlag, N.Y.

Berger, E., 1976. Heterosis and the maintenance of enzyme polymorphism, Am. Nat., 110:823-839.

Brattsten, L. B., 1979. The ecological significance of mixed function oxidations, Drug Metabolism Reviews, 10:35-58.

Brower, L. P., 1958. Larval foodplant specificity in butterflies of the Papilio glaucus group, Lepid. News, 12:103-114.

Bryant, J. P., 1981. Phytochemical deterrence of snowshoe hare browsing by adventitious shoots of four Alaskan trees, Science, 213:889-890.

Burton, R. S., and M. W. Feldman, 1983. Physiological effects of an allozyme polymorphism: Glutamate-pyruvate transaminase and response to hyperosmotic stress in the copepod, Tigriopus californicus, Biochem. Genet., 21:239-251.

Collins, M. M., 1984. Genetics and ecology of a hybrid zone in Hyalophora (Lepidoptera: Saturniidae), Univ. of California Publications in Entomology, Vol. 104, Univ. California Press, Berkeley.

Denno, R. F., and M. S. McClure, eds., 1983. "Variable Plants and Herbivores in Natural and Managed Systems", Academic Press, New York. 717 pp.

Diehl, S. R., and G. L. Bush, 1984. An evolutionary and applied perspective of insect biotypes, Annu. Rev. Entomol., 29:471-504.

Dingle, H., W. S. Blau, C. K. Brown, and J. P. Hegmann, 1982. Population crosses and the genetic structure of milkweed long life histories, in: "Evolution and Genetics of Life Histories", H. Dingle and J. P. Hegman, eds., pp. 209-229, Springer Verlag, New York.

Doskotch, R. W., S. L. Keely, and C. D. Hufford, 1972. Lipiferolide, a cytotoxic germacranolide, and γ-liriodenolide, two new sesquiterpene lactones from Liriodendron tulipifera, J.C.S. Chem. Commun., 1137.

Doskotch, R. W., S. L. Keely, C. D. Hufford, and F. S. El-Feraly, 1975. New sesquiterpene lactones from Liriodendron tulipifera, Phytochem. 14:769-773.

Doskotch, R. W., T. M. Odell and P. A. Godwin, 1977a. Feeding responses of gypsy moth larvae, Lymantria dispar, to extracts of plant leaves, Environ. Entomol., 6:563-566.

Doskotch, R. W., F. S. El-Feraly, E. H. Fairchild, and C. T. Huang, 1977b. Isolation and characterization of peroxyferolide, a hydroperoxy sesquiterpene lactone from Liriodendron tulipifera, J. Org. Chem., 42:3614-3618.

Doskotch, R. W., E. H. Fairchild, C. T. Huang, J. H. Wilton, M. A. Beno, and G. G. Christoph, 1980. Tulirinol, an antifeedant lactone for the gypsy moth larvae from Liriodendron tulipifera, J. Org. Chem., 45:1441-1446.

Fox, L. R., and R. A. Morrow, 1981. Specialization: Species property or local phenomenon? Science, 211:877-883.

Futuyma, D. J., 1983. Evolutionary interactions among herbivorous insects and plants, in: "Coevolution", D. J. Futuyma and M. Slatkin, eds., pp. 207-231, Sinauer Associates, Sunderland.

Futuyma, D. J., and S. C. Peterson, 1985. Genetic variation in use of resources by insects, Annu. Rev. Entomol., 30:217-238.

Gould, F., C. R. Carroll and D. J. Futuyma, 1982. Cross-resistance to pesticides and plant defenses: A study of the two-spotted spider mite, Ent. Exp. Appl., 31:175-180.

Gould, F., 1983. Genetics of plant-herbivore systems, in: "Variable Plants and Herbivores in Natural and Managed Systems", Denno, R. F. and M. S. McClure, eds, pp. 599-654, Academic Press, New York.

Gould, F., 1984. Mixed function oxidases and herbivore polyphagy: the devil's advocate position, Ecol. Ent., 9:29-34.

Grabstein, E. M., and J. M. Scriber, 1982a. The relationship between restriction of host plant consumption, and post-ingestive utilization of biomass and nitrogen in Hyalophora cecropia, Ent. Exp. Appl., 31:202-210.

Grabstein, E. M., and J. M. Scriber, 1982b. Hostplant utilization as affected by prior feeding experience, Ent. Exp. Appl., 32:262-268.

Hagen, R., 1986. Host plant use and electrophoretic studies of the tiger butterfly, Papilio glaucus. Ph.D. Thesis, Cornell Univ. Ithaca, NY.

Hagen, R. H., and R. C. Lederhouse, 1985. Polymodal emergence of the tiger swallowtail, Papilio glaucus (Lepidoptera: Papilionidae): Source of a false second generation in New York state, Ecol. Entomol., 10:19-28.

Hainze, J. H., and J. M. Scriber, 1985. The influence of environmental factors on diapause and voltinism in Papilio glaucus subspecies, Bull. Ecol. Soc. Amer., 66:185.

Hancock, P. L., 1983. Classification of the Papilionidae (Lepidoptera): A phylogenetic approach, Smithersia, 2:1-48.

Hilbish, T. J., and R. K. Koehn, 1985. Dominance in physiological phenotypes and fitness at an enzyme locus, Science, 299:52-54.

Koehn, R. K., A. J. Zera, and J. G. Hall, 1983. Enzyme polymorphism and natural selection, in: "Evolution of Genes and Proteins", M. Nei and R. K. Koehn, eds., pp. 115-136, Sinauer, Sunderland.

Lande, R., 1982. A quantitative genetic theory of life-history evolution, Ecology, 63:607-615.

Lechowicz, M. J., 1983. Leaf quality and host preferences of gypsy moth in the northern deciduous forest, in: "Proceedings, Forest Defoliator - Host Interactions: A Comparison Between Gypsy Moth and Spruce Budworms", pp. 67-86, General Technical Report NE-85, USDA Forest Service, Washington, DC.

Leslie, J. F., and H. Dingle, 1983. A genetic basis of oviposition preference in the large milkweed bug, Oncopeltus fasciatus, Ent. Exp. Appl., 34:215-220.

Luebke, H. J., 1985. Differentiation of hybrids in a Wisconsin blend zone; morphometric analysis with Papilio glaucus glaucus and P. g. canadensis. M.S. thesis. Univ. Wisconsin, Madison.

Manuwoto, S., J. M. Scriber, M. T. Hsia, and P. Sunarjo, 1985. Phytochemical mechanisms of antibiosis/antixenosis in tulip tree and quaking aspen leaves against the southern armyworm, Spodoptera eridania, Oecologia, 67:1-7.

Mattson, W. J., and J. M. Scriber, 1985. The nutritional ecology of folivores of woody plants: water, nitrogen, fiber and mineral considerations, in: "The Nutritional Ecology of Insects, Spiders, and Mites", F. Slansky and J. G. Rodriquez, eds., John Wiley, New York (in press).

Miller, J. S., and P. P. Feeny, 1983. Effects of benzylisoquinoline alkaloids on the larvae of polyphagous Lepidoptera, Oecologia, 58: 332-339.

Miller, J. R., and K. Strickler, 1984. Insect plant interactions; finding and accepting host plants, in: "Chemical Ecology of Insects", W. J. Bell and R. T. Carde, eds., pp. 127-157, Sinauer Associates, Sunderland.

Mitter, C., and D. J. Futuyma, 1983. An evolutionary genetic view of hostplant utilization by insects, in: "Variable Plants and Herbivores in Natural and Managed Systems", R. F. Denno and M. S. McClure, eds., pp. 427-460, Academic Press, New York.

Mitton, J. B., and M. C. Grant, 1984. Associations among protein heterozygosity, growth rate, and developmental homeostasis, Annu. Rev. Ecol. & Syst., 15:479-499.

Munroe, E., 1960. The generic classification of the Papilionidae, Can. Entomol., (Suppl. 17), 51 pp.

Nagle, J. J., and L. E. Mettler, 1969. Relative fitness of introgressed and parental populations of Drosophila mojavensis and D. arizonensis, Evolution, 23:519-524.

Nei, M., and R. K. Koehn, eds., 1983, "Evolution of genes and proteins", Sinauer Associates, Sunderland.

Oliver, C. G., 1979. Genetic differentiation and hybrid viability within and between some Lepidoptera species, Am. Nat., 114:681-694.

Palo, R. T., 1984. Distribution of birch (Betula spp.), willow (Salix spp.), and poplar (Populus spp.) secondary metabolites and their potential role as chemical defense against herbivores, J. Chem. Ecol., 10:499-520.

Pearl, I. A., and S. F. Darling, 1968. Studies on the leaves of the family Salicaceae. Hot water extracts of P. balsamifera, Phytochem., 7:1845-1849

Plapp, F. W., and T. C. Wang, 1983. Genetic origins of insecticide resistance, in: "Pest Resistance to Pesticides", G. P. Georghiou and T. Saito, eds., pp. 47-70, Plenum Publ. Corp., New York.

Rausher, M. D., 1984. Trade-offs in performance on different hosts: Evidence from within- and between-site variation in the beetle Deloyala guttata, Evolution, 38:582-595.

Ritland, D. B., and J. M. Scriber, 1985. Larval developmental rates of three putative subspecies of tiger swallowtail butterflies, Papilio glaucus, and their hybrids in relation to temperature, Oecologia, 65:185-193.

Santamour, F. S., and J. S. Treese, 1971. Cyanide production in Magnolia, Morris Arboretum Bull., 22:58-59.

Schoonhoven, L. M., and J. Meerman, 1978. Metabolic cost of changes in diet and neutralization of allelochemics, Ent. Exp. Appl., 24:689-693.

Schultz, J. C., P. J. Nothnagle, and I. T. Baldwin, 1982. Seasonal and individual variation in leaf quality of two northern hardwood tree species, Amer. J. Bot., 69:753-759.

Schwartz, D., and W. J. Laughner, 1969. A molecular basis for heterosis, Science, 166:626-627.

Scriber, J. M. 1975. Comparative nutritional ecology of herbivorous insects; Generalized and specialized feeding strategies in the Papilionidae and Saturniidae (Lepidoptera), Ph.D. Thesis, Cornell University, Ithaca, N.Y.

Scriber, J. M., 1977. Limiting effects of low leaf-water content on the nitrogen utilization, energy budget, and larval growth of Hyalophora cecropia (Lepidoptera: Saturniidae), Oecologia, 28:269-287.

Scriber, J. M., 1979. The effects of sequentially switching foodplants upon the biomass and nitrogen utilization by polyphagous and stenophagous Papilio larvae, Ent. Exp. Appl., 25:203-215.

Scriber, J. M., 1981. Sequential diets, metabolic costs, and growth of Spodoptera eridania (Lepidoptera: Noctuidae) feeding upon dill, lima bean, and cabbage, Oecologia, 51:175-180.

Scriber, J. M., 1982a. Foodplants and speciation in the Papilio glaucus group, in: "Proc. 5th Intern. Symp. Insect and Host Plant, pp. 307-314, PUDOC, Wageningen.

Scriber, J. M., 1982b. The behavior and nutritional physiology of southern armyworm larvae as a function of plant species consumed in earlier instars, Ent. Exp. Appl., 31:359-369.

Scriber, J. M., 1983. The evolution of feeding specialization, physiological efficiency, and host races, in: "Variable Plants and Herbivores in Natural and Managed Systems", R. F. Denno and M. S. McClure, eds., pp. 373-412, Academic Press, New York.

Scriber, J. M., 1984a. Host plant suitability, in: "Chemical Ecology of Insects", W. J. Bell and R. T. Carde, eds., pp. 159-202, Sinauer Associates, Sunderland.

Scriber, J. M., 1984b. Larval foodplant utilization by the world Papilionidae (Lep.): Latitudinal gradients reappraised, Tokurana (Acta Rhopalocerologica), 6/7:1-50.

Scriber, J. M., and P. P. Feeny, 1979. The growth of herbivorous caterpillars in relation to degree of feeding specialization and to growth form of their foodplants (Lepidoptera: Papilionidae and Bombycoidea), Ecology, 60:829-850

Scriber, J. M., and J. H. Hainze, 1986. Geographic invasion and abundance as facilitated by differential host utilization abilities, in: "Insect Outbreaks: Ecological and Evolutionary Processes", P. Barbosa and J. C. Schultz, eds., Academic Press, New York, in press.

Scriber, J. M., R. C. Lederhouse, and L. Contardo, 1975. Spicebush, Lindera benzoin (L.), a little known foodplant of Papilio glaucus (Papilionidae), J. Lepid. Soc., 29:10-14.

Scriber, J. M., G. L. Lintereur, and M. H. Evans, 1982. Foodplant utilization and a new oviposition record for Papilio glaucus canadensis R & J (Papilionidae:Lepidoptera) in northern Wisconsin and Michigan, Great Lakes Ent., 15:39-46

Scriber, J. M., M. H. Evans and D. Ritland, 1985. Hybridization as a causal mechanism of mixed color broods and unusual color morphs of female offspring in the eastern tiger swallowtail butterflies, Papilio glaucus, in: "Evolutionary Genetics of Invertebrate Behavior", M. Huettel, ed., Univ. Florida Press, Gainesville, in press.

Scriber, J. M. and F. Slansky, Jr., 1981. The nutritional ecology of immature insects, Annu. Rev. Entomol. 26:183-211.

Slansky, F., and J. M. Scriber, 1985. Food consumption and utilization in: "Comprehensive Insect Physiology, Biochemistry, and

Pharmacology", G.A. Kerkut and L. I. Gilbert, eds., Vol. 1, pp. 87-163, Pergamon Press, Oxford.

Sunarjo, P. I., 1985. Phytochemical studies of quaking aspen (Populus tremuloides) and tulip tree (Liriodendron tulipifera) leaves and metabolism of precocene II in rats via the mercapturic acid pathway, Ph.D. Thesis, Univ. Wisconsin, Madison.

Sunarjo, P. I., and M. T. S. Hsia, 1984. Bioactive components of quaking aspen (Populus tremuloides) and tulip tree (Liriodendron tulipfera) leaves against southern armyworm larvae (Spodoptera eridania), Abstract No. 108, Div. Pesticide Chem., 188th Amer. Chem. Soc. National Meeting, Philadelphia, Aug. 26-31.

Tabashnik, B. E., 1983. Host range evolution: The shift from native legumes to alfalfa by the butterfly Colias philodice eripyle, Evolution, 37:150-152.

Taylor, W. I., 1962. The structure and synthesis of liriodenine, a new type of isoquinoline alkaloid, Tetrahedron Letters, 14:42-45.

Thieme, H., and R. Benecke, 1971. Glycosides of native or cultivated central European Populus species and phenol glycosides of Salicaceae, Pharmazi, 25:780-788.

Waldbauer, G. P., 1968. The consumption and utilization of food by insects, Adv. Insect Physiol., 5:229-289.

Watt, W. B., 1977. Adaptation at specific loci. I. Natural selection on phosphoglucose isomerase of Colias butterflies: biochemical and population aspects, Genetics, 87:177-194.

Watt, W. B., 1985. Bioenergetics and evolutionary genetics: opportunities for new synthesis, Am. Nat., 125:118-143.

Whitham, T. G., A. G. Williams and A. M. Robinson, 1984. The variation principal: Individual plants as temporal and spatial mosaics of resistance to rapidly evolving pests, in: "A New Ecology: Novel Approaches to Interactive Systems", P.W. Price, C.N. Slobodchikoff, and W.S. Gaud, eds., pp. 15-51, J. Wiley & Sons, New York.

Zucker, W. V., 1982. How aphids choose leaves: the roles of phenolics in host selection by a galling aphid, Ecology, 63:972-981.

Appendix 1. Penultimate instar performance of individual backcross larvae fed tulip tree leaves (see also Fig. 3).

Parental Derivation (Lab Code #)	Relative Growth Rate (R.G.R.)	Relative Consumption Rate (R.C.R.)	Approximate Digestibility[2] (A.D.)	Efficiency of Conversion Digested Food[2] (E.C.D.)	Overall Efficiency[2] (E.C.I.)
(Pgg x Pgc) x Pgg (#540):	.272	1.83	57.8	25.7	(14.9)
	*.250	2.31	64.0	16.9	(10.8)
	.242	1.68	47.1	30.6	(14.4)
	*.237	1.92	31.9	81.1	(25.9
	*.232	1.79	68.1	19.0	(13.0)
	*.228	1.49	30.7	49.8	(15.3)
	*.223	1.30	26.5	65.1	(17.2)
	.216	1.42	42.6	35.7	(15.2)
	.204	1.32	35.5	43.3	(15.4)
	*.198	2.00	51.8	19.1	(9.9)
	*.197	2.06	67.6	14.1	(9.6)
	.196	1.46	42.8	31.3	(13.4)
	.192	1.27	31.7	47.9	(15.2)
	.191	1.43	43.8	30.5	(13.4)
	.190	2.02	56.6	16.7	(9.4)
	*.181	1.64	47.9	23.1	(11.1)
	.171	1.44	42.2	28.2	(11.9)
	.171	2.00	45.4	18.9	(8.6)
	.128	1.11	25.5	45.3	(11.6)
	.098	1.39	28.6	24.6	(7.0)
(Pgg x Pgc) x Pgc (#555):	.234	1.98	54.3	21.8	(11.8)
	.210	1.83	40.1	28.6	(11.5)
	.204	1.32	48.1	32.2	(15.5)
	.200	2.14	54.8	17.0	(9.3)
	*.196	1.46	42.8	31.4	(13.4)

	.193	1.97	43.0	22.7	(9.8)
	*.176	0.91	32.9	59.0	(19.4)
	.152	1.51	54.7	18.5	(10.1)
	*.145	1.49	33.6	29.0	(9.7)
	.069	1.65	74.4	5.6	(4.2)
	.053	1.13	62.1	7.5	(4.6)
	.047	1.34	75.7	4.6	(3.5)
	.006	1.58	79.2	0.5	(0.4)
	-.006	1.57	84.7	0.4	(0.4)
	-.013	1.61	52.6	1.6	(0.8)
	-.014	0.88	70.4	2.3	(1.6)
	-.016	0.98	60.3	2.7	(1.6)
	-.041	0.76	61.4	8.5	(5.2)
		--	too little eaten	--	--
		--	too little eaten	--	--
Pgc x (Pgg x Pgc) (#500):	*.295	1.98	32.0	46.6	(14.9)
	.291	2.01	44.5	32.6	(14.5)
	*.270	1.95	42.3	32.8	(13.9)
	*.270	2.46	56.5	19.4	(11.0)
	.236	1.69	63.1	22.1	(14.0)
	.203	1.16	24.1	72.7	(17.5)
	.193	2.31	62.9	13.7	(8.6)
	*.186	1.45	48.2	26.5	(12.8)
	.183	1.38	50.5	26.4	(13.3)
	.151	1.45	58.1	17.9	(10.4)
	*.126	1.47	58.3	14.6	(8.5)
	.123	1.70	59.4	12.2	(7.2)
	.108	2.88	76.9	4.9	(3.7)
	.105	1.13	51.2	18.2	(9.3)
	.102	2.46	93.0	4.5	(4.2)
	.101	1.30	50.9	15.4	(7.8)
	.100	1.24	56.5	14.2	(8.0)
	.086	1.52	76.7	7.4	(5.7)
	.054	2.06	79.7	3.3	(2.6)
	.053	2.87	90.9	2.0	(1.8)

	.042	2.10	29.2	6.8	(2.0)
		--	too little eaten	--	--
		--	too little eaten	--	--
		--	too little eaten	--	--
		--	too little eaten	--	--

[1] R.G.R. mg larval weight gain/mg mean weight during the instar/day.

[2] Efficiencies (A.D. x E.C.D. = E.C.I.) are expressed as a percent.

* = "non-switched" individuals (i.e. reared on tulip tree from neonate stage).

Appendix 2. Penultimate instar performance of individual backcross larvae fed quaking aspen (see also Fig. 4).

Parental Derivation (Lab Code #)	Relative Growth Rate (R.G.R.)	Relative Consumption Rate (R.C.R.)	Approximate Digestibility[2] (A.D.)	Efficiency of Conversion Digested Food[2] (E.C.D.)	Overall Efficiency[2] (E.C.I.)
(Pgg x Pgc) x Pgg (#540):	.152	4.88	76.5	4.1	(3.1)
	.143	2.08	51.3	13.4	(6.9)
	.139	1.59	50.2	17.4	(8.7)
	.138	1.51	27.3	33.6	(9.2)
	.130	1.97	65.8	10.1	(6.6)
	.123	1.54	48.1	16.6	(8.0)
	.119	1.94	65.2	9.4	(6.1)
	.116	1.88	63.3	9.8	(6.2)
	.109	1.60	51.6	13.2	(6.8)
	.104	1.46	53.6	13.3	(7.1)
	.100	1.35	39.2	18.9	(7.4)
	.096	1.10	34.8	24.9	(8.7)
	.094	2.28	71.8	5.7	(4.1)
	.085	1.54	58.5	9.5	(5.5)
	.082	0.91	25.8	35.2	(9.1)
	.081	1.35	48.2	12.5	(6.0)
	.078	2.59	74.6	4.0	(3.0)
	.074	1.50	53.5	9.3	(4.9)
	.070	1.41	62.2	8.0	(5.0)
	.069	1.78	61.9	6.3	(3.9)
	.064	2.11	63.2	4.8	(3.0)
	-.013	1.66	74.9	1.1	(0.8)
	-.022	2.37	80.2	1.2	(0.9)
(Pgg x Pgc) x Pgc (#555):	.217	2.53	53.8	16.0	(8.6)
	*.210	2.81	67.3	11.1	(7.5)

	*.206	2.16	49.8	19.1	(9.5)
	.205	2.42	60.6	14.0	(8.5)
	*.202	2.27	31.7	28.0	(8.9)
	.199	1.68	36.4	32.7	(11.9)
	*.199	1.91	43.4	24.1	(10.4)
	*.182	3.09	66.2	8.9	(5.9)
	*.175	2.60	60.1	11.2	(6.7)
	.169	2.31	61.6	11.9	(7.3)
	.169	2.19	64.2	12.0	(7.7)
	*.167	1.89	43.6	20.3	(8.9)
	*.166	2.53	65.0	10.1	(6.6)
	.163	1.77	50.6	18.2	(9.2)
	.156	2.12	51.1	14.4	(7.4)
	.154	2.27	46.2	14.7	(6.8)
	.130	1.92	68.6	9.9	(6.8)
	.099	1.89	69.5	7.6	(5.3)
Pgc x (Pgg x Pgc) (#500):	*.251	2.01	50.3	21.3	(10.7)
	*.223	1.86	60.1	20.0	(12.0)
	*.208	1.79	64.8	17.9	(11.6)
	*.193	2.03	56.7	16.8	(9.5)
	*.191	1.55	50.5	24.3	(12.3)
	*.186	1.51	42.5	29.0	(12.3)
	*.185	1.87	38.9	25.4	(9.9)
	*.177	1.67	46.4	22.9	(10.6)
	*.175	1.83	54.5	17.6	(9.6)
	*.175	1.93	58.4	15.6	(9.1)
	*.166	1.81	60.6	15.1	(9.2)
	.163	2.41	64.0	10.6	(6.8)
	.160	1.42	42.8	26.4	(11.3)
	.157	2.31	63.5	10.7	(6.8)
	*.146	1.39	59.7	17.7	(10.6)
	.141	1.52	50.5	18.4	(9.3)
	.131	1.65	54.5	14.6	(7.9)
	.127	1.91	54.1	12.3	(6.7)
	.115	1.80	62.1	10.2	(6.4)

.109	2.15	65.2	7.8	(5.1)
.101	2.77	33.3	10.9	(3.6)
.089	1.98	69.1	6.5	(4.5)
.081	1.09	80.0	9.2	(7.4)
.057	1.22	52.7	8.8	(4.6)
.047	1.95	68.7	3.5	(2.4)
.038	2.12	74.3	2.4	(1.8)
.036	1.54	89.3	2.7	(2.4)
.021	1.75	81.2	1.5	(1.2)
-.010	3.18	86.9	0.4	(0.3)
-.025	.30	91.2	0.8	(0.4)
	--	too little eaten	--	--
	--	too little eaten	--	--

ENZYMES INVOLVED IN THE METABOLISM OF PLANT ALLELOCHEMICALS

S. Ahmad, L. B. Brattsten[1], C. A. Mullin[2], and S. J. Yu[3]

University of Nevada
Department of Biochemistry
Reno, NV 89557

1. INTRODUCTION

Herbivorous insects have available to them two major mechanisms by which they avoid adverse effects of defensive plant allelochemicals. One is comprised of behavioral adaptations in host-seeking (Ahmad, 1983a) and feeding (Frazier, Chapter 1 and Tallamy, Chapter 8 in this text). The other is a complex of several enzymes which together spare few ingested plant chemicals from being transformed into one or more metabolites which may be utilized or eliminated. We shall, in this chapter, provide an account of these enzymes.

In addition, herbivorous insects have many special adaptations such as insensitive target sites (Berenbaum, Chapter 7 in this text) or special excretory and storage mechanisms (Berenbaum, Loc. cit.; Mullin, Chapter 5 and Brattsten, Chapter 6 in this text).

There are four enzyme categories, Table 1, of major importance in the metabolism of lipophilic foreign compounds, those that tend to be most hazardous because of their abundance in plants and the facility with which they penetrate lipid-containing cell membranes and accumulate in lipid-rich tissues such as the nervous system. These groups of enzymes are represented in all investigated living organisms to various extents.

With few exceptions, the concerted action of the enzymes results in conversion of the compounds to increasingly polar metabolites, Fig. 1. The overall metabolic process is, for convenience, organized in two phases, not always reflecting realities, Fig. 1. In phase I, reactions are catalyzed by those enzymes capable of making a direct attack on a lipophilic, non-nutrient molecule. In phase II reactions, the primary metabolite is conjugated to a highly water-soluble, endogenous group by group transfer-

[1]E.I. Du Pont de Nemours and Co., Inc., Experimental Station, Agricultural Products Department, Bldg. 402, Wilmington, DE 19898.

[2]Pennsylvania State University, Department of Entomology, Pesticide Research Laboratory and Graduate Study Center, University Park, PA 16802.

[3]University of Florida, Department of Entomology and Nematology, Gainesville, FL 32611.

Table 1. Enzymes important in the metabolism of plant allelochemicals and other foreign compounds in insects (International Union of Biochemists, 1984).

Oxidases

Cytochrome P-450-dependent monooxygenases, E.C.1.14.14.1; also called unspecific monooxygenases, polysubstrate monooxygenases (PSMO) or mixed-function oxidases (MFO).

Hydrolases

Carboxylesterases, E.C.3.1.1.1, also called aliesterases, B-esterases or carboxylic ester hydrolases.
Aryl esterases, E.C.3.1.1.2, also called A-esterases
Epoxide hydrolases, E.C.3.3.2.3, also called epoxide hydratases or arene oxide hydrolases.

Transferases

Glutathione transferases, E.C.2.5.1.18
UDP glucosyl transferases E.C.2.4.1.35, also called phenol β-glucosyl transferases.

Reductases

Carbonyl reductases (NADPH), E.C.1.1.1.184 also called AK reductases.

---

ases (Williams, 1974), in some cases after further conversion of cytochrome P-450 products, e.g. epoxides to dihydrodiol derivatives.

The majority of phase I reactions are oxidations catalyzed in most cases by cytochrome P-450-dependent monooxygenases, henceforth called PSMOs (see section 2.2). Other oxidases, the role of which in insects is

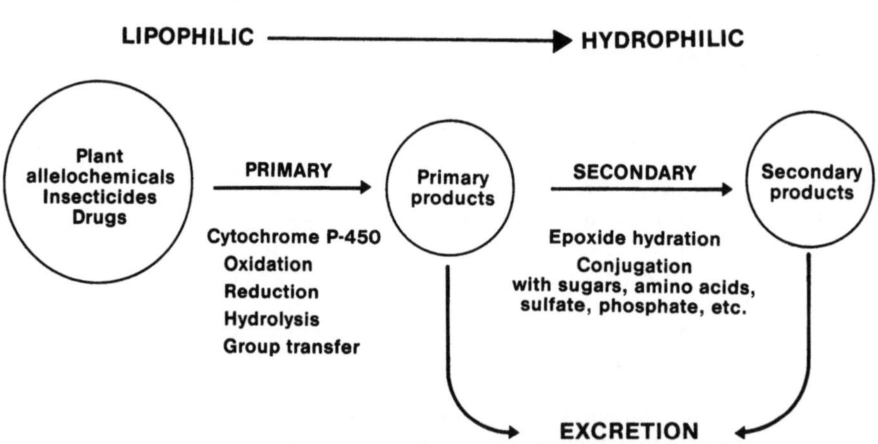

Fig. 1. Scheme for the metabolic conversion of potentially toxic xenobiotics to water soluble products that can be easily eliminated or, in some cases, stored and/or utilized.

not understood, may also be involved. Hydrolysis of ester bonds by carboxylesterases, hydrolysis of epoxide ether bonds by epoxide hydrolases, and conjugation to glutathione catalyzed by glutathione transferases are also important phase I reactions. In some or possibly many cases, the primary metabolites are excreted, but more often they are converted to secondary metabolites by phase II conjugations. Glutathione transferases are important phase II enzymes as are many other conjugating enzymes. Epoxide hydrolases play a major detoxifying role in converting primary epoxide metabolites to secondary metabolites which can be conjugated.

Phase I oxidations often result in products with less biological activity, that is, ability to interfere with vital biochemical or physiological processes. However, with appreciable frequency, oxidation products are more toxic than the parent molecule. Phase II reactions, with few exceptions, lead to harmless metabolites.

2. CYTOCHROME P-450

2.1. Occurrence

Cytochrome P-450 is absent from obligate anaerobic bacteria but present in all other organisms investigated. It has been found in bacteria, fungi, plants, many different forms of invertebrates, mostly arthropods, and in all classes of vertebrates.

2.1.1. Occurrence in Species. PSMO activity has been measured with some accuracy in at least 160 species of insects (Hodgson, 1985; Rose, 1985), a respectable number and a good beginning for an understanding of the importance of the enzyme system, but still a very small fraction of the more than 750,000 described species of insects.

Most of the insect species investigated are of economic or public health importance. The PSMOs of the housefly, *Musca domestica* L. (Diptera: Muscidae), have received more attention than those of any other species. Considerable work has also been done with lepidopterous larvae and orthopterans. This work has mostly been concerned with understanding the metabolic mechanisms by which pest insects resist poisoning by synthetic or botanical insecticides. As a consequence, there is not much information on species or developmental stages that do not directly interfere with food and fiber production or public health. Insecticide toxicity data, synergistic ratios (see section 2.4.), and a few direct measurements indicate that, in general, parasitoids and predators have less PSMO activities than generalist herbivores (Mullin, 1985), Table 2.

Microsomal cytochrome P-450 in insects and vertebrates is one of the most intensively studied enzyme systems because of its involvement in the inactivation of drugs and insecticides. The duration and effectiveness of these agents are largely dependent on the activity of the PSMOs (Gillette et al., 1969; Hodgson, 1979, 1983). The one property of the system to emerge clearly from this work is its eminent suitability for metabolizing lipophilic foreign compounds (xenobiotics). The synthetic insecticides are very recent evolutionary additions to the environment, whereas cytochrome P-450 is likely of ancient evolutionary origin (Wickramasinghee and Villee, 1975; Wilkinson, 1980). It became obvious relatively early in the history of insecticide use that polyphagous species develop high levels of insecticide resistance rather rapidly. Gordon (1961) suggested that the natural exposure of polyphagous species to a wide variety of plant allelochemicals had resulted in high capacities for their detoxification which would now enable the insects to develop resistance to synthetic insecticides. This idea directly implied plant allelochemicals as the natural

Table 2. Insect parasitoids and predators in which PSMO activities have been demonstrated or inferred.

Species	Method of assessment of activity	Reference
Hymenoptera		
<u>Vespula flavopilosa</u> (Vespidae)	NADPH cytochrome P-450 reductase	Crankshaw et al., 1981
<u>Oncophanes americanus</u> (Weed) (Braconidae)	aldrin epoxidation	Croft & Mullin, 1984
<u>Campoletis sonorensis</u> (Carlson) (Ichneumonidae)	insecticide toxicity	Plapp & Vinson, 1977
<u>Pediobius foveolatus</u> (Crawford) (Eulophidae)	aldrin epoxidation	Mullin, 1985
Neuroptera		
<u>Chrysopa carnea</u> (Stephens) (Chrysopidae)	insecticide toxicity synergistic ratio	Plapp & Bull, 1978 Brattsten & Metcalf, 1970
Coleoptera		
<u>Cicindela oregona</u> (LeConte) (Cicindelidae)	synergistic ratio	Brattsten & Metcalf, 1970
<u>Hippodamia convergens</u> (Guerin-Meneville) (Coccinellidae)	synergistic ratio	Brattsten & Metcalf, 1970
<u>Photinus pyralis</u> L. (Lampyridae)	insecticide toxicity	Brattsten & Metcalf, 1970
<u>Cantharis consors</u> (LeConte) (Cantharidae)	synergistic ratio	Brattsten & Metcalf, 1970
Diptera		
<u>Stomoxys calcitrans</u> (L.) (Muscidae)	synergistic ratio	Brattsten & Metcalf, 1973

substrates for insecticide-detoxifying enzymes for the first time. A short time thereafter, Ehrlich and Raven (1964) proposed in more general terms the theory of plant-insect coevolution whereby natural communities may remain in evolutionary equilibria by mutual biochemical and behavioral adaptations in the participants. Experimental support for these ideas was presented in 1971 (Krieger et al.) by the discovery that a cytochrome P-450-catalyzed activity, epoxidation of the insecticide aldrin, was significantly higher in polyphagous lepidopterous larvae than in oligophagous ones. These, in turn, had higher activity than monophagous species. Experiments with the Japanese beetle, <u>Popillia japonica</u> Newman (Coleoptera: Scarabaeidae) showed that naturally polyphagous, field collected beetles had higher activity than beetles fed on three different plants in the laboratory; these, in turn, had higher activity than beetles fed one plant only (Ahmad, 1983b). Subsequent experimental work indicates that the activity is better correlated with the phytochemical profile of the host plant(s) than with the degree of polyphagy, which can change considerably during the development of a species (Fox and Morrow, 1981). For instance, the specialist feeder Utetheisa bella L. (Lepidoptera: Arctiidae) has very low cytochrome P-450 content and activity (Brattsten, 1979a) possibly because it is inhibited by the pyrrolizidine alkaloids in the host plants (Brattsten, Chapter 6 in this text) but not necessarily because it is oligophagous. On the other hand, another specialist feeder, the black swallow-tail, Papilio polyxenes Fabr. (Lepidoptera: Papilionidae) has very high larval midgut cytochrome P-450 content and activity (Brattsten, 1979a). In general, lepidopterous larvae that feed on plants containing monoterpenes tend to have higher aldrin epoxidation activity than those that feed on plants containing other allelochemicals (Rose, 1985). Data of this nature are available for other xenobiotic-metabolizing enzymes in extremely few cases.

The effects of food plants on insect cytochrome P-450 are very complicated and can include direct effects such as induction (see section 2.6.) and inhibition (see section 2.5.) and indirect effects via the plant nutrient and water content, interferences by plant allelochemicals with nutrient utilization, and quite possibly, allelochemical-induced mutations.

2.1.2. Age-related Distribution of Activity. High PSMO activities seem to coincide with actively feeding life stages, reinforcing the impression that cytochrome P-450 plays an important role in the metabolism of ingested plant allelochemicals.

In lepidopterans, the larval stages have the highest activities. Only few attempts have been made to measure activities in other life stages; no activity was found in the southern armyworm moth (Krieger and Wilkinson, 1969) and low activity was measured in adults of the orange tortrix, Argyrotaenia citrana (Fernald) (Lepidoptera: Tortricidae) compared to whole larvae of the same species (Croft and Mullin, 1984); activity was high only in larval midgut preparations. In the southern armyworm, Spodoptera eridania (Cramer) (Lepidoptera: Noctuidae), the PSMO activity is variable during the larval stages and correlates well with growth and feeding activity, Fig. 2. The activity drops sharply, shortly before the molts and increases rapidly after each molt to a maximum in the most vigorously feeding last instar; it then drops to a very low level in larvae which have ceased to feed and are preparing to pupate (Krieger and Wilkinson, 1969). A similar pattern of activity variation with age was observed also in gypsy moth, Lymantria dispar (L.) (Lepidoptera: Lymantriidae) larvae (Ahmad, 1986), the variegated cutworm, Peridroma saucia (Hubner) (Lepidoptera: Noctuidae) (Berry et al., 1980), the fall armyworm, S. frugiperda (J. E. Smith) (Ibid.) larvae (Yu, 1983), and in the tobacco budworm, Heliothis virescens (Fabr.) (Ibid.) larvae (Gould and Hodgson, 1980) but not in cabbage looper, Trichoplusia ni (Hubner) (Ibid.) larvae (Kuhr, 1970) in which highest activity was measured in the prepupal stages.

Insecticides are commonly more toxic to early instar lepidopterous larvae than to fully grown last instars. This is the case with methomyl-, diazinon-, and permethrin-toxicity to fall armyworm larvae (Yu, 1983), and correlates well with the known, but not well documented, lower activity levels of cytochrome P-450 in young larvae. No PSMO activity data from first instars of any lepidopterans are available. For second instars, data are available only from the gypsy moth (Ahmad, 1986) and show low activity compared to third and subsequent instars. A careful study of larval development in Acrolepiopsis assectella Zell. (Lepidoptera: Acrolepiidae) revealed that last instars but not first or second ones could eat diets containing pulverized flowers of it's host plant, the leek, Allium porrum (Arnault, 1979). Pulverized leek foliage sustained growth and development of all instars. This was ascribed to metabolic changes during larval development although metabolism was not studied. This is the only case known where plant parts, possibly containing different allelochemicals, are toxic to young larvae but not to old ones and, thus, provides a sort of parallel with insecticide toxicities. In nature, selective feeding may contribute to the survival of young larvae on host plants containing potentially toxic allelochemicals.

There are scattered pieces of information about age-related variations in activity among other insect orders. Benke and Wilkinson (1971) found a rapidly increasing aldrin epoxidation rate in newly emerged female house crickets, Acheta domesticus (L.) (Orthoptera: Gryllidae) which remained high over a five-week period. The activity in the male house cricket increased only slightly after emergence and remained low. In the Madagascar cockroach, Gromphadorhina portentosa Schaum (Orthoptera: Blattidae), the aldrin epoxidation rate increased rapidly after adult emergence in both sexes and exceeded that in the nymphal instars. It

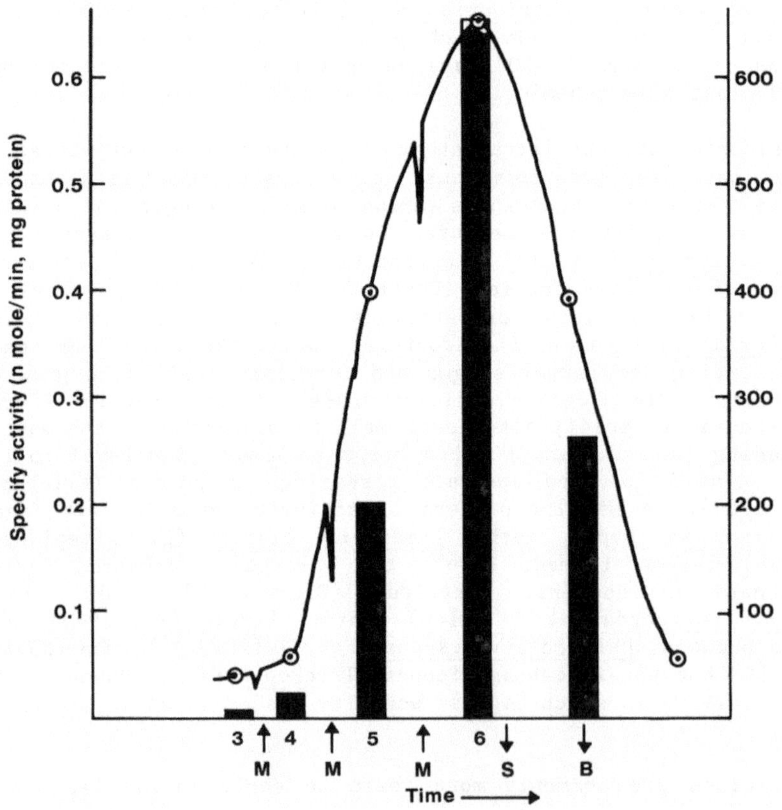

Fig. 2. Covariation of cytochrome P-450-catalyzed epoxidation of aldrin (data from Krieger & Wilkinson, 1969) and weight increase (bars) in southern armyworm larvae. M denotes molts; at S, larvae stop feeding and clear their guts in preparation for pupation; at B, larvae burrow. The numbers 3-6 indicate third through sixth (last) larval instars.

reached a plateau and remained high in the male but decreased to a low level in five-week-old females coinciding with the maturation of the ovoviviparous offspring (Benke et al., 1972). In the Pacific beetle cockroach, Diploptera punctata (Eschsholtz) (Orthoptera: Blaberidae), the adults have higher activities than the nymphs (Feyereisen and Farnsworth, 1985). House flies, the black blow fly, Phormia regina (Meigen) (Diptera: Calliphoridae) and the flesh fly, Sarcophaga bullata Parker (Diptera: Sarcophagidae) all have higher activities as adults than in the larval stages (Yu and Terriere, 1971; Perry and Buckner, 1970; Terriere and Yu, 1976).

The Japanese beetle adults have higher cytochrome P-450 contents than their larvae (Ahmad, 1983b) whereas adult boll weevils, Anthonomus grandis Boheman (Coleoptera: Curculionidae) have lower activities than boll weevil larvae (Brattsten, 1986a).

2.1.3. Tissue Distribution and Subcellular Localization. Due to their small size, careful characterization of the tissue distribution of cytochrome P-450 in insects presents problems. The most successful work has been done with lepidopterous larvae, orthopterans, and dictyopterans such as crickets, locusts, and cockroaches which have high activities in fatbody, Malpighian tubules, midgut tissues, and gastric caecae and lower activities in hindgut, nerve tissue (Nakatsugawa and Dahm, 1962; Benke and Wilkinson, 1971; Benke et al., 1972), and corpora allata (Feyereisen et al., 1981). Activities have been found in guts and fatbodies of lepidopterous larvae (Krieger and Wilkinson, 1969; Brattsten 1979b; Tate et al., 1982, Marty et al., 1982) and, at lower levels, in other tissues including Malpighian tubules, nerve tissue, integument (Krieger and Wilkinson, 1969; Morris, 1983; Collins, 1985), and silkglands (Burt et al., 1978; see Hodgson, 1985).

Apparently, in most lepidopterans, the gut tissue has the highest specific activity (Wilkinson and Brattsten, 1972; Brattsten, 1979a), but this, to some extent, depends on which cytochrome P-450-catalyzed reaction is measured. Kuhr (1970) measuring epoxidation found higher activity in cabbage looper larval fatbodies than in their midguts; this activity is usually highest in midguts. Another activity, O-demethylation of methoxyresorufin, is, however, higher in fatbody than in midgut tissues both in a lepidopterous larva, the southern armyworm (Brattsten et al., 1980), and in the nymphal Pacific beetle cockroach (Feyereisen and Farnsworth, 1985). It is thus possible to make only rather vague generalizations as to which organ is most important in protecting an insect from a certain toxicant based on work with model substrates. If possible, it would be best to determine the rate of metabolic degradation in each tissue, of the plant allelochemicals of interest; for instance, even though the gut of tobacco hornworm, Manduca sexta (L.) (Lepidoptera: Sphingidae) larvae have the highest specific activities towards selected model substrates, their nerve tissue may be of primary importance in detoxifying nicotine (Morris, 1983).

In addition to cytochrome P-450 characterized by activity towards a number of model substrates (see section 2.4.) and residing in the endoplasmic reticulum membrane of cells, there are also cytochrome P-450s with restricted substrate specificities (Ahmad, 1982). These enzymes are sometimes located in the mitochondrial membranes and sometimes in the endoplasmic reticulum membranes; they are found in a variety of tissues. Cytochrome P-450 forms participating in steroid biosynthesis and metabolism have been identified in mammals and fungi and are associated with the mitochondrial membrane. A mitochondrial cytochrome P-450 occurs in the desert locust, Schistocerca gregaria L. (Orthoptera: Acrididae) Malpighian tubules (Greenwood and Rees, 1984) and the tobacco hornworm fatbody

(Bollenbacher et al., 1977). This enzyme has also been found in the midgut of the tobacco hornworm (Mayer et al., 1978; Weirich et al., 1985). It is called ecdysone 20-monooxygenase (E.C.1.14.99.22) and is responsible for the conversion of ecdysone to 20-hydroxyecdysone; it is not known if this enzyme attacks other compounds as well. A microsomal cytochrome P-450 in fatbody, Malpighian tubules, mid-, and hindgut in the migratory locust, Locusta migratoria L. (Orthoptera: Acrididae), catalyzes the 20-hydroxylation of ecdysone (Feyereisen and Durst, 1980). Microsomal enzymes effecting epoxidation of fucosterol in midguts of the Egyptian armyworm, Spodoptera littoralis (Boisduval) (Lepidoptera: Noctuidae) larvae and other insects are likely cytochrome P-450s (Fujimoto et al., 1985). The fucosterol-24,28-epoxide is dealkylated by a complex process, at least in part catalyzed by the enzyme fucosterol epoxide lyase (Prestwich et al., 1985; Brattsten, Chapter 6 in this text). The corpora allata cytochrome P-450 which oxidizes methyl farnesoate to a juvenile hormone is microsomal and also accepts xenobiotics, notably the precocenes, as substrates (Feyereisen et al., 1981).

There is usually very little activity towards model substrates in subcellular fractions of insect tissues other than the one containing the endoplasmic reticulum membranes (Cassidy et al., 1969; Krieger and Wilkinson, 1969; Benke et al., 1972; Hansen and Hodgson, 1971, Wilkinson and Brattsten, 1972; Brattsten et al., 1976, 1980, Brattsten and Gunderson, 1981). This is usually the case also in subcellular fractions of vertebrate tissues. It is therefore the consensus that the cytochrome P-450s important in the metabolism of xenobiotics are those associated with the endoplasmic reticulum membranes of cells.

2.2. Biochemical Characteristics of Microsomal Cytochrome P-450

The cytochrome P-450s of importance in foreign compound metabolism are deeply embedded in the endoplasmic reticulum membranes of cells, Fig. 3. These membranes are fragmented and vesiculated during homogenization

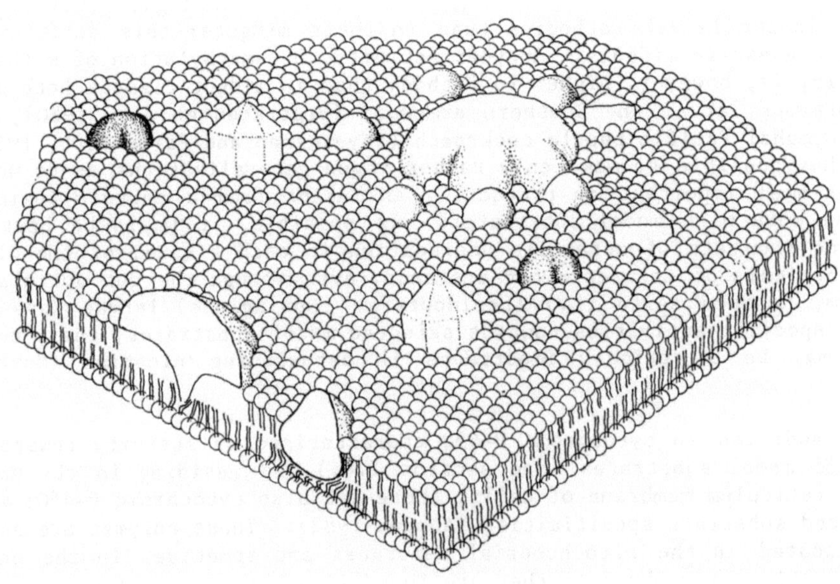

Fig. 3. Cartoon of endoplasmic reticulum membrane. A large molecule of NADPH reductase is thought to be surrounded by several smaller cytochrome P-450 molecules; epoxide hydrolase, glucose transferase, or cytochrome b_5 molecules may occur in the vicinity of the reductase-cytochrome P-450 cluster.

of the tissue and sediment only after high-speed centrifugation; they are known as microsomes (Palade and Siekevitz, 1956) and, consequently, the enzymes are also called microsomal oxidases.

Cytochrome P-450 was first detected in pig (Garfinkel, 1958) and rat liver (Klingenberg, 1958) microsomes, and characterized in 1964 by Omura and Sato who "provisionally called [it] P-450". This name stands for: pigment that absorbs light at 450 nm (when in complex with carbon monoxide). It was later discovered (Sladek and Mannering, 1966) and confirmed (Ryan et al., 1975) that the cytochrome occurs in different isoenzymic forms. At least eight different forms have been isolated from rat liver (Ryan et al., 1982). There is, thus, a family of microsomal cytochrome P-450s all of which absorb light around 450 nm when in complex with carbon monoxide, and all of which have broadly overlapping substrate specificities but show distinct substrate preferences. The isoenzyme distribution is unclear but the implication is that each cytochrome P-450-containing tissue, or perhaps even cell may have a unique composition. The composition may change with the age of the organism and in response to inducers (see section 2.6.). This complexity prompted a new name and "polysubstrate monooxygenases" (PSMOs) was recommended at the Third International Conference on the Biochemistry, Biophysics and Regulation of cytochrome P-450 in 1980.

PSMO activities as well as glutathione transferase and epoxide hydrolase activities are very low compared to those of the intermediary metabolism enzymes. This may be due to the low requirements for binding to the active site, which is, largely, a lipophilic interaction with elements of affinity to electron-rich areas of the substrate.

The microsomal PSMO system, Fig. 4, is composed of the terminal cytochrome and a flavoprotein reductase and typically depends on the membrane phosphatidylcholine for activity. The cytochrome is of an aberrant b-type; it contains iron-protoporphyrin IX as prosthetic group and reacts with molecular oxygen and carbon monoxide but not with cyanide. A lipophilic substrate binds to the active site of the oxidized cytochrome. Next, oxygen is incorporated into the complex and it is reduced by one electron transported from NADPH by the flavoprotein, NADPH hemoprotein reductase (E.C.1.6.2.4, also called NADPH cytochrome P-450 reductase or NADPH cytochrome c reductase). The complex is now reduced by a second electron of unclear origin. In this process an oxygen free radical is

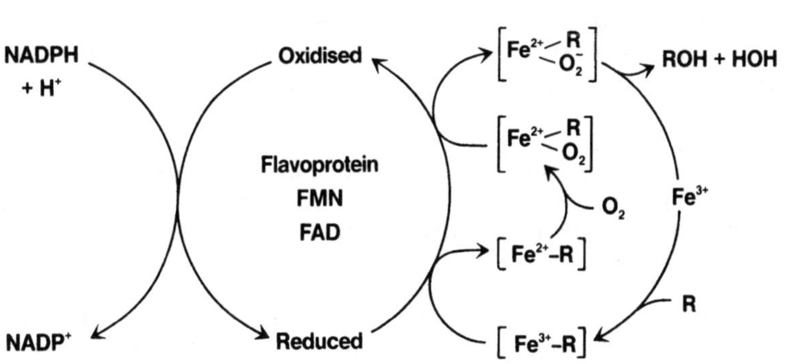

Fig. 4. Cytochrome P-450 electron transport and substrate (R) interactions. Reprinted with permission from ACS, copyright 1983.

formed and one oxygen atom is inserted into the substrate molecule. The superoxide anion free radical, $\cdot O_2^-$, is the one most likely formed (Ullrich, 1977). The other oxygen atom combines with hydrogen, H^+, to form water. The insertion of only one of the oxygen atoms into the substrate forms the basis for the term "monooxygenase" in distinction from the dioxygenases where both oxygen atoms are inserted into the substrate; this was also indicated in the name "mixed-function (of the oxygen) oxidase". In some cases, the second electron may originate from another reductase-cytochrome system, NADH cytochrome b_5, also located in endoplasmic reticulum membranes in the vicinity of the PSMO system. Cytochrome b_5, first discovered in cecropia, Hyalophora cecropia L. (Lepidoptera: Saturniidae), midgut homogenates (Chance and Pappenheimer, 1954) is normally involved in the desaturation of fatty acids. The end result is an oxygenated product, water, and reversion of the cytochrome to the oxidized (Fe^{2+}) state. Several of the interactions in this scheme can be measured (see section 2.4.).

The cytochrome P-450s are unstable proteins with molecular masses of 45,000-60,000 daltons. Their purification was difficult to accomplish with mammalian liver microsomes as the source and has not yet been achieved with an insect tissue as source. Nevertheless, partial purifications have successfully established the existence of isoenzymic forms of cytochrome P-450 also in adult house fly tissues (Capdevila et al., 1975; Yu and Terrier, 1979; Fisher and Mayer, 1984; Moldenke et al., 1984).

Studies of PSMO activity in systems established with components isolated from vertebrate liver have shown that the cytochrome b_5 system is not obligatory for full catalytic activity (Lu and West, 1978). Such studies have also shown that the system depends on phosphatidylcholine; best activity is obtained with the fully saturated (12:0) dilauroyl-substituted phosphatidylcholine (Lu et al., 1976). The lipid is thought to be important in facilitating the electron transport by assisting the binding of the cytochrome to the reductase. Nonionic detergents can, to some extent and over a narrow concentration range, substitute for the lipid. The lipid may also increase the solubility of highly lipophilic substrates which otherwise would be present in too low concentration near the active site.

Studies of cytochrome P-450 activity in systems reconstituted with partially purified components from house flies indicated that the phosphatidylcholine is not an absolute requirement for activity (Capdevila et al., 1975; Yu and Terriere, 1979, Moldenke et al., 1984). In these studies, the lack of absolute dependence on a phospholipid could be due to incomplete removal of detergent used for solubilization, which could substitute for the phospholipid. However, in a study with larvae of the southern armyworm, the microsomal phospholipid fraction did not change when activities toward aldrin, p-chloro N-methylaniline, or methoxyresorufin were increased by induction (Brattsten et al., 1986) and no significant changes in Arrhenius activation energies were found. The only difference in membrane components between control and induced tissues was a significant increase in protein with cytochrome P-450 activity. This also indicates a lower dependence on the phospholipid in insects than generally expected, based on work with liver microsomal PSMO systems, although insect PSMOs are capable of catalyzing the same apparent reactions as the vertebrate ones.

Most of the insect work on reconstitution of the PSMO system has been done with components partially purified from house fly tissues. The concentration of phosphatidylethanolamine in the membranes of flies, beetles, and some other insects is higher than that of phosphatidylcholine (Bridges, 1983), whereas in mammalian tissues the latter predominates;

flies are, moreover, unable to synthesize choline. It is, therefore, conceivable that the apparent lack of absolute phosphatidylcholine dependence may be related to its lower abundance in fly membranes. Some evolutionary adaptation may have occurred in the cytochrome P-450 or reductase protein, or in both, such that the phosphatidylcholine is no longer necessary to facilitate the electron flow between them.

Another possibility is that the insect PSMO system is similar to the ancestral, prokaryotic form which is soluble, and that the mammalian system may have acquired the phosphatidylcholine dependence. A great deal of experimental work is needed to evaluate these ideas.

The overall cytochrome P-450-catalyzed reaction can be summarized as follows:

$$RH + O_2 + NADPH + H^+ \longrightarrow ROH + NADP^+ + H_2O$$

Characteristically and importantly this reaction does not proceed in the reverse direction except under extraordinary conditions (see section 3.1.).

2.3. Apparent Reactions Catalyzed by Cytochrome P-450

All reactions catalyzed by cytochrome P-450 are monooxygenations, i.e., one oxygen atom from molecular oxygen is inserted in the substrate molecule. The cytochrome P-450s are, with few exceptions, carbon oxidizers. Occasionally a phosphate or sulfur atom can serve as the recipient of the oxygen. The apparent reactions have been categorized (Williams and Millburn, 1975; Nakatsugawa and Morello, 1976) into the following groups:

CH hydroxylations,
- N-dealkylation
- O-dealkylation
- S-dealkylation

π-bond oxygenations,
- Aromatic hydroxylation
- Epoxidation
- Oxidative desulfuration

Oxygenations at an unshared electron pair,
- S-oxidation
- N-oxidation

With the exception of plant allelochemicals that have found commercial use as insecticides, it is difficult to find well-documented examples of metabolism of these compounds in insects where the enzyme involved as well as the metabolites produced are identified. We shall therefore illustrate the above listed reactions with examples from both plant-derived and synthetic insecticide metabolism in insects, with the justification that an enzyme protein molecule can not "know" the synthetic origin of a molecule presented to it as a potential substrate in vivo or in vitro and, thus, can not discriminate between man-made and plant-made chemicals.

There are four general criteria which should be met to ascertain the involvement of cytochrome P-450 in a biotransformation (Wislocki et al., 1980). First, the in vivo treatment of the insects with known inducers or inhibitors should be reflected in the rate of metabolism of an administered compound. Second, metabolism can be inhibited by using specific inhibitors in an in vitro assay optimized for cytochrome P-450 activity. Third, for a compound to be metabolized, it first has to bind to the active site of the enzyme. Cytochrome P-450, being a "pigment", allows visualization of substrate binding by spectral perturbations (see section 2.5.) that tend to be characteristic of broad classes of compounds. An established spectral perturbation is a good, but not infallible, indicator that a cytochrome P-450 is, indeed, involved in the reaction. An alternative good, but not infallible, indication for cytochrome P-450 involvement is the in vitro measurement of substrate-enhanced rates of oxidation of NADPH which is consumed during cytochrome P-450 catalysis (see section 2.5.). The fourth criterion for cytochrome P-450 involvement is the demonstration of product formation in an incubation system with purified, reconstituted components. This would provide indisputable proof, due to absence of any other enzymes in the incubation, but is not yet practically available in insect work. We have tried to use examples in which at least one of these criteria is satisfied.

2.3.1. CH-hydroxylation. Aliphatic hydroxylations are very common, with the conversion of DDT to dicofol

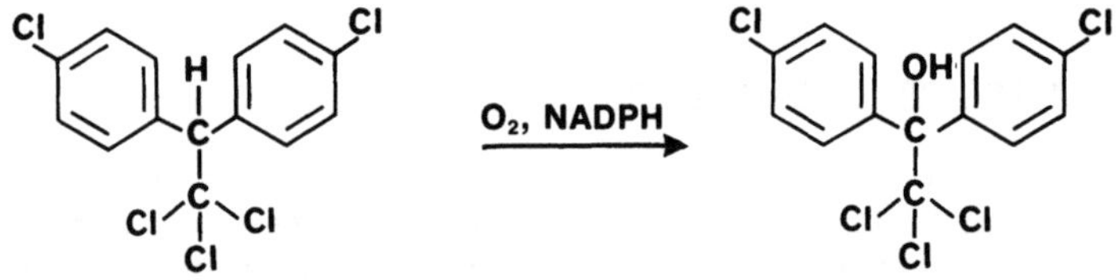

as a well-documented example in several insect species including the gypsy moth (Respicio, 1975) and the house fly (Tsukamoto and Casida, 1967). Monoterpenes often undergo this kind of biotransformation. For example, pulegone is hydroxylated to 9- and 10-hydroxy-pulegone in vitro by cytochrome P-450 from southern armyworm larval midguts (Trammell, 1982):

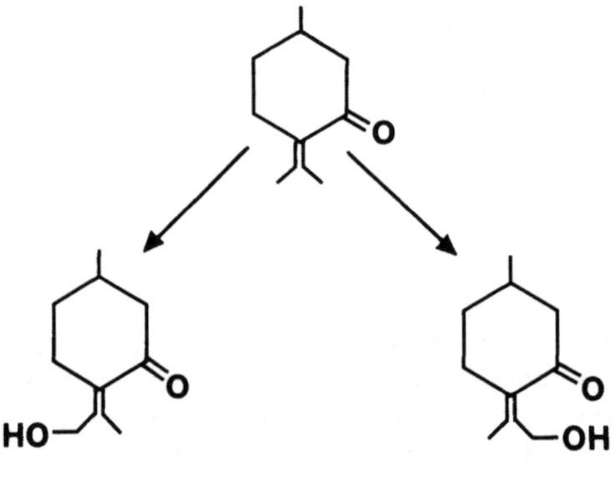

Oxidation of monoterpenes in bark beetles (Coleoptera: Scolytidae) in some cases leads to products with pheromonal activity (Brattsten, Chapter 6 in this text).

N-, O-, and S-dealkylations occur by oxygenation of a carbon atom adjacent to the heteroatom. The oxygenation intermediate is unstable and, in many cases, breaks up into a dealkylated part and an aldehyde or ketone:

$$R{-}N{<}^{H}_{CH_3} \xrightarrow{O_2,\ NADPH} [R{-}NH{-}CH_2OH] \longrightarrow R{-}NH_2 + HCHO$$

N- and O-dealkylations are important in the detoxification of many carbamate and organophosphorous insecticides (Nakatsugawa and Morelli, 1976) and therefore well documented in several species of insects. N-demethylation of p-chloro-N-methylaniline and O-demethylation of methoxy-resorufin are standard methods of estimating cytochrome P-450 activity (see section 2.5.). Because of the abundance of N- and O-alkylated natural products, these reactions are probably important in herbivorous insects (Brattsten, Chapter 6 in this text). S-dealkylation has only been indicated in the case of house flies S-demethylating the insecticide aldicarb (Metcalf et al., 1967) and may not be widespread in the metabolism of plant allelochemicals.

2.3.2. π-Bond oxygenations. Aromatic hydroxylation can take place by two different mechanisms. A form of cytochrome P-450 can directly attach an oxygen atom to a carbon double-bonded to another carbon. This mechanism is identical to the one by which aliphatic carbons are hydroxylated and appears to be of minor occurrence, at least with compounds containing aromatic rings or ring systems. The predominant mechanism is hydroxylation via an intermediate, unstable epoxide. This is well illustrated with chlorobenzene oxidations by rat liver microsomes (Wislocki et al., 1980 and references therein):

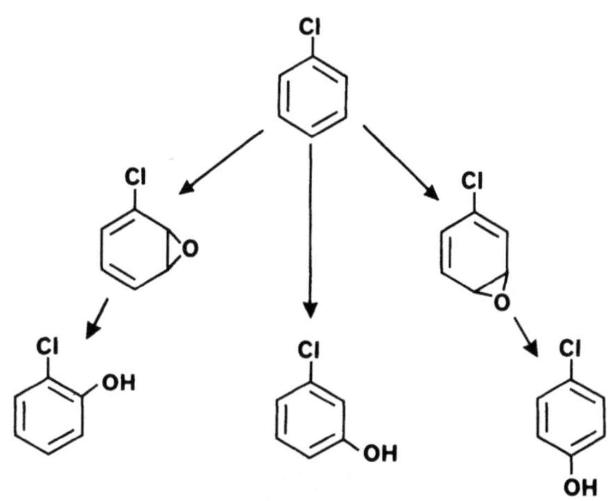

(2-OH) (3-OH) (4-OH)

Here oxygen can be directly attached to the 3- (meta) position, or inserted as a 2,3- or 3,4-arene oxide as a result of a π-bond interaction. The arene oxides are short lived and destined to one of three possible end products. They can spontaneously rearrange to the phenol. In this case, the 2,3-epoxide rearranges to the 2-(ortho) phenol and the 3,4-epoxide to the 4-(para) phenol. The latter two metabolites predominate over the 3-phenol. The formation of all three phenols was inhibited by cytochrome P-450 inhibitors and all were formed in reconstituted systems. The two arene oxides can also be converted either to the respective trans-dihydrodiols by epoxide hydrolases (see section 4.2.) or conjugated to glutathione (see section 5.1.).

Insect cytochrome P-450s can perform aromatic hydroxylations by the arene oxide intermediate mechanism. Major primary metabolites of the insecticide carbaryl in several insects (Menzie, 1969; Ahmad et al., 1980) are 4-hydroxy-, 5-hydroxy-, and trans-5,6-dihydro-5,6-dihydroxycarbaryl, the latter formed via epoxide hydrolase action on the 5,6-arene oxide (see also section 5.4.):

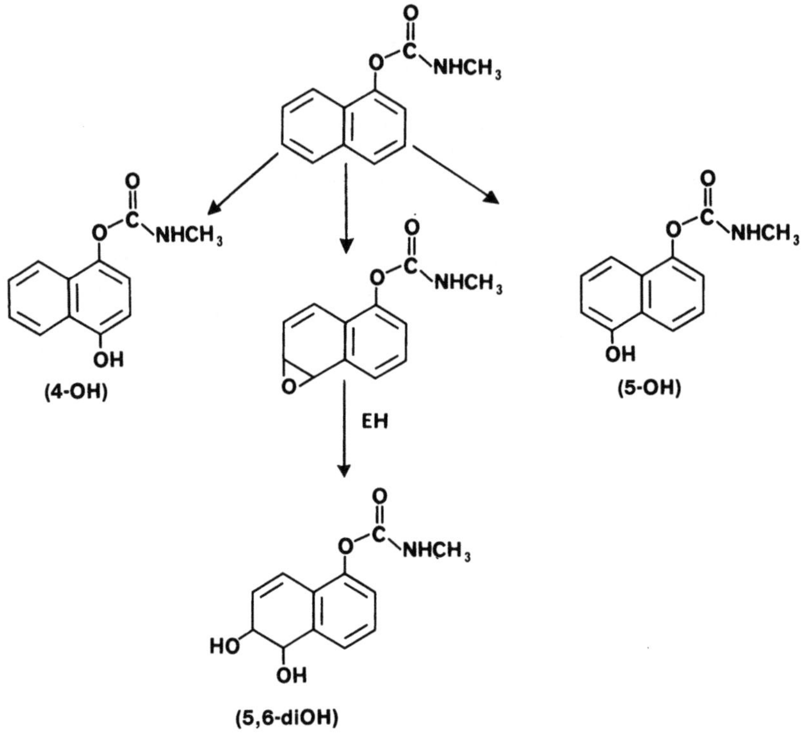

Epoxidations also occur frequently with unsaturated aliphatic compounds or side groups. An example is provided by the dihydrodiol metabolite of pyrethrin I formed in vivo in rats (Elliott et al., 1972):

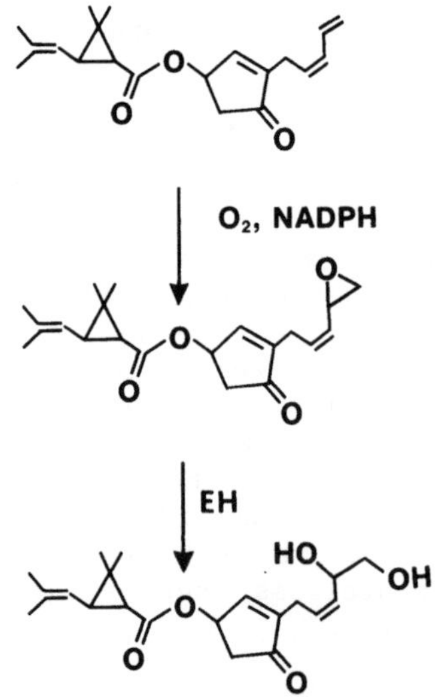

Epoxidations are an important intermediate step in the conversion of plant sterols to cholesterol in insects (Brattsten, Chapter 6 in this text). Sitosterol and other plant sterols are converted to fucosterol epoxide (Downer, 1978) by a process which requires NADPH and oxygen and therefore, likely, is catalyzed by cytochrome P-450 (Ahmad, 1979). The gypsy moth pheromone, disparlure, is an epoxide probably formed by a cytochrome P-450. Southern armyworm midgut microsomes produce disparlure from the precursor olefin (Brattsten, 1979a):

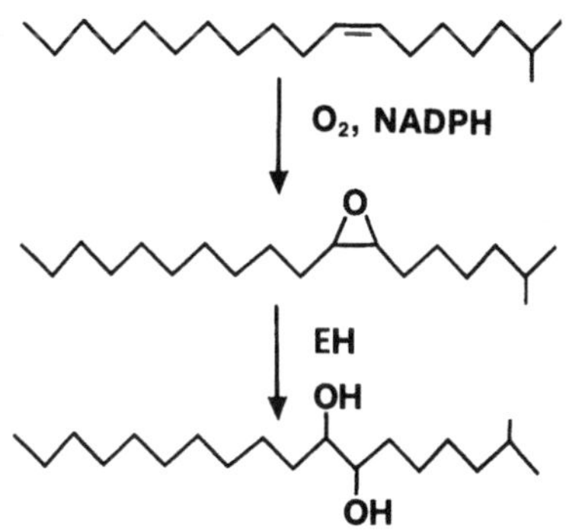

This pheromone is inactivated in the gypsy moth (Prestwick, pers. comm.) when no longer needed by an epoxide hydrolase-catalyzed conversion to the dihydroxy alkane (see section 5.4.). Trans- and cis-epoxide hydrolase (see section 5.4.) have been detected in antennal extracts of gypsy moth males and females; the lack of activity differences between the sexes suggests that the enzyme is not specialized for hydrolysis of this male-specific pheromone (Mullin and Cardé, unpubl.). In the house fly, the sex pheromone consists of a C23 olefin or alkene, (Z)-9-tricosene, (Z)-14-epoxytricosane, (Z)-14-tricosen-10-one, and a series of C23-30 methyl-branched alkanes (Adams and Holt, 1986). Blomquist et al. (1987) have confirmed that cytochrome P-450 converts the C23 olefin to the C23 epoxide and ketone, the latter through an intermediate PSMO-catalyzed hydroxylated product, (Z)-14-tricosen-10-ol. The 9,10-epoxidation is predominant over the 10-hydroxylation, and piperonyl butoxide strongly inhibits the in vivo formation of both reactions.

Experiments using two microsomal preparations, one containing gut tissues, the other containing peripheral tissues indicated that the gut enzyme was very efficient in catalyzing both reactions of (Z)-9-tricosene. Ahmad et al. (1987) demonstrated that intact abdominal cuticular epithelium and non-gut microsomal PSMO from the house fly are both active in the oxidative biotransformation of the C23 olefin. Cuticle-associated abdominal tissues are sites for pheromone synthesis in the house fly (Dillwith and Blomquist, 1982). The cuticular PSMO associated with the site of pheromone synthesis is therefore more likely involved in the synthesis of the oxygenated pheromone components than the gut enzyme.

(Z)-9-Tricosene forms a type I spectrum (see section 2.4.) with oxidized cytochrome P-450 from both males and females (Ahmad et al., 1987), direct evidence of the olefin's interaction with the cytochrome P-450. Distinctly differential inhibition by piperonyl butoxide indicates that the peripheral PSMO is directly involved in pheromone production and is qualitatively different from the gut enzyme.

Oxidative desulfuration of organophosphate insecticides occurs in tobacco budworm larvae (Whitten and Bull, 1978) and other insects. This cytochrome P-450-catalyzed biotransformation is probably not important in nature but is of major importance in organophosphorous insecticide toxicity. In this process, a phosphothionate is converted to an organophosphate as in, for example, the conversion of methyl parathion to methyl paraoxon:

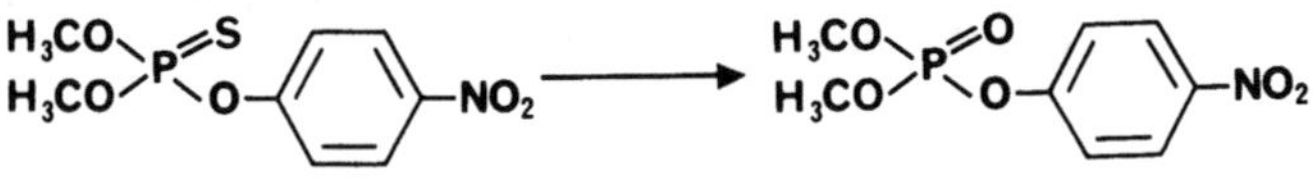

The paraoxon metabolite is a better inhibitor of acetylcholinesterase than the parent insecticide. The mechanism of this reaction has been studied with liver microsomes and is analogous to an epoxide formation followed by spontaneous rearrangement whereby elemental sulfur may become covalently bound to the active site of the enzyme (Nakatsugawa and Dahm, 1967; Nakatsugawa and Morelli, 1976). The reaction has been illustrated as follows:

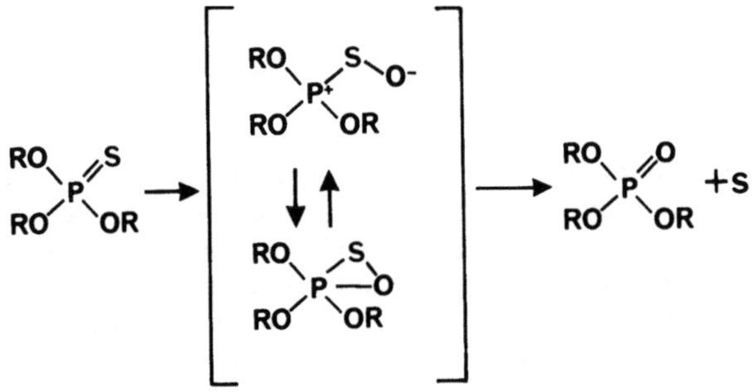

The reaction is remarkable because it produces a more toxic metabolite, cytochrome P-450 attacks a π-electron cloud generated between two heteroatoms, and it has the potential of mechanistically inactivating the enzyme.

2.3.3. Oxygenation at an Unshared Electron Pair. S-oxidations of thioethers to sulfoxides and sulfones occur in insects as shown with insecticides as substrates. Sulfoxidation of phorate with fall armyworm larval midgut microsomes (Yu, 1985) showed that cytochrome P-450 catalyzes this reaction which produced phorate sulfoxide only, although phoratoxon, the P=O analog, was produced concurrently:

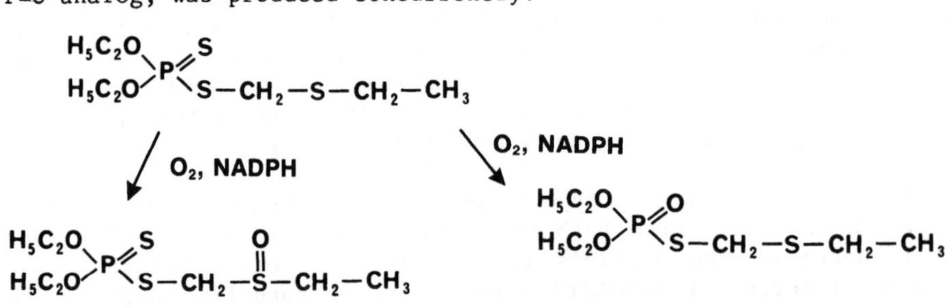

Both reactions were inhibited by carbon monoxide and piperonyl butoxide, cytochrome P-450 inhibitors, and required NADPH. Sulfoxidation usually results in increased acetylcholinesterase inhibition due to facilitated hydrolysis of the leaving group from the phosphorylated acetylcholinesterase.

Sulfoxidations can also be catalyzed by an oxygenase that does not depend on cytochrome P-450, the FAD-containing microsomal monooxygenase, E.C.1.14.13.8 (Hodgson, 1983). This enzyme has not been found in insects. Since plants produce many thioether-containing compounds the investigation of this reaction in insects, in particular, the extent of cytochrome P-450 involvement, merits attention.

N-oxidations occur frequently with alkaloids, the classical example being nicotine N-oxidation:

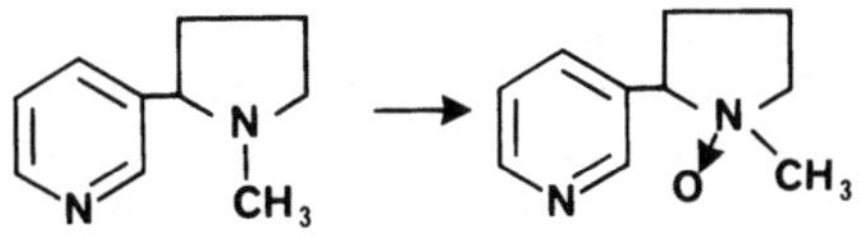

However, this reaction may not be catalyzed by cytochrome P-450 either. The FAD-containing monooxygenase is known to catalyze this reaction in liver microsomes (Tynes and Hodgson, 1985). Gorrod (1973) categorized the nitrogen-containing compounds into three groups: those with a basic amino group with pKa values between eight and eleven; those with weakly basic amino groups with pKa values between one and seven; and those with a non-basic nitrogen. The first group would preferentially be N-oxidized by oxidases other than cytochrome P-450. Those in the second group, e.g. nicotine, may undergo N-oxidation by non-P-450-dependent oxidases or N-dealkylation by cytochrome P-450 depending on the overall steric and electrochemical properties of the molecule. The compounds in the third group would undergo cytochrome P-450-catalyzed N-dealkylations (Brattsten, 1979b). This reasoning is consistent with the preference of cytochrome P-450 to oxygenate carbon atoms.

2.4. Measurements of Cytochrome P-450-catalyzed Reactions in Insects

The only way to ascertain whether cytochrome P-450 catalyzes the biotransformation of a certain compound is to incubate it with as pure a fraction of the enzyme as practically possible in a buffer that contains the required cofactors, and to identify the metabolite(s). It is difficult to obtain active cytochrome P-450 in sufficient quantity purified beyond the microsomal fraction of insect tissues. Cytochrome P-450 may be up to six-fold enriched in microsomes compared to tissue homogenates (Brattsten et al., 1980). Since microsomes may also contain other oxidases, such as the FAD-monooxygenase, it is important to quantify metabolite production in the absence and presence of specific cytochrome P-450 inhibitors, especially if heteroatoms are oxidized. An example of this is the sulfoxidation of phorate. This reaction is catalyzed to a large extent by the FAD-monooxygenase in mouse liver microsomes (Tynes and Hodgson, 1985) but by cytochrome P-450 in microsomes from fall armyworm tissues (Yu, 1985), because metabolite formation was inhibited by piperonyl butoxide and carbon monoxide in the insect microsomes. This complete investigation of the metabolism of a given compound depends on the availability of pure, putative metabolite standards and/or on access to high-resolution analytical instruments.

The fact that NADPH is oxidized to NADP during cytochrome P-450 catalysis offers an indirect method for evaluating whether a certain compound is a cytochrome P-450 substrate. NADPH absorbs strongly at 340 nm and it is, thus, possible to measure the disappearance of this absorbance with time in a buffered incubation also containing the putative substrate. This method only provides quantitative information and none about the nature of the metabolite(s) formed. Since there is always endogenous NADPH-oxidation in microsomes, i.e. in the absence of an exogenous substrate, it is important to quantify this and subtract it from the reaction rate obtained with the exogenous substrate present. Endogenous oxidation of NADPH in insect microsomes has been documented by Folsom and Hodgson (1970), Ahmad and Forgash 91973, 1978), and Brattsten et al., (1984). This method is not infallible because certain compounds such as nicotine and several steroids did not cause measurable NADPH oxidation in southern armyworm midgut microsomes (Brattsten et al., 1984) although they are known cytochrome P-450 substrates.

Another indirect evaluation of a compound's potential for being a cytochrome P-450 substrate is the measurement of the optical difference spectrum of cytochrome P-450 in its presence. Like all hemoproteins, cytochrome P-450 absorbs radiation in the visible wavelength range. It is, however, difficult to measure absolute absorbances of cytochrome P-450 in microsomal particles due to wave-length-dependent light scattering by

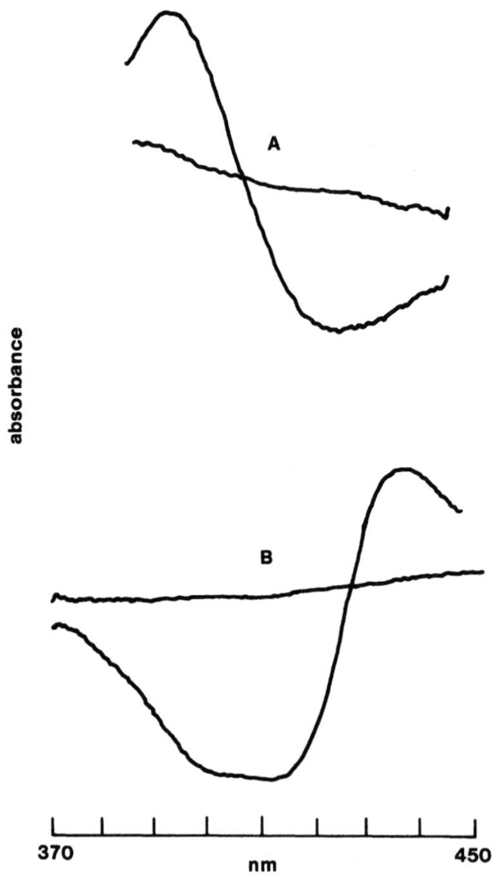

Fig. 5. Spectral perturbations resulting from substrate interaction with oxidized (Fe^{3+}) cytochrome P-450.

A. Type I spectrum with a peak at 392 nm and a trough at 421.5 nm caused by binding of 10 μM pulegone to cytochrome P-450 in 2.0 mg microsomal protein from southern armyworm larval midguts.

B. Type II spectrum with absorption maximum at 435 nm and minimum at 414 nm caused by indole binding.

the membrane fragments. Instead, the spectral perturbation, measured as the difference in absorbance in the presence and absence of a compound (ligand) that binds to the active site of the cytochrome, can be quantified in relation to ligand concentration (Brattsten and Gunderson, 1981). This provides an apparent spectral dissociation constant that indicates the affinity of the cytochrome for a compound. Fig. 5 shows the two major difference spectra with oxidized microsomes. Nitrogen-containing compounds usually produce the type II difference spectrum, thought to reflect binding to the heme iron in addition to the lipophilic interaction. Most other compounds give rise to the type I spectrum, presumably, reflecting the lipophilic binding only to the protein portion of the cytochrome (Dauterman and Hodgson, 1978). Other types of spectral perturbation with oxidized and reduced microsomes have proven useful in insecticide resistance studies with genetically isolated strains of the house fly (Hodgson and Kulkarni, 1979), but have not yet been used for studying interactions with plant allelochemicals.

The spectral method does not reveal the rate by which the compound can be metabolized since it only illustrates the first step, binding, in a sequence of events leading to an oxidized product. Piperonyl butoxide binds with high affinity to the cytochrome and produces a quantifiable spectral perturbation, but is metabolized extremely slowly. It is therefore, an excellent inhibitor of cytochrome P-450.

The most direct measurement of cytochrome P-450 is by the spectral perturbation caused by complex formation with carbon monoxide of the reduced cytochrome. This difference spectrum shown in Fig. 6, allows quantification of the cytochrome content of the microsomes provided that there is not substantial degradation to the inactive form, cytochrome P-420. The latter also binds carbon monoxide but has no catalytic activity. It can form during microsome preparation by many factors including unsuitable pH and ionic strength of the buffer, presence of proteolytic enzyme activity in the homogenate, rough mechanical treatment during homogenization and resuspension of the microsomes, or unsuitable storage conditions. The presence of glycerol in the buffer after centrifugation has a considerable stabilizing effect on the cytochrome and protects it from damage by overdoses of sodium dithionite, the reducing agent. Fig. 6 shows a small conversion to cytochrome P-420. This method was first described in 1964 by Omura and Sato who quantified the heme in a sample of rabbit liver cytochrome P-450 completely converted to P-420 and thereby were able to assign a molar extinction coefficient, 91 mM^{-1}, cm^{-1}, to cytochrome P-450. This extinction coefficient is validly used for quantifying cytochrome P-450 from all types of sources inasmuch as all cytochromes P-450 are hemoproteins.

The carbon monoxide complex-generated difference spectrum also allows some discrimination between different isoenzymic forms without further purification. Rodent liver microsomes induced (see section 2.6.) with phenobarbital have an absorbance maximum at 450 nm, whereas those induced with 3-methylcholanthrene absorb maximally at 448 nm. Southern armyworm midgut microsomal cytochrome P-450 absorbs at 450 nm, but when induced with pentamethylbenzene shows an absorbance peak at 449 nm (Brattsten et al., 1980). A large cytochrome P-420 peak can displace the absorbance maximum of cytochrome P-450. It is therefore best to reduce the microsomes with NADPH, which does not reduce cytochrome P-420, for accurate location of the absorbance maximum (Hodgson and Kulkarni, 1979; Hodgson, 1985).

There are many compounds which can be used as substrates to measure the catalytic activity of cytochrome P-450. Ideally the substrate should undergo only one reaction, the product of which can be conveniently and

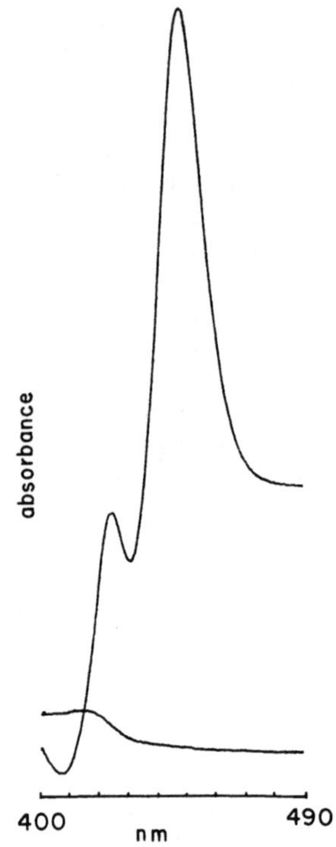

Fig. 6. Spectral perturbation caused by carbon monoxide binding to dithionite-reduced (Fe^{2+}) cytochrome P-450 in 2.0 mg microsomal protein from southern armyworm larval midguts.

accurately quantified at very low concentrations. Substrates commonly used with insect tissues include aldrin, heptachlor, dihydroisodrin, p-chloro N-methylaniline, aniline, p-nitroanisole and methoxyresorufin. In addition, biphenyl, benzo(a)pyrene, N, N-dimethyl-p-nitrophenyl carbamate, and others have been used. Some of the substrates used in insect work, reactions, and methods of product quantification are shown in Table 3.

None of the model substrates used is a natural product. The synthetic origin of the compound is, as pointed out before, not recognized by the enzyme. It is important to use more than one single model substrate in evaluating the metabolic capacity of an insect because of the presence of isoenzymic forms (Ahmad, 1986; Gould, 1984). This is particularly important in studies of induction (see section 2.6.) because the different forms can respond differently to any one inducer (Brattsten, 1979b, Yu, 1985; Yu, Chapter 4 in this text).

Work with model substrates provides a good starting point for assessing the potential for metabolism of a plant allelochemical of interest but cannot replace direct metabolic studies of the compound in question. Model substrates will remain useful, however, to ascertain the activity of the enzyme preparation used for metabolic studies with the intrinsically unstable cytochrome P-450, particularly in homogenates from insect tissues that often contain endogenous inhibitors.

2.5. Inhibition of cytochrome P-450

Inhibitory interactions have received considerable attention because of the potential practical importance in pest control of inhibition of insect PSMO-activity.

Table 3. Model substrates commonly used for measurement of insect cytochrome P-450 activity.

Reaction		Quantification	Reference
Aldrin epoxidation:	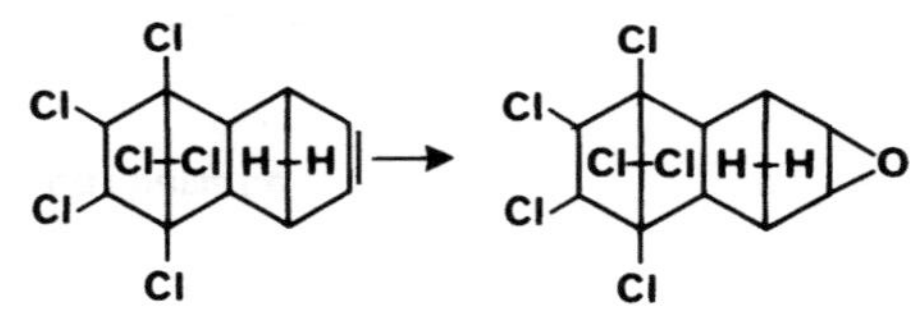	GLC	Lewis et al., 1967
Heptachlor epoxidation:	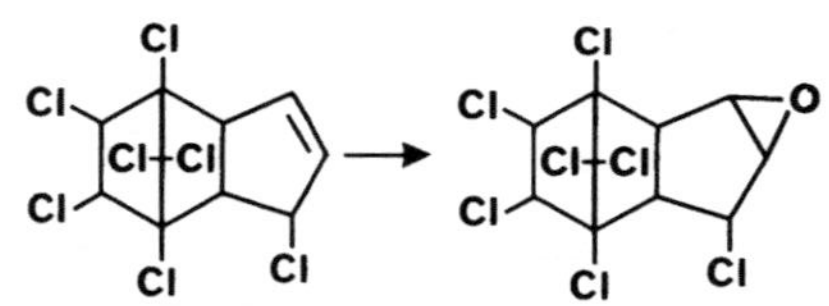	GLC	Nakatsugawa et al., 1965
Dihydroisodrin hydroxylation:	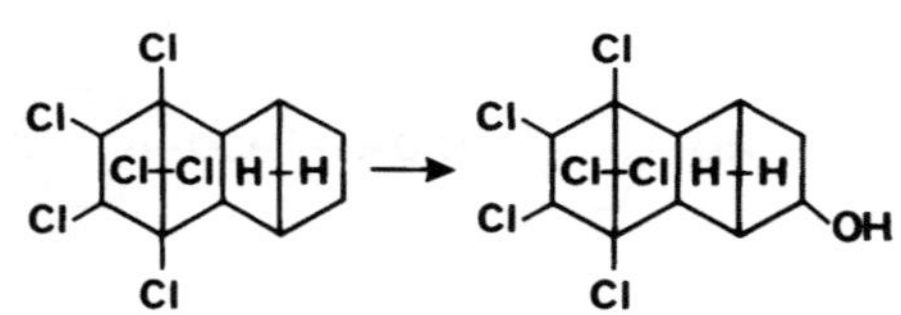	GLC	Krieger & Wilkinson, 1971
p-chloro N-methylaniline N-demethylation:	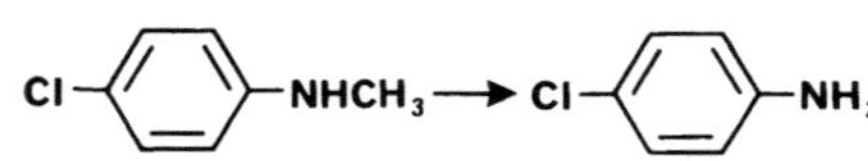	diazotization, spectrophotometry	Kupfer & Bruggeman, 1966

(continued)

Table 3 (cont.)

Reaction	Quantification	Reference
p-nitroanisole O-demethylation: 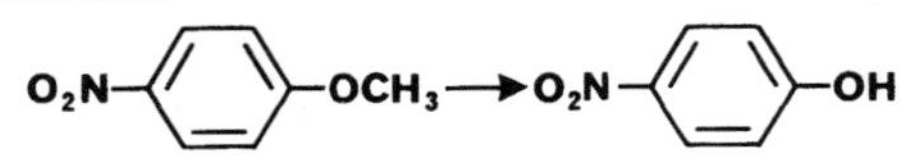	spectrophotometry	Hansen & Hodgson, 1971
Aniline hydroxylation: 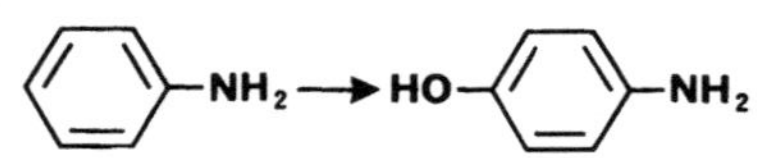	spectrophotometry	Schenkman et al., 1967
Methoxyresorufin O-demethylation: 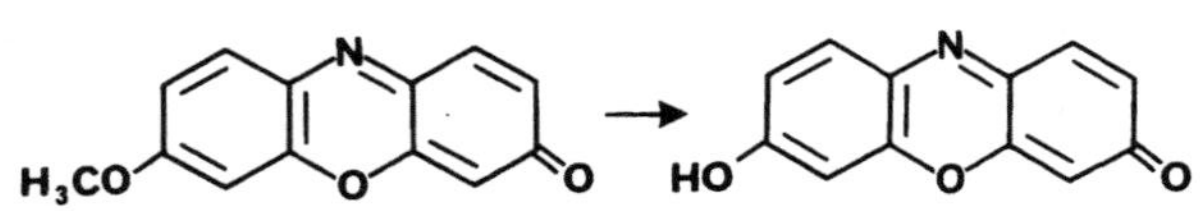	direct fluorometry	Mayer et al., 1977
N, N-dimethyl-p-nitrophenyl carbamate N-demethylation: 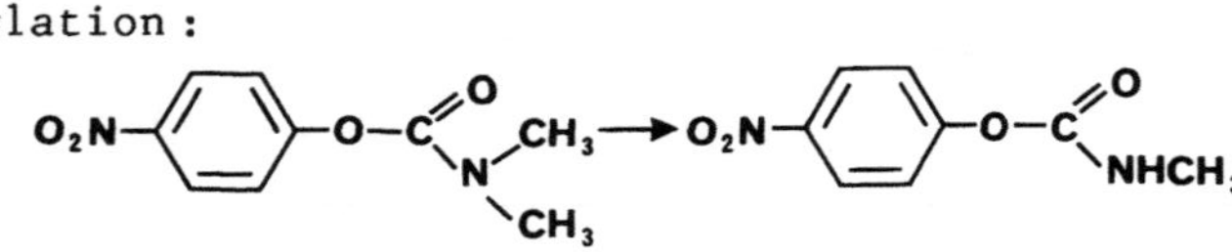	spectrophotometry	Hodgson & Casida, 1961

2.5.1. Endogenous Inhibitors. In vitro measurement of insect PSMO activities is often hampered by the release, during enzyme preparation, of proteases that destroy activity. Two digestive proteases that can interfere have been reported (Krieger and Wilkinson, 1970; Orrenius et al., 1971; Wilkinson and Brattsten, 1972; Hodgson, 1985). In addition to proteases, an eye pigment, xanthommatin and a small species of RNA may produce misleading interferences in in vitro activity assays (Wilkinson and Brattsten, 1972; Hodgson, 1985). None of these is likely to have any importance in vivo.

2.5.2. Synergists as Inhibitors. Insecticide synergists are compounds that enhance the toxicity of an insecticide when applied together with it at non-lethal concentrations, by inhibiting the enzyme that would normally detoxify the insecticide (Metcalf, 1967). The most successful commercially used synergist, piperonyl butoxide, contains the methylenedioxyphenyl group and was developed with the plant allelochemical sesamin as model (see section 2.5.3.).

The methylenedioxyphenyl compounds bind to the active site of cytochrome P-450 and undergo metabolism very slowly (Wilkinson and Hicks, 1969; Casida, 1970); they thereby act as alternate substrates. However, the interaction is rather complex and may involve formation of one of the several free radical species of carbon including a carboxonium ion, a carbanion, or a carbene (Hodgson, 1985). The interactions have been studied extensively by spectral perturbations of the cytochrome (Hodgson, 1985).

Other cytochrome P-450 inhibitors with a potential use as synergists include the 1,2,3-benzothiadiazoles (Gil and Wilkinson, 1977) and several substituted imidazoles (Wilkinson et al., 1974a, 1974b).

2.5.3. Plant Allelochemicals as Inhibitors. The methylenedioxyphenyl synergists owe their existence to sesamin, a natural product present in sesame oil. The latter, when tested with other vegetable oils as carriers for pyrethrin insecticides, caused a significantly increased toxicity of the pyrethrins (Eagleson, 1940), later associated with sesamin (Haller et al., 1942). Natural methylenedioxyphenyl-containing compounds, including lignans and other phenylalanine derivatives occur widely in plants (Newman, 1962; Vickery and Vickery, 1981) including *Chrysanthemum cinerariaefolium*, the commercial source of the pyrethrin insecticides (Doskotch and el-Feraly, 1969).

Myristicin, a major lignan in parsnip, *Pastinaca sativa* and other umbellifers, and an insecticide synergist (Lichtenstein and Casida, 1963), synergizes the toxicity of the co-occurring linear furanocoumarin xanthotoxin better than piperonyl butoxide in the corn earworm, *Heliothis zea* (Boddie) (Lepidoptera: Noctuidae) (Berenbaum and Neal, 1985). The methylenedioxyphenyl-containing amides pipercide, dihydropipercide, and guineensine from black pepper, *Piper nigrum*, fruits are moderately toxic to the adzuki bean weevils, *Callosobruchus chinensis* L. (Coleoptera: Bruchidae), when each is administered alone but synergize each other two to three-fold when applied together in a 1:1:1 ratio (Miyakado et al., 1983). Gypsy moth larvae may encounter cytochrome P-450 inhibitors in their normal food plants; they have higher activities when reared on a wheat germ diet compared to larvae fed oak leaves (Ahmad and Forgash, 1978), although nutritional factors may also influence activities.

These and other cases have inspired ideas that plants may elaborate on their defensive allelochemicals by employing synergistic mixtures (Berenbaum, 1985; Raffa and Priester, 1985). Experimental work in this area should help our understanding of insect-plant co-evolution.

Several kinds of chemicals are known to undergo activation by cytochrome P-450 to highly reactive metabolites, usually epoxides, that can form adducts with cellular macromolecules and thus cause cell destruction, or they may instantly adduct to the cytochrome protein itself and thus inhibit it (Ortiz de Montellano and Correia, 1983). Acetylenic compounds are known to do this. Pyrrolizidine alkaloids are cytotoxic in vertebrate liver (Bull et al., 1968) and destroy cytochrome P-450 in rat liver (Miranda et al., 1980). Chromenes are cytotoxic in mammalian liver (Halpin et al., 1984) and also damage insect cells with high cytochrome P-450 content (Ellis-Pratt, 1983). Since sulfhydryl group inhibitors such as p-chloromercuriphenyl sulfonate are known cytochrome P-450 inhibitors (Folsom and Hodgson, 1970), compounds such as parthenin, coronopilin and other sesquiterpene lactones may also prove to be cytochrome P-450 inhibitors.

2.6. Induction of cytochrome P-450

Cytochrome P-450 and several other xenobiotic-metabolizing enzymes undergo a rapid increase in activity when exposed to a large number of environmental chemicals. This phenomenon is termed induction and depends on de novo synthesis of enzyme protein. The inducing compound may stimulate the synthesis of more protein of the kind already present, or of closely related, but sufficiently different isoenzymic forms, to allow metabolism of a different kind of chemical. Studies of cytochrome P-450 induction provide the strongest indications that insects have different isoenzymic forms (see Yu, Chapter 4 in this text). Induction is a strictly temporary phenomenon, the enhanced activity level lasting only as long as a sufficiently high concentration of the inducing chemical remains in the tissue. Thus, the effect of highly biodegradable inducing chemicals, such as most plant allelochemicals, is of short duration unless the insect constantly ingests a continuing dose of the compound.

The precise molecular mechanism(s) of induction in insects is unknown. It may proceed similarly to the classic operon model in bacteria via a soluble protein receptor. This is the case in mouse liver cells which contain a receptor that specifically binds polycyclic aromatic hydrocarbons and dioxins (Nebert and Jensen, 1979; Okey et al., 1979). Induction by barbiturates and other compounds may proceed via some different mechanism in rodents since no receptor protein specific for these compounds has been found. There are some indications that the induction mechanism in insects may be different. The polycyclic aromatic hydrocarbons, although metabolized by insect cytochrome P-450 (Anderson, 1978; Chang et al., 1983) are, at best, weak inducers of cytochrome P-450-catalyzed reactions in insects (Brattsten et al., 1976; Chang et al., 1983; Terriere and Yu, 1974; Tate et al., 1982). No cytosolic protein receptor from an insect tissue has been found (Denison et al., 1985) or unambiguously characterized (Plapp, 1984).

There are no obvious structural similarities among the many chemicals capable of inducing cytochrome P-450. However it appears that binding to the cytochrome's active site may be a prerequisite for induction. Piperonyl butoxide and other methylenedioxyphenyl compounds which inhibit the cytochrome by binding to its active site, also induce activity (Thongsinthusak and Krieger, 1974; Dickins et al., 1978; Cook and Hodgson, 1983). Substitution on the methylene carbon to prevent it from binding to the active side of cytochrome P-450 produces compounds without ability to induce (Cook and Hodgson, 1983).

2.7. Regulation of Cytochrome P-450 Activity

From what limited information is available about the variation of

PSMO activity with age in insects, it is obvious that the activity is highly variable. Generally, activity is high in active life stages, actively feeding life stages including immature stages bewteen molts, and typically low in inactive, resting or protected stages such as eggs, molting immatures, pupae, and non-feeding adults. This seems to indicate a strong influence by external factors. Considering the important roles cytochrome P-450 plays in the metabolism of endogenous substances, notably hormones, it seems more realistic to assume some basic internal control upon which the external factors may be superimposed to modulate the general level of activity. It is also apparent that PSMO activity differs between tissues both in general activity levels (Tate et al., 1982; Brattsten et al., 1980; Feyereisen and Farnsworth, 1985) and in response to inducing chemicals (Brattsten et al., 1980). This indicates independent control between tissues and therefore intrinsic control. The few cases in which activity has been monitored in sexed adults (Benke et al., 1972; Benke and Wilkinson, 1971; Feyereisen and Farnsworth, 1985) clearly show sex-related differences in activity levels and different degrees of such differences depending on which reaction is studied. This also implies intrinsic or internal control of cytochrome P-450 activities.

The many differences in inducibility of different cytochrome P-450-catalyzed reactions (Yu, Chapter 4 in this text) indicate a direct influence on separate DNA segments coding for different cytochrome P-450 isozymes or, at least, multiple regulatory genes, rather than an effect on one single regulatory gene.

Non-genetic internal factors can also influence cytochrome P-450 activities. Being dependent on reducing equivalents from the high-energy intermediary metabolite NADPH, the general nutritional condition of the insect may affect activity. Cytochrome P-450 is also dependent on the intact endoplasmic reticulum membrane (see section 2.2.). The membrane is susceptible to lipid peroxidation which, in mammals, depends on iron. If the organism is exposed to a foreign chemical normally conjugated to glutathione (see section 6.1.) to the extent of depleting endogenous glutathione levels, iron, normally bound by glutathione and other endogenous factors, may be released and stimulate microsomal lipid peroxidation. This would cause a depression of cytochrome P-450 activity (Levine, 1982). This possibility has not been explored in insects.

3. OTHER BIOLOGICAL OXIDATIONS

Cytochrome P-450 is not the only mechanism capable of catalyzing oxidation of foreign chemicals, altough it is, by far, the most important one. Over 200 oxidases, many of which act on more than one substrate, are known from living organisms (Keevil and Mason, 1978). The participation of copper-, non-heme iron-, and flavin-dependent oxygenases, in contrast to iron-porphyrin monooxygenases such as cytochrome P-450, needs experimental work in foreign compound metabolism. Dioxygenases, enzymes which insert both oxygen atoms from molecular oxygen into the substrate molecule, occur in higher organisms but seem to be most characteristic of prokaryotes where they are a major mechanism for breaking open aromatic rings.

3.1. FAD-monooxygenase

FAD-monooxygenase (E.C.1.14.14.8) is a microsomal oxidase that does not rely on the metal-containing porphyrin nucleus for activity but instead uses flavin adenine dinucleotide (FAD) as prosthetic group. FAD-monooxygenase, like cytochrome P-450, is embedded in the endoplasmic reticulum membranes and depends on NADPH for reducing equivalents. It uses only molecular oxygen and inserts only one oxygen atom into the substrate molecule. Unlike cytochrome P-450, it is not inhibited by methylenedioxyphenyl compounds and does not appear inducible by typical cytochrome P-450 inducers such as phenobarbital or 3-methylcholanthrene; it is, instead, activated by tert-octylamine (Hodgson, 1983). It has been isolated from mammalian liver cells (Poulsen, 1981; Tynes and Hodgson, 1985) and appears especially suited to oxidize nitrogen and sulfur. Many insecticides containing these heteroatoms are oxidized by FAD-monooxygenase in mammals. There is no published information about this enzyme in insects. The insecticide phorate readily undergoes sulfoxidation catalyzed by FAD-monooxygenase from mammalian liver. There are no indications that this enzyme is involved in phorate sulfoxidation in midgut microsomes from the generalist feeder, the fall armyworm (Yu, 1985; section 2.3.3.). It may be of interest to search for this enzyme in insect herbivores that specialize to feed on plants with sulfur-containig allelochemicals. Oxidations of tertiary amines may be catalyzed by this enzyme (section 2.3.3.) or by some other amine monooxygenase.

3.2. Dehydrogenases

Human liver alcohol dehydrogenase (E.C.1.1.1.1, alcohol:NAD oxidoreductase) has been intensively investigated because of its importance in alcoholism. The mystery with this enzyme is its occurrence in very high concentrations in the mammalian liver although ethanol is neither a normal nutrient nor a normal intermediary metabolite in mammals. Up to 16 different molecular forms of the enzyme are known (Bosron and Li, 1980). Between them, they readily accept primary and secondary alcohols of widely dissimilar structures as substrates and, therefore, may be of importance in foreign compound metabolism. Being oxidoreductases they, however, also catalyze the reduction of the aldehydes formed; more often, the enzymes involved in foreign compound-metabolism catalyze a reaction only in one direction. The alcohol dehydrogenases are cytosolic metalloenzymes which rely on NAD/NADH. Alcohol dehydrogenases have been extensively investigated in *Drosophila* and a few other dipterans (Moxon et al., 1982; Delden, 1982; Mercot, 1985) but are otherwise unexplored in insects. Nevertheless, metabolic studies clearly imply their presence and involvement in the metabolism of plant allelochemicals in insects. The major metabolite of nicotine in several insects is cotinine:

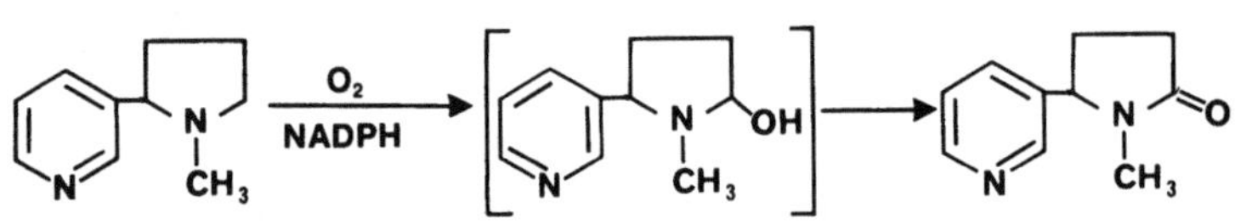

It is formed by initial oxidation of the carbon next to the pyrrole heteroatom by cytochrome P-450 followed by further oxidation, presumably, catalyzed by an alcohol dehydrogenases, of the secondary alcohol formed. It is unlikely that the further oxidation occurs spontaneously. Many allelochemicals could be oxidized by a similar mechanism; the conversion of the ingested phenyl glycoside salicin to salicylaldehyde in the defensive glands of some chrysomelid beetle larvae (Pasteels et al., 1983) may be another case.

4. BIOLOGICAL REDUCTIONS

Whereas higher organisms including insects certainly employ reductive reactions in their metabolism of foreign compounds, reductions appear characteristic of and more frequently utilized by microorganisms; it is probably often gut microorganisms that are responsible for the production of reduced metabolites observed in animals (Goldman, 1982; Rowland, 1986).

4.1. Nitro-, Azo-, and Tertiary Amine Oxide Reduction

In vitro experiments indicate that cytochrome P-450 can catalyze reductions of nitro- and azo- groups and of tertiary amine oxides in the absence of oxygen. All three groups, as well as others, are capable of forming free radicals (Kalyanaraman, 1982). In the absence of oxygen, the reactive group may bind directly to the site otherwise occupied by molecular oxygen and undergo radical formation through some direct interaction. Reductions of this kind have been documented with microsomes from mammalian liver and synthetic nitro- and azo- compounds (Wislocki et al., 1980); nitro reduction has been demonstrated only in one case with an insect. Nitrobenzene was reduced to aniline by extracts of several tissues of the Madagascar cockroach under anaerobic in vitro conditions (Rose and Young, 1973), and flavin (FMN) added to the incubations greatly stimulated the rate of conversion. Nitro and azo compounds may not be common among plant allelochemicals. This, and the strict requirement for anaerobic conditions imply that the reaction may not be of great importance in vivo.

Many alkaloids occur as tertiary N-oxides, a more water-soluble form than the free base and, therefore, potentially less toxic. Insects feeding on alkaloid-containing plants, therefore, ought not to have an enzyme that readily reduces tertiary N-oxides. Cytochrome P-450 operating in an anaerobic atmosphere in vitro can catalyze this reaction (Kato et al., 1978) as well as the reduction of arene oxides such as benz(a)anthracene (Booth et al., 1975). No work other than that with the Madagascar cockroach has been done to elucidate the occurrence of enzyme-catalyzed reductions of these kinds in insects.

4.2. Carbonyl Reductases

Many carbonyl-group-containing drugs and other foreign compounds are reduced by a group of apparently specialized enzymes related to alcohol dehydrogenases (E.C.1.1.1.2, alcohol $NADP^+$ oxidoreductase). They have been found in many mammalian, reptile, and amphibian liver tissues as well as in yeast and *Drosophila* flies (von Wartburg and Wermuth, 1980; Davidson et al., 1978; Felsted and Bachur, 1980). They convert a large variety of structurally dissimilar lipophilic aldehydes and ketones to alcohols which can be conjugated and therefore more easily excreted. Almost all work to date with these enzymes has been done to elucidate their role in drug metabolism. Being cytosolic and fairly stable proteins, these enzymes can be purified by standard techniques and are usually monitored by measurement of NADPH absorbance at 340 nm with time in the presence of a candidate substrate, assuming a one-to-one stoichiometry. Whereas the aldehyde reductases in general do not catalyze the reduction of the less reactive organic ketones, ketoreductases appear to catalyze the reduction of both aldehydes and ketones. Daunorubicin, an antibiotic from *Streptomyceus peucetius*, adriamycin, and aflatoxin (Ahmed et al., 1979) are some of the few natural foreign compounds known to undergo metabolic inactivation catalyzed by these enzymes which are inhibited by flavonoids such as quercetin, rutin, and dicoumarol (Wermuth, 1981). They are not inducible (Bachur, 1976) by cytochrome P-450 inducers.

The milkweed cardenolide uscharidin is converted in vitro to the reduced enantiomeres calotropin and calactin by gut and fatbody homogenates of Monarch butterfly, *Danaus plexippus* L. (Lepidoptera: Danaidae), larvae (Marty and Krieger, 1984):

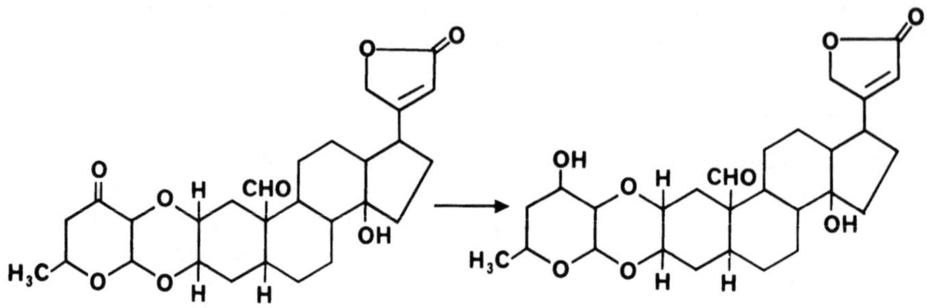

This reaction proceeds only in one direction, towards metabolites with increased polarity, as seems characteristic for foreign compound-metabolizing enzymes.

5. HYDROLASES

The hydrolases are a very large class of enzymes all of which basically split a covalent bond with the help of a water molecule. Most of the enzymes in this class (IUB, 1984) hydrolyze endogenous substances and are important in intermediary metabolism. The class also contains some enzymes of known importance in foreign compound metabolism such as esterases (E.C.3.1.1), glycosidases (E.C.3.2), and ether hydrolases represented by epoxide hydrolases (E.C.3.3.2.3).

5.1. Esterases

An esterase is an enzyme that catalyzes the hydrolysis of an ester bond without any requirement of high-energy cofactors:

$$R-C(=O)OR^{1} + H_2O \longrightarrow R-C(=O)OH + HOR^{1}$$

However, of the known esterases there are many that can also catalyze the hydrolysis of phosphoesters, sulfuric acid esters, and amide bonds. Many of these esterases have very broad substrate specificities, and may, while fulfilling some important but unidentified endogenous function, also be important in the metabolism of plant allelochemicals. The terminology and identification of these esterases is hampered by their apparent, great versatility (Walker and Mackness, 1983); an esterase may be able to hydrolyze both a carboxylester bond and a phosphoester bond, or an amide bond as well as a carboxylester bond. There are, in other words, carboxylesterases with phosphatase or amidase activity. In addition, any carboxylester bond may be hydrolyzed by several different esterases. The present IUB classification of esterases into carboxylesterases (E.C.3.1.1.1, also called aliesterases or B-esterases) and arylesterases (E.C.3.1.1.2, also called A-esterases or paraoxonases) is based on the pioneering work by Aldridge (1953) and Augustinson (1959, 1961) who suggested two general groups of esterases based on their sensitivity towards inhibition by organophosphorous insecticides. Arylesterases are not inhibited by paraoxon, whereas carboxylesterases are. The inhibition is non-competitive and results from irreversible (or nearly so) phosphorylation of the seryl residue in the active site. Carboxylesterases are thus serine hydrolases which can be further distinguished from the cholinesterases (E.C.3.1.1.8), some of which are important as insecticide targets, by the inhibition of the latter, but not the former, by 10^{-5}M physostigmine (Kapin and Ahmad, 1980). In general, the carboxylesterases can be seen as representing one end of a, probably continuous, series of enzymes catalyzing hydrolysis, with the lipases (E.C.3.1.1.3) at the other end. Esterases metabolize relatively small, polar molecules with an ester bond at least 1000 times faster than lipases, which prefer long-chain fatty acid esters as substrates. Because of their importance in insecticide toxicity, considerable attention has been paid to esterases in insects and several comprehensive reviews are available (Oosterbaan and Jansz, 1965; Dauterman, 1976, 1985; Ahmad and Forgash, 1976; Brattsten, 1979a; Junge, 1984; Junge and Klees, 1984; Heymann, 1980).

In many cases, no attempt is made to distinguish carboxylesterase activity from that of other esterases such as aryl esterases, cholinesterases or even lipases. This results in an overall measurement of the ability of the insect to hydrolyze a model substrate, often 1-naphthyl acetate, and may not reveal the relative importance of esterases in the detoxification of an allelochemical. The resultant activity is often referred to a "general esterase" or "esterase" activity.

5.1.1. Carboxylesterases. Carboxylesterase activity has been studied in many insect species, Table 4, and appears to be universally present in all insect tissues, although different forms may reside in different tissues. The presence in all tissues of lepidopterous larvae has been shown in the southern armyworm (Abdel-Aal and Soderlund, 1980) and the gypsy moth by Kapin and Ahmad (1980). In both species, the midgut had the highest activity. A similar distribution of esterase activity in different tissues including fore-, mid-, and hindgut, and muscles is present in adult house flies (Ahmad, 1976) where different mixtures of isozymes predominate in the different tissues. The temporal variation in activity in gypsy moth larval guts indicates a coincidence of high enzyme activity and feeding activity (Kapin and Ahmad, 1980). In a study of two generalist feeders, the corn earworm and the cabbage looper larvae, Turunen and Chippendale (1977) found more isoenzymic esterase forms in their midgut than in two specialized feeders, the larvae of the southwestern corn borer, Diatraea grandiosella (Dyar) (Lepidoptera: Pyralidae), and the imported cabbageworm, Pieris rapae (L.) (Lepidoptera: Pieridae). These studies support the idea that carboxylesterases are important in foreign compound metabolism.

Carboxylesterases occur in many isoenzymic forms with broadly overlapping substrate preferences. There are both microsomal and cytosolic forms. Turunen and Chippendale (1977) found at least four different forms in the lepidopterous larval guts. At least 15 bands with sensitivity towards 1-naphthyl acetate were obtained in the cattle tick, Boophilus microplus (Acari: Ixodidia) larvae by isoelectric focusing (De Jersey et al., 1985). Ahmad (1976) demonstrated at least ten isoenzymic forms in whole adult house flies by vertical slab polyacrylamide gel electrophoresis. Kao et al. (1985), showed by chromatofocusing that four of the adult house fly esterases hydrolyze both p-nitrophenyl butyrate and paraoxon, and that their activities correlated with resistance to organophosphorous insecticides. In another fly, the sheep blow fly, Lucilia cuprina (Wiedemann) (Diptera: Calliphoridae), Hughes and Raftos (1985), demonstrated 16 esterase bands by electrophoresis; only one of these bands, E3, was associated with organophosphate resistance. E3 was present in susceptible flies but, presumably, occurs in a mutant form in the resistant flies where it hydrolyzes organophosphates more effectively than the original E3 but has lost the ability to hydrolyze 1- and 2-naphthyl acetate. Six carboxylesterases were identified in the blood sucking bug, Triatoma infestans Klug (Heteroptera: Reduviidae) none of which hydroylzed paraoxon (Casabe and Zerba, 1981); instead, an arylesterase was responsible for paraoxon hydrolysis in this bug.

Esterases clearly have the capacity to respond to external factors;

Table 4. Some of the insects in which carboxylesterase activity has been documented.

ORDER Species	Life Stage	Tissues	Substrate	Reference
THYSANURA				
Thermobia domestica	larva	whole	juvenile hormone I	Ajami & Riddiford, 1973
ORTHOPTERA				
Periplaneta americana	adult	nerve cord, brain fatbody, integument	1-naphthyl acetate	Cook & Forgash, 1965
	adult	in vitro	trans-permethrin cis-permethrin	Shono et al., 1978
	adult	proventricular, midgut, gastric caeca	1-naphthyl acetate	Hipps & Nelson, 1974
	adult	whole	malathion	Matsumura & Sakai, 1968
	adult	whole	1-naphthyl acetate indophenyl acetate 5-bromoindoxyl acetate	Afsharpour & O'Brien, 1963
	nymph	fatbody, gut	juvenile hormone I	Fox & Masare, 1976
	adult	whole, all tissues	O-nitrophenyl acetate	Casida, 1955
	adult	all tissues	phenyl acetate, isoamyl acetate	Metcalf et al., 1955; 1956
Teleogryllus commodus	eggs	whole, caeca	1-naphthyl acetate	Jameson et al., 1976
Blattella germanica	adult	head	O-nitrophenyl acetate	Casida, 1955

	adult	whole	trans-resmethrin	Jao & Casida, 1974
	nymph	whole	juvenile hormone I	Ajami & Riddiford, 1973
Gromphadorphina portentosa	adult	midgut	juvenile hormone I	Slade & Wilkinson, 1974
Locusta migratoria	adult	nerve cord	O-nitrophenyl acetate	Hopf, 1954
Schistocerca gregaria	nymph	hemolymph	juvenile hormone III	Pratt, 1975
Blaberus giganteus	nymph	hemolymph	juvenile hormones I & III 1-naphthyl acetate, p-nitrophenyl acetate	Hammock et al., 1977
HETEROPTERA Triatoma infestans	adult	integument	phenylthioacetate	Fontan & Zerba, 1987
	adult	thorax, abdomen	1-naphthyl acetate	Casabe & Zerba, 1981
Oncopeltus fasciatus	adult	whole	trans-resmethrin	Jao & Casida, 1974
	adult	head	O-nitrophenyl acetate	Casida, 1955
	adult	whole	1-naphthyl acetate indophenyl acetate 5-bromoindoxyl acetate	Afsharpour & O'Brien, 1963
	nymph	whole	juvenile hormone I	Ajami & Riddiford, 1973
Pyrrhocoris apterus	nymph adult	whole nerve, fatbody,	juvenile hormone I 1-naphthyl acetate	Ajami & Riddiford, 1973
Rhodnius prolixus	adult	hemolymph, epidermis	5-bromoindoxyl acetate	Wigglesworth, 1958

Table 4 (cont.)

ORDER Species	Life Stage	Tissues	Substrate	Reference
Dysdercus cingulatus	adult	hemolymph	juvenoid esters	Slama & Jarolim, 1980
HOMOPTERA				
Myzus persicae	adult	whole	immunoassay	Devonshire & Moores, 1984
	adult	whole	1-naphthyl acetate	Devonshire & Sawicki, 1979
Aphis nerii	adult	whole	1-naphthyl acetate	Mullin, 1985
Macrosiphum euphorbiae	adult	whole	1-naphthyl acetate	Mullin, 1985
	adult	whole	O-nitrophenyl acetate	Lord & Potter, 1951
Schizaphis graminum	adult	whole	1-naphthyl acetate p-nitrophenyl acetate	Volkova & Titova, 1984
Nephotettix cincticeps	adult	whole	p-nitrophenyl acetate 2-naphthyl acetate	Motoyama et al., 1984 Ozaki & Koike, 1965
Laodelphax striatellus	adult	whole	2-naphthyl acetate	Miyata, 1983
Nilaparvata lugens	adult	whole	2-naphthyl acetate	Miyata, 1983
Dactynotus ambrosiae	nymph	whole	juvenile hormone I	Ajami & Riddiford, 1973
Acrythosiphon pisum	adult	whole	O-nitrophenyl acetate	Lord & Potter, 1951
Trialeurodes vaporariorum	adult	whole	O-nitrophenyl acetate	Casida, 1955

Paraphlepsius irroratus	adult	whole	O-nitrophenyl acetate	Casida, 1955
NEUROPTERA				
Chrysopa carnea	larva	whole	trans-permethrin cis-permethrin	Bashir & Crowder, 1983
	larva	whole	pyrethroid esters	Ishaaya & Casida, 1981
LEPIDOPTERA				
Spodoptera eridania	larva	epidermis, gut, fatbody, Malpighian tubules, head, silk gland	transpermethrin cis-permethrin	Abdel-Aal & Soderlund, 1980
	larva	head	O-nitrophenyl acetate	Casida, 1955
		midgut, fatbody	1-naphthyl acetate 2-naphthyl butyrate	Brattsten, 1986c
	larva	hemolymph, midgut, fatbody, integument, Malpighian tubules	juvenile hormone I	Slade & Wilkinson, 1974
S. frugiperda	larva	midgut, fatbody	1-naphthyl acetate 2-naphthyl butyrate	Brattsten, 1986c
	larva	midgut	1-naphthyl acetate	Yu & Hsu, 1985
S. littoralis	larva	whole homogenate	1-naphthyl acetate 2-naphthyl butyrate p-nitrophenyl acetate	Riskallah, 1983
	larva	hemolymph	juvenoid esters	Slama & Jarolim, 1980
Trichoplusia ni	larva	gut, integument	trans-permethrin cypermethrin	Ishaaya & Casida, 1980

Table 4 (cont.)

ORDER Species	Life Stage	Tissues	Substrate	Reference
	larva	midgut	p-nitrophenyl acetate	Dowd et al., 1983a
	larva	midgut	1-naphthyl acetate	Turunen & Chippendale, 1977
	larva	hemolymph	juvenile hormones	Sparks & Hammock, 1979
Pseudoplusia includens	larva	midgut	p-nitrophenyl acetate trans-permethrin	Dowd et al., 1983a
Pseudaletia unipuncta	larva	mdigut	p-nitrophenylthiol acetate p-nitrophenylthiol butyrate	Tanada et al., 1980
Wiseana cervinata	larva	gut, fatbody, carcass	trans-permethrin cis-permethrin	Chang & Jordan, 1983
Heliothis zea	larva	midgut	trans-permethrin cis-permethrin	Bigley & Plapp, 1978
	larva	midgut	1-naphthyl acetate	Turunen & Chippendale, 1977
Heliothis virescens	larva	midgut	trans-permethrin cis-permethrin	Bigley & Plapp, 1978
	larva	midgut, fatbody	1-naphthyl acetate	Brattsten, 1986a

Argyrotaenia citrana	larva adult	midgut whole	1-naphthyl acetate	Croft & Mullin, 1984
Colias eurytheme and 13 other spp.	larva	hemolymph	juvenile hormone III	Wing et al., 1984
Lymantria dispar	larva	midgut, Malpighian tubules, nerve cord, fat-body, hindgut, foregut, testes, muscles, brain integument, hemolymph	1-naphthyl acetate	Kapin & Ahmad, 1980
Manduca sexta	larva	hemolymph	1-naphthyl acetate	Wongkrobat & Dahlman 1976
	larva	hemolymph, fatbody, integument	juvenile hormone I	Slade & Wilkinson, 1974
Diatraea grandiosella	larva	midgut	1-naphthyl acetate	Turunen & Chippendale, 1977
Diatraea saccharalis	larva	hemolymph 1-naphthyl acetate	juvenile hormones,	Roe et al., 1983
Pieris rapae	larva	midgut	1-naphthyl acetate	Turunen & Chippendale, 1977 Clements, 1967
Hyalophora gloveri	larva	hemolymph, fatbody	juvenile hormone I	Whitmore et al., 1972
Hyalophora cecropia	larva	hemolymph, fatbody, midgut, Malpighian tubules, integument	juvenile hormone I	Slade & Wilkinson, 1974
Antheraea pernyi	pupa	whole	juvenile hormone I	Ajami & Riddiford, 1973

Table 4 (cont.)

ORDER Species	Life Stage	Tissues	Substrate	Reference
Samia cynthia	pupa	whole	juvenile hormone I	Ajami & Riddiford, 1973
Diataraxia oleracea	larva, eggs	whole	O-nitrophenyl acetate	Lord & Potter, 1951
Ephestia kuhniella	eggs	whole	O-nitrophenyl acetate	Lord & Potter, 1951
Plutella maculipennis	larva	whole	O-nitrophenyl acetate	Lord & Potter, 1951
Plodia interpunctella	larva	gut, fatbody, carcass, hemolymph	1-naphthyl acetate malathion	Beeman & Schmidt, 1982
Galleria mellonella	adult, larva	head	O-nitrophenyl acetate	Casida, 1955
	larva, pupa	hemolymph	1-naphthyl acetate juvenile hormone III	Hwang-Hsu et al., 1979
COLEOPTERA				
Callosobruchus maculatus	adult larva	gut	1-naphthyl acetate	Gatehouse et al., 1985
Anthonomus grandis	adult larva	head-thorax, abdomen whole	1-naphthyl acetate 1-naphthyl acetate	Brattsten, 1986a Brattsten, 1986a
Tenebrio molitor	pupa	whole	juvenile hormoneI	Ajami & Riddiford, 1973
	larva	whole	O-nitrophenyl acetate ethyl butyrate	Lord & Potter, 1951
	larva	whole	*trans*-resmethrin *trans*-tetramethrin	Jao and Casida, 1974

<u>Epilachna varivestis</u>	adult	head	O-nitrophenyl acetate	Casida, 1955
	adult	whole	1-naphthyl acetate indophenyl acetate 5-bromoindoxyl acetate	Afsharpour & O'brien, 1963
<u>Hippodamia convergens</u>	adult	midgut	1-naphthyl acetate	Mullin, 1985
	larva	whole	juvenile hormone I	Ajami & Riddiford, 1973
<u>Tribolium castaneum</u>	larva	whole	1-naphthyl acetate p-nitrophenyl acetate phenyl acetate	Mackness et al., 1983
	adult	whole	O-nitrophenyl acetate	Lord & Potter, 1951
	adult	whole	malathion	Dyte & Rowlands, 1968
<u>Leptinotarsa decemlineata</u>	adult	whole	1-naphthyl acetate indophenyl acetate 5-bromoindoxyl acetate	Afsharpour & O'Brien, 1963
	adult	hemolymph, fatbody abdomen	juvenile hormone I	Kramer et al., 1977
<u>Attagenus piceus</u>	larva	whole	O-nitrophenyl acetate	Casida, 1955
<u>Sitophilus granarius</u>	adult	whole	O-nitrophenyl acetate	Casida, 1955
HYMENOPTERA <u>Apis mellifera</u>	adult	head, abdomen	phenyl acetate, 27 substrates	Metcalf et al., 1955, 1956
	adult	head	O-nitrophenyl acetate	Casida, 1955

Table 4 (cont.)

ORDER Species	Life Stage	Tissues	Substrate	Reference
	adult	midgut	1-naphthyl acetate 5-bromoindoxyl acetate indophenyl acetate	Afsharpour & O'Brien, 1963
	adult	whole	juvenile hormone I	Ajami & Riddiford, 1973
Oncophanes americanus	larva adult	whole	1-naphthyl acetate	Croft & Mullin, 1984
Pediobius foveolatus	adult	whole	1-naphthyl acetate	Mullin, 1985
Atta texana	pupa	whole	juvenile hormone I	Ajami & Riddiford, 1973
Solenopsis invicta	pupa	whole	juvenile hormone I	Ajami & Riddiford, 1973
DIPTERA Anopheles albimanus	adult	whole	1-naphthyl acetate 2-naphthyl acetate	Georghiou & Pasteur, 1978
Anopheles arabiensis	larva adult	whole whole	1-naphthyl acetate 2-naphthyl acetate malathion	Hemingway, 1985
Culex quinquefasciatus	larva adult	whole whole	4-methylumbelliferyl acetate	Pasteur et al., 1984
Culex tarsalis	larva	whole	1-naphthyl acetate, malathion and 6 other substrates	Matsumura & Brown, 1963

<u>Culex pipiens complex</u>	adult	whole	4-methylumbelliferyl acetate and other substrates	Pasteur et al., 1981
	adult	whole	juvenoid esters, malathion, 1-naphthyl acetate	Brown & Hooper, 1979
	adult	whole	1-naphthyl acetate	Villani et al., 1983
<u>Culex triaeniorhynchus</u>	adult	whole	2-naphthyl acetate	Yasutomi, 1971
<u>Lucilia cuprina</u>	adult	whole	1-naphthyl acetate 2-naphthyl acetate	Hughes & Raftos, 1985
<u>Musca domestica</u>	adult	whole less head	ethyl butyrate triacetin trimyristin, fats, oils	Ahmad, 1970
	adult	foregut, mid- and hindgut, thoracic muscles, gonads	1-naphthyl acetate	Ahmad, 1976
	larva	whole gut, muscles, fatbody	1-napthyl acetate	Ahmad, 1976
	adult	whole	2-naphthyl acetate	Kao et al., 1985
	adult	whole	1-naphthyl acetate 2-naphthyl acetate 1-naphthyl butyrate	Kao et al., 1985
	all stages	whole, head	O-nitrophenyl acetate	Casida, 1955
	adult	whole	trans-resmethrin	Jao & Casida, 1974

Table 4 (cont.)

ORDER Species	Life Stage	Tissues	Substrate	Reference
			trans-tetramethrin	
	adult	head	phenyl acetate, 26 other substrates	Metcalf et al., 1955 1956
	adult	whole	1-naphthyl acetate, 2-naphthyl acetate	van Asperen, 1962
	adult	whole less head	hydroprene	Terriere & Yu, 1973
	adult	whole less head	permethrin	Shono et al., 1979
	adult	whole	malathion	Motoyama et al., 1980
	pupa	whole	juvenile hormone I	Ajami & Riddiford, 1973
Drosophila melanogaster	pupa	whole	juvenile hormone I	Ajami & Riddiford, 1973
	adult	whole	p-nitrophenyl acetate	Morton & Holwerda, 1985
Drosophila hydrei	larva, pupa	hemolymph, fatbody, body wall	juvenile hormone I	Klages & Emmerich, 1979
Ceratitis capitata	adult	whole	p-nitrophenyl acetate	Koren et al., 1984
Chrysomya putoria	adult	whole	methyl propionate	Townsend & Busvine, 1969
	larva	fatbody	malathion	

Chaoborus americanus	pupa	whole	juvenile hormone I	Ajami & Riddiford, 1973
Sarcophaga bullata	adult	hemolymph whole less head	1-naphthyl acetate	Maa & Terriere, 1983a
	adult pupa, larva	abdomen whole	6 juvenoid esters	Terriere & Yu, 1977
Phormia regina	adult	hemolymph whole less head	1-naphthyl acetate	Maa & Terriere, 1983a
	adult pupa, larva	abdomen whole	6 juvenoid esters	Terriere & Yu, 1977

in addition to spot mutations causing mutant forms as in the sheep blow fly (Hughes and Raftos, 1985) and in house flies (Oppenoorth and van Asperen, 1960), the amount of protein with esterase activity in an insect can increase due to amplification of the allele coding for it. This appears to be the case in insecticide-resistant clones of the green peach aphid, Myzus persicae (Sulzer) (Homoptera: Aphididae) where the amount of an esterase, E4, increases in a geometric fashion in 16 clones with increasing resistance (Devonshire and Sawicki, 1979; Devonshire and Moores, 1982).

On the other hand, the esterases appear to be only marginally affected by allelochemicals in the food plants that induce other detoxifying enzymes. Very moderate increases in activity were observed in the fall armyworm fed several dietary allelochemicals and host plants (Yu, Chapter 4 in this text). There were no increases in midgut esterase activity in tobacco budworm larvae fed dietary (+)-α-pinene, β-caryophyllene, gossypol, umbelliferone or scopoletin (Brattsten, 1986b) or in adult boll weevil esterase activities; boll weevil larvae fed the same compounds had modestly increased microsomal esterase activity only when eating a 0.2% gossypol diet. These compounds were completely without effect on soluble esterase activities in the tobacco budworm and boll weevil. A similarly weak response to phenobarbital has been observed with mammalian liver carboxylesterases (Kaneko et al., 1979; Raftell et al., 1977). Somewhat larger host plant effects on esterase activities have been seen with sucking type arthropods such as aphids and spider mites (Mullin, Chapter 5 in this text). It is possible that a different picture might be obtained if, instead of a model substrate such as 1-naphthyl acetate, an appropriate allelochemical was used as substrate. Otherwise, it appears that modification in carboxylesterase activity to meet an environmental chemical challenge is often accomplished by a permanent genetic change. This may be related to the endogenous roles of the esterases and the substrate specificities of the various isoenzymic forms.

The big mystery about the esterases is why there are so many different forms. Other than involvement in digestion, few endogenous roles for esterases have been identified. An esterase in the antennae, legs and wings of the cabbage looper moth inactivates their pheromone (Ferkovich et al., 1982), and a similar pheromone-degrading esterase has recently been isolated from the sensory hairs of the silkmoth, Antheraea polyphemus (Cramer) (Lepidoptera: Saturnidae) (Vogt et al., 1985). The hemolymph of the cabbage looper contains an esterase which is highly specific for juvenile hormones (Sparks and Hammock, 1979; Hammock and Quistad, 1981; Sparks and Rose, 1983). The juvenile hormone-specific esterases also occur in the hemolymph of many other insect species including the bugs Pyrrhocoris apterus L. (Heteroptera: Pyrrhocoridae), Dysdercus cingulatus Fabr. (Ibid.), the mealworm, Tenebrio molitor L. (Coleoptera: Tenebrionidae), and in at least 14 species of Lepidoptera (Wing et al., 1984) including the greater wax moth Galleria mellonella L. (Lepidoptera: Pyralidae) and the Egyptian armyworm (Slama and Jarolim, 1980). This carboxylesterase activity can be measured fluorometrically by using as substrate 4-methyl umbelliferyl esters of juvenile hormone acids and related acids, or radiometrically with the appropriate tritium-labeled compound.

5.2. Glycosidases

Glycosidases (E.C.3.2) are a group of hydrolytic enzymes specialized to break glycosidic bonds to O-, N-, or S- groups on the aglycone. Many different sugars are attached to allelochemicals with a suitable functional group by plants, the most common sugar being glucose. Often, more than one sugar is attached in a different place of the same aglycone or in tandem. The combined aglycone and sugar moiety is called a glycoside in general and, if the sugar is known, it may be called accordingly. The glycosides represent the water soluble forms of the aglycone and are usually non-toxic whereas many of the aglycones are highly toxic. The α-glycosidases may be viewed as digestive enzymes whose function is to release useful carbohydrates from sugars with α-linkages. The β-glycosidases, on the other hand, hydrolyze sugar residues from β-glycosidic allelochemicals, the major storage form in plants, and from structural carbohydrates such as cellulose and chitin. This is clearly done at the risk of also releasing toxic aglycones. Although α-glycosidases are found in all insects, β-glycosidases are less common. The latter have been found in the tissues of a few insect species including the migratory locust with highest activity in the crop, the American cockroach, *Periplaneta americana* (L.) (Orthoptera: Blattidae), the mealworm, a water boatman (*Notonecta* sp.), the bean aphid, *Aphis fabae* Scop. (Homoptera: Aphididae) (Robinson, 1956) and other aphid species (Adams and Drew, 1965), the cowpea weevil, *Callosobruchus maculatus* (Fabr.) (Coleoptera: Bruchidae) (Gatehouse et al., 1985), *Rhagium inquisitor* L. (Coleoptera: Cerambycidae) (Chipoulet and Chararas, 1985), in larvae of *Seirarctia echo* (Abbot and Smith) (Lepidoptera: Arctiidae) (Teas, 1967), and in many wood- and litter-feeding arthropods (Martin, 1983).

Since most glycosidic plant allelochemicals are O-glycosides with a β-linkage, most of the enzymes hydrolyzing these bonds are O-glycosyl hydrolases (E.C.3.2.1). These enzymes occur in several different isoenzymic forms and may be able to split off several different sugars, although the locust crop contained a β-D-glucosidase as well as β-D-glucuronidase (Robinson, 1956). The enzymes are probably not specific for the aglycone although enzymes are occasionally named as if they were, for instance, quercitrinase (E.C.3.2.1.66) found in a microorganism (Westlake, 1963). The enzyme hydrolyzing amygdalin, a β-D-glucosidase (E.C.3.2.1.21) also hydrolyzes β-D-galactosides, β-D-xylosides, and β-D-arabinosides (IUB, 1984). However, considerable specificity is indicated for plant β-glucosidases (Hosel and Conn, 1982).

Although these soluble and sometimes extracellular enzymes occur in the tissues of vertebrates (Sanchez-Bernal et al., 1984) the most important source of these hydrolytic activities in mammals is the intestinal microflora as shown in experiments with germ-free rats (Goldman, 1982; Rafter et al., 1983). One can not exclude the possibility that insect alimentary microflora provides the major source in insects; in none of the studies quoted had any extra precaution in addition to "rinsing" been taken to exlude gut microbes.

Being hydrolytic enzymes, the β-glycosidases, if present, have a high catalytic activity compared to cytochrome P-450 and many group transferases (see section 6). This activity is not necessarily desirable and can in many cases be hazardous.

5.3. S-Glycosyl Hydrolases

An enzyme, myrosinase (E.C.3.2.3.1, also called thioglucosidase or glucosinolase) catalyzes the hydrolysis of β-D-thioglucosides such as those typically found in the glucosinolates. This enzyme is well known in plants (Ettlinger et al., 1961; Cole, 1975) and microorganisms (Oginsky et al., 1965; Marangos and Hill, 1974), but unknown in higher organisms except for the cabbage aphid, *Brevicoryne brassicae* (L.) (Homoptera: Aphidae), where the glucosinolase of the aphid and its host plant were electrophoretically different (MacGibbon and Allison, 1971). Glucosinolates, characteristic of cruciferous and a few other plants, are hydrolyzed to mustard oils. When ingesting glucosinolate-containing plant tissue the insect may also ingest the plant enzyme responsible for the conversion, or, myrosinase in gut microorganisms may hydrolyze them. The free mustard oils spontaneously rearrange to one of several products depending on the microenvironment. One of the possible products, isothiocyanate, is an irritant. Glucosinolates are deterrents for most insects except a few adapted species (Blau et al., 1978); the mechanism(s) whereby the adapted species avoid the post-ingestive mustard oil effects are unknown.

5.4. Ether hydrolases

The epoxide hydrolases (E.C.3.3.2.3 also called epoxide hydratase or arene-oxide hydrolase) can split a cyclic ether bond in lipophilic oxiranes by stereospecific addition of water. These enzymes were at one time classified as lyases (E.C.4.2.1.63) but have been reclassified as hydrolases since they use only a water molecule and no high-energy cofactors for the catalytic action.

The epoxide hydrolases were first discovered in the metabolism of trans-L-epoxysuccinic acid by a cell-free extract from a Flavobacterium sp. (Martin and Foster, 1955) and of epoxyestrogens in rat liver slices (Breuer and Knuppen, 1961). They have subsequently been found in all living organisms investigated. They occur as membrane-bound (microsomal) and cytosolic (soluble) proteins in multiple isoenzymic forms.

Work with mouse liver indicates that the cytosolic forms are predominantly trans-epoxide hydrolases, i.e. they preferentially attack disubstituted epoxides with a trans configuration, whereas the microsomal forms preferentially attack substrates with a sterically hindered cis configuration, i.e. they are cis-epoxide hydrolases. In both cases, there is a trans addition of water resulting in the respective 1,2-diol (Mumby and Hammock, 1979; Hammock et al., 1980; Mullin and Hammock, 1982; Hammock and Hasagawa, 1983):

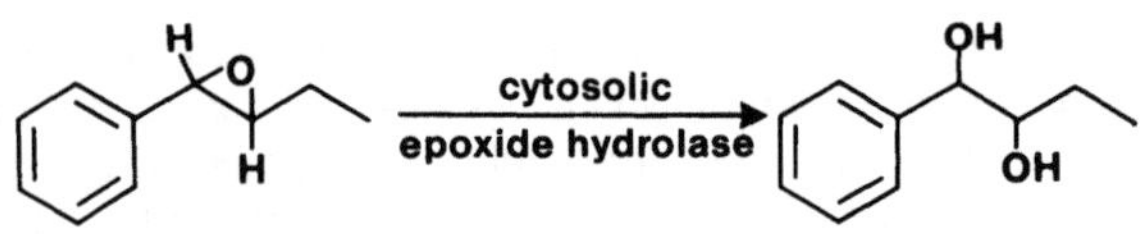

trans-β-ethylstyrene oxide β-ethylstyrene diol

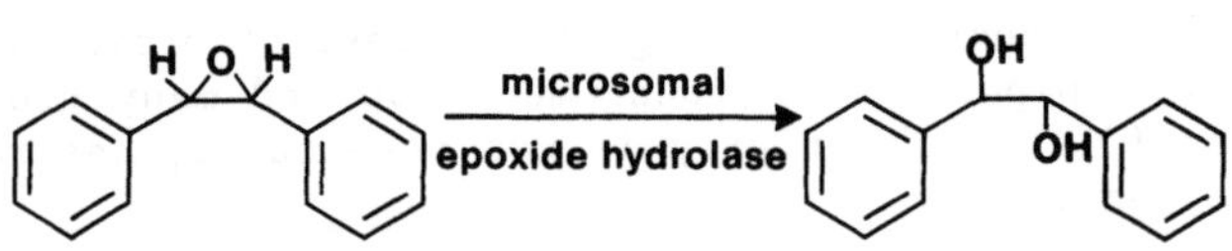

cis-stilbene oxide stilbene glycol

Much less is known about the subcellular distribution and specificity of epoxide hydrolases in arthropod tissues, although both trans and cis epoxides are hydrolyzed in cytosolic and microsomal fractions (Mullin et al., 1984; Ottea and Hammock, unpubl.). A microsomal enzyme was purified from midguts of the southern armyworm (Mullin and Wilkinson, 1980) and occurs in many other insects including the blow fly, Calliphora erythrocephala Meigen (Diptera: Calliphoridae), the house fly (Brooks et al., 1970); the fruit fly (Baars, 1980), the mealworm (Brooks, 1973), the flour beetle, Tribolium castaneum (Herbst) (Coleoptera: Tenebrionidae) (Cohen, 1981), the Madagascar cockroach (Slade et al., 1975), the American cockroach (Nelson and Matsumura, 1973), the fall armyworm (Mullin and Croft, 1984; Yu and Hsu, 1985), the orange tortrix and its ectoparasite, Oncophanes americanus (Weed) (Hymenoptera: Braconidae) (Croft and Mullin, 1984), the honey bee, Apis mellifera L. (Hymenoptera: Apidae) (Yu et al., 1984), and in addition in 33 other insect species (Mullin, 1985; Mullin, Chapter 5 in this text). It also occurs in the twospotted spider mite, Tetranychus urticae Koch (Acari: Tetranychidae), and the predatory mite Amblyseius fallacis (Garman) (Acari: Phytoseiidae) (Mullin et al., 1982).

Epoxide hydrolases are important in the detoxification of cytochrome P-450-generated epoxide metabolites which often are highly reactive, forming adducts with cellular macromolecules including DNA, RNA and proteins (see section 2.3.2.). As with most other enzymes involved in foreign compound metabolism, the epoxide hydrolases characteristically catalyze a one-directional reaction, are induced by exogenous chemicals (Yu, Chapter 4 in this text) and lack a clearly established endogenous function. A specialized epoxide hydrolase is of critical importance in regulating the level of juvenile hormone in insects (Slade et al., 1976; Hammock and Quistad, 1976, 1981; Yu and Terriere, 1978). Epoxide hydrolases are also involved in the deactivation of pheromones (see section 2.3.2.).

Epoxide hydrolase may also be of critical importance in herbivore digestion; the epicuticular waxes of plants contain unsaturated long-chain fatty acids that may undergo cytochrome P-450-catalyzed epoxidation followed by conversion to diols. Mullin and Croft (1984) found that the ratio of trans-epoxide hydrolases to cis-epoxide hydrolases is usually large in insect herbivores but tends to be low in predators and parasites (Mullin and Croft, 1985); this is true also for mites (Mullin et al., 1982). Coincidentally, the plant allelochemicals that can undergo epoxidation tend to have a trans configuration, whereas constitutive substances common to both animals and plants tend to have a cis configuration (Mullin, 1985).

Epoxide hydrolases, in particular the trans-selective ones are potently inhibited by chalcone oxides (Mullin and Hammock, 1982). Chalcones are common in plants, usually occur in hydroxylated form (Vickery and Vickery, 1981), and are regarded as universal flavonoid precursors; they might have a fortuitous role in plant defenses if oxidized to epoxide hydrolase inhibitors (pro-synergists) by insect cytochrome P-450, which would leave the insect exposed to potentially dangerous epoxides formed from co-occuring allelochemicals. In view of the current debate of the insect-plant co-evolution theory, this idea deserves investigation.

5.5 Other hydrolases

Several cases are known where insects utilize potentially toxic plant allelochemicals for their own physiological or defensive purposes; having an insensitive target site is perhaps a necessary prerequisite which may form the evolutionary basis for subsequent adaptations for utilization (Berenbaum, Chapter 7 in this text). The hydrolytic enzyme urease (E.C.3.5.1.5) enables the seed predator Caryedes brasiliensis (Coleoptera: Bruchidae) to utilize the non-protein amino acid L-canavanine as a nitrogen source (Rosenthal et al., 1982). This enzyme, which is regarded absent in most insects, occurs in a highly active form in Caryedes brasiliensis (Rosenthal et al., 1976). The beetle converts L-canavanine to L-canaline and urea in a hydrolytic reaction catalyzed by arginase; the urea is then converted to carbon dioxide and ammonia and the latter is incorporated into L-glutamate to form L-glutamine (Rosenthal and Janzen, 1985; see Brattsten, Chapter 6 in this text).

Hydrolyzable tannins are hydrolyzed by certain tree-feeding locusts to gallic acid (Bernays, 1978) which is utilized in the cross-linking of cuticular proteins (Bernays and Woodhead, 1982). The hydrolytic enzyme(s) involved in this process have not been characterized, but may be specialized through evolutionary adaptation.

6. GROUP TRANSFER ENZYMES

Group transferases are a group of enzymes catalyzing the addition, by a covalent bond, of an endogenous, water-soluble group such as a sugar, an amino acid, acetate, sulfate, phosphate and others to the metabolite of a compound or the compound itself, that often is a foreign compound; group transferases also participate in the endogenous metabolism of organisms. These enzymes catalyze synthetic reactions and the products are known as conjugates. Because a covalent bond is established, the reactions require energy input. The energy is in the form of a high energy intermediary cofactor such as ATP, GTP, an activated endogenous group, an activated compound or metabolite, or in some cases a high-energy, reactive group such as the sulfhydryl goup in glutathione. The reactions are viewed as group transfers between donors, usually the endogenous groups, and acceptors, normally the foreign compounds or their primary metabolites.

Group transfer enzymes are known to play an important role in Phase II metabolism (see section 1) and also to participate in Phase I metabolism, for example, in the detoxification of organophosphorous insecticides. Very little is known about the participation of group transferases in the primary metabolism of plant allelochemicals except for work with plant-derived drugs in mammals. Conjugations usually result in detoxified, highly water-soluble and excretable products, but some exceptions are known where a conjugate has more biological activity than its parent compound. For instance, one of the glucuronides of morphine is as potent an analgesic as morphine itself (Caldwell, 1982).

Several different types of conjugation occur; the most common ones are listed in Table 5. Glucuronidations are probably uncommon in insects which, instead, utilize glucose and other hexoses without prior oxidation. Phosphorylations are characteristic and important in vertebrate intermediary metabolism, but appear to be used also in foreign compound metabolism in insects. Macromolecular adduct formation and lipid conjugation both of which can have toxic and lethal consequences, are with few exceptions not well known in insects. Neither is the occurrence of methylations which unlike most other conjugations decrease the water solubility of the conjugate. Several authors have reviewed conjugation reactions and the group transfer enzymes in general (Williams, 1974; Smith and Litwack, 1980; Jakoby, 1980; Jakoby and Habig, 1982; Jakoby et al., 1982; Caldwell, 1982; Rafter et al., 1983; Paulson et al., 1986) and in insects (Smith, 1968; Yang, 1976; Ahmad and Forgash, 1976; Dauterman and Hodgson, 1978; Dauterman, 1980, 1985; Motoyama and Dauterman, 1980; Wilkinson, 1986).

6.1. Glutathione Transferases

The conjugation of reduced glutatione (GSH) to a lipophilic foreign compound or its metabolite, bearing an electrophilic site preferably on or adjacent to an aromatic ring, is catalyzed by a group of cytosolic glutathione transferase isoenzymes. The reaction does not require participation of a high-energy intermediate such as ATP or NADPH but is, indirectly, energy-requiring. Energy is required to form the two peptide bonds in the tripeptide glutatione, γ-glutamylcysteinylglycine, and to keep the cysteinyl sulfhydryl group in the reduced state; glutathione can thus be regarded as both the conjugating agent and an atypical high-energy intermediate.

Table 5. Types of conjugation reactions.

Reaction	Donor Group	Acceptor Groups
Glutathione conjugation	Glutathione	Electrophilic part of lipophlic molecules, e.g. epoxide, halide
Amino acid conjugation	Glycine, others	-COOH
Glucuronidation	UDP-glucuronic acid	-COOH, -OH, $-NH_2$, -SH
Glycoside formation	UDP-glucose	-COOH, -OH, -SH
Sulfation	PAPS	-OH, $-NH_2$, -SH
Acetylation	Acetyl coenzyme A	-OH, $-NH_2$
Methylation	S-adenosyl methionine	-OH, $-NH_2$,
Thiocyanate formation	Sulfane sulfur	CN-
Phosphorylation	Phosphate	-OH
Macromolecular adduct formation	RNA, DNA, Proteins	Epoxide
Lipid conjugates		

Many sufficiently reactive foreign compounds and/or their metabolites readily combine spontaneously with glutathione; the importance of the enzyme in the reaction is thought to be mainly the providing of proximity between the substrate and glutathione and to promote the conversion of GSH to the reactive nucleophilic glutathione thiolate ion, GS^-, which combines with the substrate (Jakoby and Habig, 1980). To compensate for this rather vague catalytic function, glutathione transferases occur at high concentration in major tissues; up to three percent of soluble liver proteins in vertebrates are glutathione transferases; likewise, tissue concentrations of glutathione are high, ranging from three to ten mM in vertebrates (Jakoby and Habig, 1980).

Cytosolic glutathione transferase activity occurs in midgut and fatbodies of lepidopterous larvae including the southern (Brattsten et al., 1984) and fall armyworms (Yu, 1982; Gunderson et al., 1986), the tobacco budworm (Gould and Hodgson, 1980; Brattsten, 1986a), and the greater wax moth (Chang et al., 1981). It also occurs in Malpighian tubules and testes of the southern armyworm (Gunderson et al., 1986). The activities have been found in the tufted apple budmoth, Platynota idaeusalis (Walker) (Lepidoptera: Tortricidae) (Wells et al., 1983), house flies (Ottea and Plapp, 1981; Oppenoorth et al., 1977; Hayaoka and Dauterman, 1983), fruit flies (Jansen et al., 1984), the Mediterranian fruit fly, Ceratitis capitata (Wiedemann) (Diptera: Tephritidae) (Yawetz and Koren, 1984), the migratory locust, Locusta migratoria cinerascens (Orthoptera: Locustidae) (Menguelle et al., 1985), the New Zealand grass grubs, Costelytra zealandica (White) (Coleoptera: Scarabaeidae) (Clark et al., 1985) and the porina moth, Wiseana cervinata (Lepidoptera: Hepialidae) (Clark and Drake, 1984), fatbodies of American cockroaches (Usui et al., 1977), other cockroaches, a flour beetle (Tenebrio), a cotton stainer (Dysdercus), a turnip beetle (Phaedon), and locusts (Schistocerca) (Cohen et al., 1964), the orange tortrix and the braconid Oncophanes americanus (Croft and Mullin, 1984), the honey bee (Yu et al., 1984), and adult and larval boll weevils (Brattsten, 1986a). The gluthatione transferases thus occur widely among insects of different orders, with different life styles and in different life stages; they also occur in several different tissues.

Glutathione transferase activities are most often measured with 1,2-dichloro-4-nitrobenzene as model substrate in a rapid and convenient direct spectrophotometric assay:

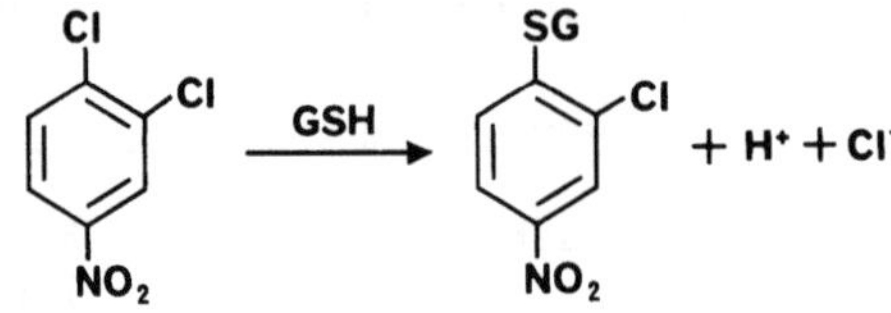

The resulting glutathione conjugate absorbs light at 340-344 nm. Several other model substrates have been used with insect enzyme preparations including 1-chloro-2,4-dinitrobenzene, methyl iodide, butyl iodide, trans-cinnamaldehyde, 1,2-epoxy-3-(p-nitrophenoxy) propane (Usui et al., 1977; Yu, 1984; Clark et al., 1984; Gunderson et al., 1986; Brattsten, 1986a). A distinct advantage in studying the glutatione transferases is their considerable storage stability; post-microsomal supernatants can be stored for many months at -20°C without loss in activity. There are also successful methods for purifying the activities (Clark et al., 1984).

The glutathione transferases participating in foreign compound metabolism in insects appear to be cytosolic (Brattsten, unpublished) although no subcellular distribution studies have been published; in rat liver, there are also microsomal glutatione transferases with catalytic properties similar to those of the cytosolic enzymes (Morgenstern et al., 1985).

The glutathione transferase isoenzymes used to be named according to the functional group of the substrate they acted on, a practice that lead to some confusion until the overlapping substrate specificities between the isoenzymic forms were established. Model substrates used with purified forms of cytosolic rat liver enzyme illustrate this, Table 6.

Glutathione transferases are important in secondary, or Phase II, metabolism (see section 1) where they inactivate reactive metabolites formed by cytochrome P-450 catalysis. They are also important in the primary, or Phase I, metabolism of organophosphorous insecticides (Motoyama and Dauterman, 1980). It is unknown to what extent they metabolize ingested plant allelochemicals in insects. Many plant allelochemicals fulfill the requirements of glutathione transferase substrates; they include α,β-unsaturated aliphatic or aromatic aldehydes such as cinnamaldehyde. Thiocyanates and isothiocyanates such as those in the glucosinolates are also conjugated to glutathione (Scheline, 1978).

It has recently been established that a glutathione transferase is identical to DDT-dehydrochlorinase which converts DDT to the non-toxic product, DDE (Clark and Shaaman, 1984); the mechanism of this reaction, which is unusual in that glutathione participates only in catalytic amounts, was described as an E2 elimination (Wilkinson, 1986) in which a benzylic hydrogen is first abstracted followed by elimination of a chlorine. In fact, some insects, for instance the red-banded leaf roller, Argyrotaenia velutiana (Walker) (Lepidoptera: Tortricidae), were resistant to DDT from the inception of its use without any use-related selection pressure (natural tolerance) (Glass and Chapman, 1952). This, together with the ability of many plant allelochemicals to serve as glutathione transferase substrates, leads to the idea that these enzymes may have survival value for insects feeding on certain plants and, indeed, participate in the primary metabolism of plant allelochemicals. Another strong indication for this is their inducibility by plant allelochemicals and synthetic compounds observed in several insect species (Yu, Chapter 4 in this text).

Table 6. Activities of purified rat liver cytosolic glutathione transferases.

SUBSTRATE	ISOENZYME AA	A	B	C	E
1-chloro-2,4-dinitrobenzene	14	62	11	10	0.01
1,2-dichloro-4-nitrobenzene	0.008	4.3	0.003	2.0	4.1
Iodomethane	1.4	0	0.6	0	8.9
Ethacrynic acid	0.30	0	0.26	0.11	0
Bromosulfophthalein	0.004	0.530	0.006	0.180	0
Trans-4-phenyl-3-butene-2-one	--	0.02	0.001	0.40	--

Data from Baars, 1979; nmole/min, mg protein

Several plant allelochemicals including ellagic acid, purpurogallin, quercetin and other phenols (Das et al., 1984) and dicarboxylic acids (Clark et al., 1967) are potent in vitro inhibitors of the glutathione transferases. The inhibition is competitive with respect to reduced glutathione and non-competitive with respect to the substrate used. These compounds are not necessarily specific inhibitors of the glutathione transferases.

Another way of inhibiting the rate of glutathione conjugation reactions is to deplete the organism of its endogenous glutathione; insects kept in an atmosphere of high carbon dioxide have reduced concentration of glutathione in their tissues (Friedlander and Navarro, 1984) apparently due to inhibition of the glutathione biosynthesis pathway.

In vertebrates, the glutathione-conjugated xenobiotics or their primary metabolites are often excreted in the form of mercapturic acid derivatives; this conversion requires the participation of several other enzymes which also occur in at least some insects (Smith, 1968):

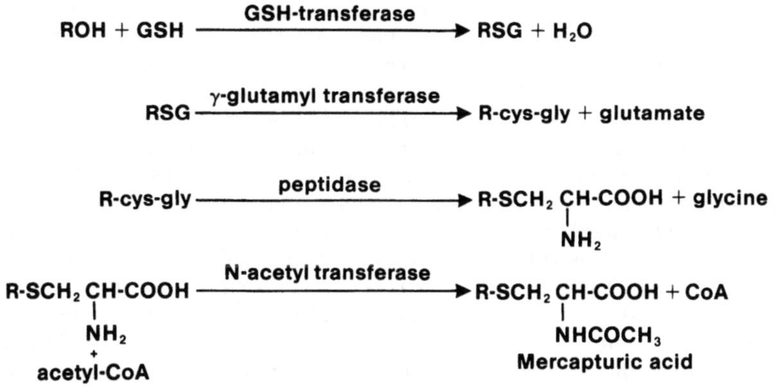

There have been no investigations of mercapturic acid formation from ingested plant allelochemicals in insects. Dykstra and Dauterman (1978) demonstrated excretion of a 2,4-dinitrophenyl mercapturic acid derivative in the American cockroach. In house flies, the mercapturic acid derivative of 2,4-dinitrophenol is increased following induction of glutathione transferase activity by phenobarbital (Abd-Elraof and Dauterman, 1981) indicating that either all enzymes involved are inducible or that the first step, catalyzed by glutathione transferase, is rate limiting.

6.2. Hexose Transferases

Conjugation to a sugar is probably the most common fate of a foreign compound. Formation of glucuronic acid conjugates (glucuronides) is the most extensively investigated reaction because mammals rely primarily on it. The most widely occurring hexose conjugation reaction is, however, glucosylation or formation of glucose conjugates (glucosides) because it is the predominant reaction in plants and insects. "Glycoside" is a term for hexose conjugates in general, formed with glucose or any of a number of other sugars; plants often use sugars other than glucose and also frequently add more than one sugar moiety in tandem or to different sites of the xenobiotic, to the first formed conjugate. Pupae of the stable fly, Stomoxys calcitrans (L) (Diptera: Muscidae) have mannosyl transferases involved in the intermediate metabolism of lipids and proteins (Mayer et al., 1983); it is not known if mannosyl transferases also participate in foreign compound metabolism in insects.

Glucose conjugation requires the activation of the glucose by a high-energy intermediate:

$$\text{UTP} + \text{D-glucose} \xrightarrow[\text{pyrophosphorylase}]{\text{UDP-glucosyl-}} \text{UDP-}\alpha\text{-D-glucose} + \text{Pi}$$

$$\text{UDP-}\alpha\text{-D-glucose} + \text{R-OH} \xrightarrow[\text{transferase}]{\text{UDP-glucosyl-}} \text{RO-}\beta\text{-D-glucose}$$

Whereas there are few and unconfirmed examples of glucuronidation, which requires the extra steps of oxidizing glucose, in insects, e.g. house flies and blow flies (Terriere et al., 1961), mammals are known to form glucosides to a small extent (Heirweigh et al., 1971).

Glucose transferase activity has been reported from many insects; Smith (1968) lists 41 species among Orthoptera, Coleoptera, Lepidoptera, Diptera and Hemiptera in which glucose conjugation has been shown. In most cases the conjugations were measured with xenobiotic phenols and aromatic acids. Glycoside formation renders a foreign compound or its primary metabolite highly water soluble, and, undoubtedly, constitutes a major elimination mechanism in insects. However, insects may also use glycosides as storage forms. The 4-O-β-D-glucoside of protocatechuic acid is thought to be a storage form in one of the colleterial glands of cockroaches for the reactive protocatechuic acid used in forming the ootheca (Kent and Brunet, 1959). Zygaenid moths biosynthesize (Wray et al., 1983) their own cyanogens, presumably for defensive purposes, and store them in the hemolymph in the form of the β-glucosides of the 2-butanone and acetone cyanohydrins lotaustralin and linamarin (Davis and Nahrstedt, 1979; Jones et al., 1962); when a moth is crushed, cyanide is liberated by the action of intracellularly located β-glycosidases followed, probably, by spontaneous decomposition of the cyanohydrin. Larvae of the moth Seirarctia echo convert the aglycone of cycasin, methylazoxymethanol, a strong alkylating agent, back to the β-glucoside, cycasin, when fed to them in an artificial diet (Teas, 1967). These larvae are adapted to feed on cycads and have a highly active β-glycosidase in their gut which apparently first converts the naturally ingested cycasin to its toxic aglycone; the latter is subsequently returned to the glucoside form by UDP-glucosyl transferase, perhaps mainly located in the fatbody (Teas, 1967).

In addition to O-glucosides, insects also biosynthesize S-glucosides, but no N-glucosides have been found in insects. Crickets and cockroaches excreted β-D-S-glucosylthiophenol and β-D-S-glucosyl-5-mercaptouracil after dosing with the respective aglucones (the non-sugar part of a glucoside) and homogenates of their fatbodies and gonads also produced these S-glucosides (Gessner and Acara, 1968). Insects are thought to readily form N-acetyl derivatives in preference to N-glycosides (Dauterman, 1985).

The fatbody appears to be the tissue with the highest concentration of UDP-glucose transferases, but activity also occurs in other tissues, such as gonads (Gessner and Acara, 1968) and midguts (Mehendale and Dorough, 1972).

The UDP-glucose transferases are relatively unknown biochemically. In some insect species they occur in the soluble fraction of the cell, whereas in others they may be associated with microsomal and/or mitochondrial fractions (Yang, 1976; Mehendale and Dorough, 1972). It appears that several isoenzymic forms may be present; two isozymes with preference for either hydroquinone or 4-nitrophenol were found in the fatbody of a locust (Trivelloni, 1964). The pH-optimum for the in vitro reaction seems to range from 6.5 to 8.5. Several model substrates have been used to measure the activity in insect tissues; 4-methyl umbelliferone glucosidation can be measured spectrofluorometrically (Smith and Turbert, 1961), or the disappearance of simple substituted phenols from the incubation medium can be quantified spectrophotometrically (Dutton, 1962).

Insects very frequently eliminate insecticides or insecticide metabolites as glucosides (Dauterman, 1985; Dorough, 1979) but because the enzymes primarily detoxifying insecticides are of more urgent interest, the glucose transferases have received relatively scant attention. It is, however, likely that they play an important role in insects that feed on phenol-rich plant tissues.

6.3. Amino Acid Conjugations

The most common alternative to glucosidation in insects is probably conjugation to amino acids, in particular glycine. Even so, very little is known about amino acid conjugation. In mammals, the reaction requires that the xenobiotic first be activated by energy-requiring reaction, after which it can be combined with the amino acid:

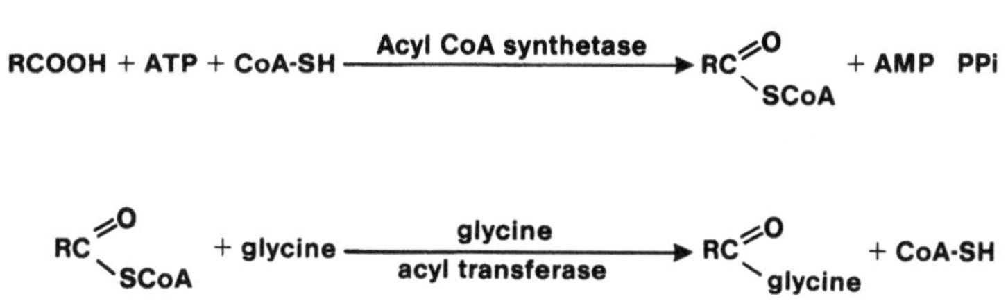

Neither the first enzyme in the sequence, one of the acid-thiol ligases (E.C.6.2.1), nor the second one, also called acyl-CoA:glycine N-acyltransferase (E.C.2.3.1.13) have been characterized in insect tissues; the second enzyme appears to be specific for glycine in mammals.

Several species of insects including a locust, two lepidopterans, house flies and *Aedes* mosquitoes produce glycine conjugates when exposed to substituted benzoic acids (Smith, 1968). The reactions appear to take place in the midgut and fatbody of silkworm, *Bombyx mori* (L.) (Lepidoptera: Bombycidae) larvae (Shyamala, 1964).

Although glycine is the most common amino acid involved in conjugation in insects, several other amino acids form conjugates with tropital, a cytochrome P-450 inhibitor, in house flies (Esaac and Casida, 1968).

Cabbage looper larvae and American cockroaches produce metabolites of the synthetic pyrethroid insecticide permethrin conjugated to serine, glutamine or glutamic acid as well as to glycine (Shono et al., 1978).

6.4. N-acetylation

Locusts, silkworm larvae, and wax moth larvae acetylate aromatic amines (Smith, 1964). Blow flies acetylate serotonin, histamine, and tyramine (Karlson and Ammon, 1963). In nerve tissue of European corn borer, *Ostrinia nubilalis* (Hubner) (Lepidoptera: Pyralidae) larvae, an arylamine acetyl-transferase is important in acetylating biogenic amines (Evans et al., 1980).

N-acetylation activity is also present in other tissues, primarily Malpighian tubules and fatbody. N-acetylation is a major mechanism in the sclerotization of the insect cuticle. This may explain the large amount of activity measured in the "carcass" of the European corn borer larvae (Evans et al., 1980). If this activity is present in suitable tissues which are exposed to amine xenobiotics, the mechanism for their N-acetylation is already in place and, thus, N-acetylation may replace glycosidation of amines in insects (Dauterman, 1985).

6.5. Phosphate conjugation

Another mechanism of great importance in intermediary metabolism which, in insects, is apparently also utilized to facilitate the elimination or storage of foreign compounds is conjugation to the phosphate group. Phosphate conjugates of xenobiotics appears to be rare in vertebrates, but has been shown in several species of insects. Exposure to phenolic compounds results in excretion of phosphorylated phenols in *Costelytra* grass grubs, house flies and blow flies (Binning et al., 1967), and may be a major mechanism of phenol detoxification in the Madagascar cockroach (Yang and Wilkinson, 1973; Gil et al., 1974; Wilkinson, 1986). Tandem conjugates of phenols containing glucose linked to phosphate have been isolated from four insect species (Ngah and Smith, 1983).

It is not known if or how the phosphotransferase in those insects, the Madagascar cockroach, tobacco hornworm larvae, and adult house flies, where it has been demonstrated by in vitro measurements (Yang and Wilkinson, 1973) differs from intermediary metabolism phosphokinases. The insect phosphotransferase was found in the soluble fraction of midgut cells of the tobacco hornworm and the Madagascar cockroach; it requires magnesium ions and ATP for activity and may transfer the phosphate group directly from ATP (Wilkinson, 1986).

6.6. Sulfotransferases

Conjugation to sulfate is an energetically expensive form of conjugation and may be susceptible to a rate limiting supply of sulfate groups; in addition to eliminations, sulfate conjugates are, likely, important storage forms. To serve as a conjugating agent the sulfate must be activated to a phosphorylated form, 3'-phosphoadenosine 5'-phosphosulfate (PAPS):

$$SO_4^{2-} + ATP \underset{\text{adenylyl transferase}}{\overset{\text{ATP-sulfate}}{\rightleftharpoons}} APS + PPi$$

$$APS + ATP \xrightarrow[\text{3'-phosphotransferase}]{\text{ATP:adenylyl sulfate}} PAPS + ADP$$

$$ROH + PAPS \xrightarrow{\text{Sulfotransferase}} ROSO_3H + PAP$$

The adenylyl sulfate (APS) formed in the first reversible reaction catalyzed by a cytosolic enzyme (E.C.2.7.7.4) is activated in the second reaction, the committing, irreversible step, possibly also the rate-limiting one, catalyzed by the cytosolic ATP:adenylylsulfate 3'-phosphotransferase, also called adenylylsulfate kinase (E.C.2.7.1.25). In vertebrates, both these enzymes are highly specific whereas the third enzyme, the sulfotransferase, also cytosolic, appears to occur in several isoenzymic forms with overlapping substrate preferences, the major two being 3'-phosphoadenylyl sulfate: phenol sulfotransferase (E.C.2.8.2.1) and 3'-phosphoadenylyl sulfate: alcohol sulfotransferase (E.C.2.8.2.2), catalyzing reactions with aromatic and aliphatic compounds, respectively. A third sulfotransferase of potential major importance is the steroid sulfotransferase (3'-phosphoadenylyl sulfate: phenolic-steroid sulfotransferase, E.C.2.8.2.15) because sulfate conjugates are strongly implied as storage forms for the insect molting hormones and/or for ingested phytosterol precursors thereof (Robbins et al., 1971; Yang et al., 1973; Wilkinson, 1986). In all, the Nomenclature Committee of the International Union of Biochemists lists 16 2.8.2-sulfotransferases (IUB, 1984) some of which may be identical proteins studied with different substrates. None of these enzymes have been purified from insect tissues.

Most sulfate conjugates have been identified in excreta and insect tissues after exposure to a suitable xenobiotic; in vitro experiments with several insect species (Yang and Wilkinson, 1973) showed that sulfoconjugation takes place in the soluble fraction of cells from midgut tissues and requires inorganic sulfate, ATP and magnesium ions, identical requirements to those of in vitro sulfoconjugations with mammalian liver high-speed supernatants (Hiltz and Lipman, 1955; Banerjee and Roy, 1966); it is therefore reasonable to assume that the catalytic mechanisms in insect tissues are similar or even identical to those in vertebrates. Recently, Isaac et al. (1982) demonstrated that PAPS-synthesizing enzymes occur at high levels in the gut cytosolic fraction of the Egyptian armyworm, and that the rate of sulfoconjugation of p-nitrophenol is limited by the amount of PAPS synthesized. Several p-nitrophenyl sulfates have been isolated from five other insect species (Ngah and Smith, 1983).

6.7. Thiosulfate Sulfur Transferase

The enzyme thiosulfate:cyanide sulfur transferase, historically called rhodanese (E.C.2.8.1), catalyzes the reaction in which cyanide or other strong nucleophiles are combined with a sulfane sulfur atom (a divalent sulfur atom covalently bonded only to another sulfur atom):

$$S\text{-}SO_3^{2-} + CN^- \longrightarrow SCN^- + SO_3^{2-}$$

This enzyme is biochemically very well known from work with mammals (Westley, 1973). It represents a conjugation mechanism where there is no need for a high-energy intermediate because of the reactivity of the cyanide group, one of the strongest ionic nucleophiles in existence.

Thiosulfate sulfur transferase is probably of ubiquitous occurrence having been demonstrated in actinomycetes, autotrophic and heterotrophic bacteria, plants, fungi, and animals (Westley, 1981). It is located in the interior of mitochondria of midgut, fatbody and Malpighian tubule cells in lepidopterous larvae (Long and Brattsten, 1982) and in mammalian liver and many other tissues.

The insensitivity to cyanide of many insect species, adapted to feed on cyanogenic plants, ascribed to thiosulfate sulfur transferase activity (Dowd et al., 1983b), is probably unrelated to this activity. The activity is present in species feeding on cyanogenic plants and also, at similar levels, in other species, not known to feed on cyanogenic plants; the distribution is apparently unrelated to cyanide exposure (Long and Brattsten, 1982; Beesley et al., 1985; see Brattsten, Chapter 6 in this test).

The location of the enzyme in the interior of mitochondria makes it an unlikely defense mechanism for cyanide; even though cyanide very readily penetrates membranes, the cosubstrate for the reaction, thiosulfate or other sulfane sulfur compounds would reach the interior of mitochondria much too slowly to counteract the very rapid poisoning effects of unmetabolized cyanide. Vertebrates are extremely sensitive to cyanide poisoning; yet, in vitro measurements of thiosulfate sulfur transferase activity in tissues of the dog (Himwich and Saunders, 1948) led to the absurd estimate that the dog has enough activity in its liver alone to detoxify four kg of cyanide in 15 minutes (Hollingworth, 1976). Clearly, this activity has very little to do with defense against cyanide, and several other functions for the enzyme have been proposed (Westley, 1973) among them involvement in the turnover of non-heme iron sulfur protein constituents in the mitochondrial respiratory pathway.

A more likely candidate for cyanide detoxification in insensitive insects may be the enzyme β-cyanoalanine synthase (E.C.4.4.1.9) known to occur widely in plants where it channels cyanide into the intermediary metabolism. This enzyme has been found in several insects (Duffey, 1981); in a heliconiine butterfly species, it is present in the actively feeding larval stage but disappears in the pupal stage (Davis and Nahrstedt, 1985), consistent with the behavior of the enzymes involved in foreign compound metabolism. The thiosulfate sulfur transferase activity, in contrast, remains constant throughout much of the insect life cycle, at least in the southern armyworm, and is not inducible by external chemicals (Long and Brattsten, 1982).

It is also possible that cyanide reacts spontaneously with endogenous sulfane sulfur compounds in insect biological fluids and tissues, if present; extremely little is known about sulfur metabolism in insects, the study of which may shed some light on cyanide metabolism, as well.

7. SUMMARY

Insects possess a rich and diversified assortment of enzymes which taken together constitutes a very powerful defense against chemical toxicants of all kinds and of both natural and synthetic origin. Along with behavioral adaptations to avoid poisoning, this assembly of detoxification mechanisms, undoubtedly, contributes to the evolutionary success of herbivorous insects.

For the detoxification and elimination of lipophilic toxicants, the microsomal cytochrome P-450-dependent monooxygenases are of major importance as well as carboxylesterases. In many cases, glutatione transferases are of major importance. These three enzyme systems constitute what is known as primary or Phase I metabolism which operates in close association with secondary or Phase II metabolism. Among Phase II enzymes, glutathione transferases and epoxide hydrolases may be the most important ones in the defense against lipophlic toxicants.

This review reflects the perceived relative importance of the enzymes; there is a wealth of information about the Phase I enzymes in insects and relatively very little about the Phase II mechanisms. The disproportionate review here of some enzymes, therefore, inevitably results from the fact that investigations in insects have focussed on cytochrome P-450, carboxylesterases, glutathione transferases, and epoxide hydrolases because of their obvious and great influence on the toxicity of commercially used insecticides. It seems quite likely that future work will reveal a "primary" importance of glucose transferases and other group transfer enzymes in the detoxification, elimination, and/or storage of plant allelochemicals in insect herbivores.

8. ACKNOWLEDGEMENT

The recent work on house fly pheromone metabolism was supported in part by National Science Foundation grant DCB-8416558. We thank G. J. Blomquist for valuable suggestions, K. E. Kirkland for critical reading, S. Ganteaume, D. Rainey, and C. Cairo for skillful typing, and T. Johns of DuPont's Lavoisier Library for many creative literature searches.

9. REFERENCES

Abdel-Aal, Y. A. I. and D. M. Soderlund, 1980. Pyrethroid-hydrolyzing esterases in southern armyworm larvae: tissue distribution, kinetic properties, and selective inhibition, Pestic. Biochem. Physiol., 14:282-289.

Abd-Elraof, T. K. and W. C. Dauterman, 1981. The effect of phenobarbital on mercapturic acid biosynthesis in the house fly, Musca domestica, Insect Biochem., 11:649-651.

Adams, J. B. and M. E. Drew, 1965. A celluose-hydrolyzing factor in aphid saliva, Can. J. Zool., 43:489-496.

Adams, T. S. and G. G. Holt, 1986. Effect of pheromone components when applied to different models on male sexual behavior in the housefly, Musca domestica, J. Insect Physiol., in press.

Afsharpour, F. and R. D. O'Brien, 1963. Column chromatography of insect esterases, J. Insect Physiol., 9:521-529.

Ahmad, S., 1970. Studies on aliesterase, lipase and peptidase in susceptible and organophosphate-resistant strains of house fly (Musca domestica L.), Comp. Biochem. Physiol., 32:465-474.

Ahmad, S., 1976. Larval and adult house fly carboxylesterase: isozymic composition and tissue pattern, Insect Biochem., 6:541-547.

Ahmad, S., 1979. The functional roles of cytochrome P-450 mediated systems: present knowledge and future areas of investigations, Drug Metab. Rev., 10:1-14.

Ahmad, S., 1982. Roles of mixed-function oxidases in insect herbivores, in: "Proceedings of the 5th Symposium on Insect-Plant Relationships", J. H. Visser and A. K. Minks, eds., pp. 41-47, PUDOC, Wageningen.

Ahmad, S., Ed. 1983a, "Herbivorous Insects: host-seeking Behavior and Mechanisms:, Academic Press, New York.

Ahmad, S., 1983b. Mixed-function oxidase activity in a generalist herbivore in relation to its biology, food plants, and feeding history, Ecology, 64:235-243.

Ahmad, S., 1986. Enzymatic adaptations of herbivorous insects and mites to phytochemicals, J. Chem. Ecol., 12:533-559.

Ahmad, S. and A. J. Forgash, 1973. NADPH oxidation by microsomal preparations of gypsy moth larval tissues, Insect Biochem., 3:263-273.

Ahmad, S. and A. J. Forgash, 1976. Non-oxidative enzymes in the metabolism of insecticides, Drug. Metab. Rev., 5:141-164.

Ahmad, S. and A. J. Forgash, 1978. Gypsy moth mixed-function oxidases: gut enzyme levels increased by rearing on a wheat germ diet, Ann. Entomol. Soc. Am., 71:449-452.

Ahmad, S., A. J. Forgash and Y. T. Das, 1980. Penetration and metabolism of [^{14}C] carbaryl in larva of the gypsy moth, Lymantria dispar (L.), Pestic. Biochem. Physiol., 14:236-248.

Ahmad, S., K. E. Kirkland and G. J. Blomquist, 1987. Evidence for a sex pheromone-metabolizing cytochrome P-450 monooxygenase in the house fly, Musca domestica L., Arch. Insect Biochem. Physiol., in press.

Ahmed, N. K., R. L. Felsted and N. R. Bachur, 1979. Comparison and characterization of mammalian xenobiotic ketone reductases, J. Pharmacol. Exp. Ther., 209:12-19.

Ajami, A. M. and L. M. Riddiford, 1973. Comparative metabolism of the cecropia juvenile hormone, J. Insect Physiol., 19:635-645.

Aldridge, W. N., 1953. Serum esterases. 1. Two types of esterase (A and B) hydrolysing p-nitrophenyl acetate, propionate and butyrate, and a method for their determination, Biochem. J., 53:110-116.

Anderson, R. S., 1978. Aryl hydrocarbon hydroxylase in an insect, Spodoptera eridania, Comp. Biochem. Physiol., 59C:87-93.

Arnault, C., 1979. Influence de substances de la plante-hote sur le developpement larvaire d'Acrolepiopsis assectella (Lepidoptera, Acrolepiidae) en alimentation artificielle, Entomol. Exp. Appl., 25:64-74.

Augustinson, K. B., 1959. Electrophoretic studies on blood plasma esterases, Acta Chem. Scand., 13:571-592.

Augustinson, K. B., 1961. Multiple forms of esterase in vertebrate blood plasma, Ann. N. Y. Acad. Sci., 94:944-870.

Baars, A. J., 1979. Xenobiotic-metabolizing enzymes in the fruit fly Drosophila melanogaster and the albino rat, with emphasis on glutathione transferase, dissertation, State University of Leiden.

Baars, A. J., 1980. Biotransformation of xenobiotics in Drosophila melanogaster and its relevance for mutagenicity testing, Drug. Metab. Rev., 11:191-221.

Bachur, N. R., 1976. Cytoplasmic aldo-keto reductases: a class of drug-metabolizing enzymes, Science, 193:595-597.

Banerjee, R. K. and A. B. Roy, 1966. The sulfotransferases of guinea pig liver, Molec. Pharmacol., 2:56-66.

Bashir, N. H. H. and L. A. Crowder, 1983. Mechanisms of permethrin tolerance in the common green lacewing, Chrysopa carnea (Neuroptera: Chrysopidae), J. Econ. Entomol., 76:407-409.

Beeman, R. W. and B. A. Schmidt, 1982. Biochemical and genetic aspects of malathion-specific resistance in the Indian meal moth (Lepidoptera: Pyralidae), J. Econ. Entomol., 75:945-949.

Beesley, S. G., S. G. Compton and D. A. Jones, 1985. Rhodanese in insects, J. Chem. Ecol., 11:45-50.

Benke, G. M. and C. F. Wilkinson, 1971. In vitro microsomal epoxidase activity and susceptibility to carbaryl and carbaryl-piperonyl butoxide combinations in house crickets of different age and sex. J. Econ. Entomol., 64:1032-1034.

Benke, G. M., C. F. Wilkinson and J. N. Telford, 1972. Microsomal oxidases in a cockroach, Gromphadorhina portentosa, J. Econ. Entomol., 65:1221-1229.

Berenbaum, M. R., 1985. Brementown revisited: interactions among allelochemicals in plants. Rec. Adv. Phytochem., 19:139-169.

Berenbaum, M. R. and J. J. Neal, 1985. Synergism between myristicin and xanthotoxin, a naturally co-occurring plant toxicant, J. Chem. Ecol., 11:1349-1358.

Bernays, E. A., 1978. Tannins: an alternative viewpoint, Entomol. Exp. Appl., 24:44-53.

Bernays, E. A. and S. Woodhead, 1982. Plant phenols utilized as nutrients by a phytophagous insect, Science, 216:201-203.

Berry, R. E., S. J. Yu and L. C. Terriere, 1980. Influence of host plant on insecticide metabolism and management of variegated cutworm, J. Econ. Entomol., 73:771-774.

Bigley, W. S. and F. W. Plapp, 1978. Metabolism of cis- and trans-[^{14}C] permethrin by the tobacco budworm and the bollworm, J. Agric. Food Chem., 26:1128-1134.

Binning, A., F. J. Barby, M. P. Heenan and J. N. Smith, 1967. The conjugation of phenols with phosphate in grass grubs and flies, Biochem. J., 103:42-48.

Blau, P. A., P. Feeny and L. Contardo, 1978. Allylglucosinolate and herbivorous caterpillars: a contrast in toxicity and tolerance, Science, 200:1296-1298.

Blomquist, G. J., J. W. Dillwith and T. S. Adams, 1987. Biosynthesis and endocrine regulation of sex pheromone production in Diptera, in: "Pheromone biochemistry", G. D. Prestwich and G. J. Blomquist, eds., Academic Press, Miami, in press.

Bollenbacher, W. E., S. L. Smith, J. J. Wielgus and L. I. Gilbert, 1977. Evidence for an α-ecdysone cytochrome P-450 mixed-function oxidase in insect fatbody mitochondria, Nature, 268:660-663.

Booth, J., A. Hewer, G. R. Keysell and P. M. Sims, 1975. Enzymatic reduction of aromatic hydrocarbon epoxides by the microsomal fraction of rat liver, Xenobiotica, 5:197-203.

Bosron, W. F. and T. K. Li, 1980. Alcohol dehydrogenase, in: "Enzymatic Basis of Detoxification", W. B. Jakoby, ed., Vol. 1, pp. 231-248, Academic Press, New York.

Brattsten, L. B., 1979a. Ecological significance of mixed-function oxidations, Drug. Metab. Rev., 10:35-58.

Brattsten, L. B., 1979b. Biochemical defense mechanisms in herbivores against plant allelochemicals, in: "Herbivores, Their Interaction with Secondary Plant Metabolites", G. A. Rosenthal and D. H. Janzen, eds., pp. 199-270, Academic Press, New York.

Brattsten, L. B., 1986a. Metabolic insecticide defenses in the boll weevil compared to those in a resistance-prone species, Pestic. Biochem. Physiol., in press.

Brattsten, L. B., 1986b. Inducibility of metabolic defenses in the boll weevil and the tobacco budworm, Pestic. Biochem. Physiol., in press.

Brattsten, L. B., 1986c. Potential role of plant allelochemicals in the development of insecticide resistance, in: "Indirect effects of plant allelochemicals", P. Barbosa, ed., J. Wiley and Sons, New York, in press.

Brattsten, L. B. and C. A. Gunderson, 1981. Isolation of insect microsomal oxidases by rapid centrifugation, Pestic. Biochem. Physiol., 16:187-198.

Brattsten, L. B. and R. L. Metcalf, 1970. The synergistic ratio of carbaryl with piperonyl butoxide as an indicator of the distribution of multifunction oxidases in the Insecta, J. Econ. Entomol., 63:101-104.

Brattsten, L. B. and R. L. Metcalf, 1973. Synergism of carbaryl toxicity in natural insect populations, J. Econ. Entomol., 66:1347-1348.

Brattsten, L. B., S. L. Price and C. A. Gunderson, 1980. Microsomal oxidases in midgut and fatbody tissues of a broadly herbivorous insect larva, Spodoptera eridania Cramer (Noctuidae). Comp. Biochem. Physiol., 66C:231-237.

Brattsten, L. B., C. F. Wilkinson and M. M. Root, 1976. Microsomal hydroxylation of aniline in the southern armyworm (Spodoptera eridania), Insect Biochem., 6:615-620.

Brattsten, L. B., C. K. Evans, S. Bonetti and L. H. Zalkow, 1984. Induction by carrot allelochemicals of insecticide-metabolizing enzymes in the southern armyworm (Spodoptera eridania), Comp. Biochem. Physiol., 77C:29-37.

Brattsten, L. B., C. A. Gunderson, J. T. Fleming and K. N. Nikbahkt, 1986.

Temperature and diet modulate cytochrome P-450 activities in southern armyworm, Spodoptera eridania Cramer, caterpillars, Pestic. Biochem. Physiol., 25:346-357.

Breuer, M. and R. Knuppen, 1961. The formation and hydrolysis of 16 , 17 -epoxyoestratriene-3-ol by rat liver tissue, Biochem. Biophys. Acta, 49:620-621.

Bridges, R. G., 1983. Insect phospholipids, in: "Metabolic Aspects of Lipid Nutrition in Insects", T. E. Mittler and R. H. Dadd, eds., pp. 159-181, Westview Press, Boulder.

Brooks, G. T., 1973. Insect epoxide hydrase inhibition by juvenile hormone analogues and metabolic inhibitors, Nature, 245:382-384.

Brooks, G. T., A. Harrison, S. E. Lewis, 1970. Cyclodiene epoxide ring hydration by microsomes from mammalian liver and house flies, Biochem. Pharmacol., 19:255-273.

Brown, T. M. and G. H. S. Hooper, 1979. Metabolic detoxication as a mechanism of methoprene resistance in Culex pipiens pipiens, Pestic. Biochem. Physiol., 12:79-86.

Bull, L. B., C. C. J. Culvenor and A. T. Dick, 1968. "The Pyrrolizidine alkaloids", J. Wiley and Sons, New York.

Burt, M. E., R. J. Kuhr and W. S. Bowers, 1978. Metabolism of precocene II in the cabbage looper and European corn borer, Pestic. Biochem. Physiol., 9:300-303.

Caldwell, J., 1982. Conjugation reactions in foreign-compound metabolism: definition, consequences, and species variations, Drug Metab. Rev., 13:745-777.

Capdevila, J., N. Ahmad and M. Agosin, 1975. Soluble cytochrome P-450 from house fly microsomes. Partial purification and characterization of two hemo-protein forms, J. Biol. Chem., 250:1048-1060.

Casabé, N. and E. Zerba, 1981. Esterases of Triatoma infestans and its relationship with the metabolism of organophosphorous insecticides, Comp. Biochem. Physiol., 68C:255-258.

Casida, J. E., 1955. Comparative enzymology of certain insect acetylesterases in relation to poisoning by organophosphorous insecticides, Biochem. J., 60:487-496.

Casida, J. E., 1970. Mixed-function oxidase involvement in the biochemistry of insecticide synergists, J. Agric. Food Chem., 18:753-772.

Cassidy, J. D., E. Smith and E. Hodgson, 1969. An ultrastructural analysis of microsomal preparations from Musca domestica and Prodenia eridania, J. Insect Physiol., 13:1573-1578.

Chance, B. and A. M. Pappenheimer, 1954. Kinetic and spectrophotometric studies of cytochrome b_5 in midgut homogenates of Cecropia, J. Biol. Chem., 209:931-943.

Chang, C. K. and T. W. Jordan, 1983. Distribution of permethrin-hydrolyzing esterases from Wiseana cervinata larvae, Pestic. Biochem. Physiol., 19:190-195.

Chang, C. K., A. G. Clark, A. Fieldes and S. Pound, 1981. Some properties of a glutathione S-transferase from the larvae of Galleria mellonella, Insect Biochem., 11:179-186.

Chang, K. M., C. F. Wilkinson, K. Hetnarski and M. Murray, 1983. Aryl hydrocarbon hydroxylase in larvae of the southern armyworm (Spodoptera eridania), Insect Biochem., 13:87-94.

Chipoulet, J. M. and C. Chararas, 1985. Survey and electrophoretical separation of the glycosidases of Rhagium inquisitor (Coleoptera: Cerambycidae) larvae, Comp. Biochem. Physiol., 80B:241-246.

Clark, A. G. and B. Drake, 1984. Purification and properties of glutathione S-transferases from larvae of Wiseana cervinata, Biochem. J., 217:41-50.

Clark, A. G. and N. A. Shaaman, 1984. Evidence that DDT-dehydrochlorinase from the house fly is a glutathione S-transferase, Pestic. Biochem. Physiol., 22:249-261.

Clark, A. G., F. J. Darby and J. N. Smith, 1967. Species differences in the inhibition of glutatione S-aryl transferase by phthaleins and dicarboxylic acids, Biochem. J., 103:49-54.

Clark, A. G., N. A. Shaaman, W. C. Dauterman and T. Hayaoka, 1984. Characterization of multiple glutatione transferases frρm the house fly, Musca domestica (L.), Pestic. Biochem. Physiol., 22:51-59.

Clark, A. G., G. L. Dick. S. M. Martindale and J. N. Smith, 1985. Glutathione S-transferases from the New Zealand grass grub, Costelytra zealandica. Their isolation and characterization and the effect on their activity of endogenous factors, Insect Biochem., 15:35-44.

Clements, A. N., 1967. A study of soluble esterases in Pieris brassicae (Lepidoptera), J. Insect Physiol., 13:1021-1030.

Cohen, A. J., J. N. Smith and H. Turbert, 1964. Comparative detoxification, 10. The enzymic conjugation of chlorocompounds with glutathione in locusts and other insects, Biochem. J., 90:457-464.

Cohen, E., 1981. Epoxide hydrase activity in the flour beetle, Tribolium castaneum (Coleoptera, Tenebrionidae), Comp. Biochem. Physiol., 69B:29-34.

Cole, R. A., 1975. 1-cyanoepithioalkanes: major products of alkenyl glucosinolate hydrolysis in certain Cruciferae, Phytochemistry, 14:2293-2294.

Collins, P. J., 1985. Induction of the polysubstrate monooxygenase system in the native budworm Heliothis punctiger (Wallengren) (Lepidoptera: Noctuidae), Insect Biochem., 15:551-555.

Cook, B. J. and A. J. Forgash, 1965. The identification and distribution of the carboxylic esterases in the American cockroach, Periplaneta americana (L.), J. Insect Physiol., 11:237-250.

Cook, J. C. and E. Hodgson, 1983. Induction of cytochrome P-450 by methylene dioxyphenyl compounds: importance of the methylene carbon, Toxicol. Appl. Pharmacol., 68:131-139.

Crankshaw, D. L., K. Hetnarski and C. F. Wilkinson, 1981. Interspecies cross-reactivity of an antibody to southern armyworm (Spodoptera eridania) midgut NADPH-cytochrome c reductase, Insect Biochem., 11:593-597.

Croft, B. A. and C. A. Mullin, 1984. Comparison of detoxification enzyme systems in Argyrotaenia citrana (Lepidoptera: Tortricidae) and the ectoparasite Oncophanes americanus (Hymenoptera: Braconidae), Environ. Entomol., 13:1330-1335.

Das, M., D. R. Bickers and H. Mukhtar, 1984. Plant phenols as in vitro inhibitors of glutathione S-transferases, Biochem. Biophys. Res. Comm., 120:427-433.

Dauterman, W. C., 1976. Extra microsomal metabolism of insecticides, in: "Insecticide Biochemistry and Physiology", C. F. Wilkinson, ed., pp. 149-176, Plenum Publ. Corp., New York.

Dauterman, W. C., 1980. Metabolism of toxicants: phase II reactions, in: "Introduction to Biochemical Toxicology", E. Hodgson and F. E. Guthrie, eds., pp. 92-105, Elsevier, New York.

Dauterman, W. C., 1985. Insect metabolism: extramicrosomal, in: "Comprehensive Insect Physiology, Biochemistry, and Pharmacology", G. A. Kerkut and L. I. Gilbert, eds., Vol. 12, pp. 713-730. Pergamon Press, New York.

Dauterman, W. C. and E. Hodgson, 1978. Detoxication mechanisms in insects, in; "Biochemistry of Insects", M. Rockstein, ed., pp. 541-577, Academic Press, New York.

Davidson, W. S., D. J. Walton and T. G. Flynn, 1978. A comparative study of the tissue and species distribution of NADPH-dependent aldehyde reductase, Comp. Biochem. Physiol., 60B:309-315.

Davis, R. H. and A. Nahrstedt, 1979. Linamarin and lotaustralin as the source of cyanide in Zygaena filipendulae L. (Lepidoptera), Comp. Biochem. Physiol., 64B:395-397.

Davis, R. H. and A. Nahrstedt, 1985. Cyanogenesis in insects, in: "Comprehensive Insect Physiology, Biochemistry, and Pharmacology", G. A. Kerkut and L. I. Gilbert, eds., Vol. 11, pp. 635-657. Pergamon Press, New York.

De Jersey, J., J. Nolan, P. A. Davey and P. W. Riddles, 1985. Separation and characterization of the pyrethroid-hydrolyzing esterases of the cattle tick, Boophilus microplus, Pestic. Biochem. Physiol., 23:349-357.

Delden, V. van Evbia, 1982. The alcohol dehydrogenase polymporhism in Drosophila melanogaster: selection at an enzyme locus, Evolutionary Biol., 15:197-222.

Denison, M. S., J. W. Hamilton and C. F. Wilkinson, 1985. Comparative studies of aryl hydrocarbon hydroxylase and the Ah receptor in non-mammalian species, Comp. Biochem. Physiol., 80C:319-324.

Devonshire, A. L. and G. D. Moores, 1982. A carboxylesterase with broad substrate specificity causes organophosphorous, carbamate, and pyrethroid resistance in peach-potato aphids (Myzus persicae), Pestic. Biochem. Physiol., 18:235-246.

Devonshire, A. L. and G. D. Moores, 1984. Immunoassay of carboxylesterase activity for identifying insecticide resistant Myzus persicae, Proc. Br. Crop Prot. Conf.: Pests, Dis., 2:515-520.

Devonshire, A. L. and R. M. Sawicki, 1979. Insecticide-resistant Myzus persicae as an example of evolution by gene duplication, Nature, 280:140-141.

Dickins, M., J. W. Bridges, C. R. Elcombe and K. J. Netter, 1978. A novel hemoprotein induced by isosafrole pre-treatment in the rat, Biochem. Biophys. Res. Commun., 80:89-96.

Dillwith, J. W. and G. J. Blomquist, 1982. Site of sex pheromone production in the housefly, Musca domestica L., Experientia, 38:471-473.

Dorough, H. W., 1979. Metabolism of insecticides by conjugation mechanisms, Pharmac. Therap., 4:433-471.

Doskotch, R. W. and F. S. El-Feraly, 1969. Isolation and characterization of (+)-sesamin and B-cyclopyrethrosin from pyrethrum flowers, Can. J. Chem., 47:1139-1142.

Dowd, P. F., C. M. Smith and T. C. Sparks, 1983a. Influence of soybean leaf extracts on ester cleavage in cabbage and soybean loopers (Lepidoptera: Noctuidae), J. Econ. Entomol., 76:700-703.

Dowd, P. F., C. M. Smith and T. C. Sparks, 1983b. Detoxification of plant toxins by insects, Insect Biochem., 13:453-468.

Downer, R. G. H., 1978. Functional role of lipids in insects, in: "Biochemistry of Insects", M. Rockstein, ed., pp. 57-92, Academic Press, New York.

Duffey, S. S., 1981. Cyanide and arthropods, in: "Cyanide in Biology", B. Vennerland, E. E. Conn, C. J. Knowles, J. Westley and F. Wissing, eds., pp. 385-414, Academic Press, New York.

Dutton, G. J., 1962. The mechanism of o-aminophenyl glucoside formation in Periplaneta americana, Comp. Biochem. Physiol., 7:39-46.

Dykstra, W. G. and W. C. Dauterman, 1978. Excretion, distribution and metabolism of S-(2,4-dinitrophenyl) glutatione in the American cockroach, Insect Biochem., 8:263-265.

Dyte, C. E. and D. G. Rowlands, 1968. The metabolism and synergism of malathion in resistant and susceptible strains of Tribolium castaneum (Herbst) (Coleoptera, Tenebrionidae), J. Stored Prod. Res., 4:157-173.

Eagleson, C., 1940, U. S. Patent No. 2,202,145.

Ehrlich, P. R. and P. H. Raven, 1964. Butterflies and plants: a study in coevolution, Evolution, 18:586-608.

Elliott, M., N. F. James, E. C. Kimmel and J. E. Casida, 1972. Metabolic fate of pyrethrin I, pyrethrin II and allethrin administered orally to rats, J. Agr. Food Chem., 20:300-313.

Ellis-Pratt, G., 1983. The mode of action of pro-allatocidins, in: "Natural Products for Innovative Pest Management", D. L. Whitehead and W. S. Bowers, eds., pp. 323-355, Pergamon Press, New York.

Esaac, E. G. and J. E. Casida, 1968. Piperonylic acid conjugates with alanine, glutamate, glutamine, and serine in living house flies, J. Insect Physiol., 14:913-925.

Ettlinger, M. G., G. P. Dateo, Jr., B. W. Harrison, T. J. Mabry and C. P. Thompson, 1961. Vitamin C as a coenzyme: the hydrolysis of mustard oil glucosides, Proc. Nat. Acad. Sci. USA, 47:1875-1880.

Evans, P. H., D. M. Soderlund and J. R. Aldrich, 1980. In vitro N-acetylation of biogenic amines by tissues of the European corn borer, Ostrinia nubilalis Hubner, Insect Biochem., 10:375-380.

Felsted, R. L. and N. R. Bachur, 1980. Ketone reductases, in: "Enzymatic Basis of Detoxication", W. B. Jakoby, ed., Vol. 1, pp. 281-293. Academic Press, New York.

Ferkovich, S. M., J. E. Oliver and C. Dillard, 1982. Pheromone hydrolysis by cuticular and interior esterases of the antennae, legs, and wings of the cabbage looper moth, Trichoplusia ni (Hubner), J. Chem. Ecol., 8:859-866.

Feyereisen, R. and F. Durst, 1980. Control of cytochrome P-450 monooxygenases in an insect by the steroid moulting hormone, in: "Microsomes, Drug Oxidations, and Chemical Carcinogens", M. J. Coon, A. H. Conney, R. W. Estabrook, H. V. Gelboin, J. R. Gillette, and P. J. O'Brien, eds., pp. 595-598, Academic Press, New York.

Feyereisen, R. and D. E. Farnsworth, 1985. Developmental changes of microsomal cytochrome P-450 monooxygenases in larval and adult Diploptera punctata, Insect Biochem., 6:755-761.

Feyereisen, R., G. E. Pratt and A. F. Hamnett, 1981. Enzymic synthesis of juvenile hormone in locust corpora allata: evidence for a microsomal cytochrome P-450 linked methylfarnesoate epoxidase, Eur. J. Biochem., 118:231-238.

Fisher, C. W., and R. T. Mayer, 1984. Partial purification and characterization of phenobarbital-induced house fly cytochrome P-450, Arch. Insect Biochem. Physiol., 1:127-138.

Folsom, M. D. and E. Hodgson, 1970. Biochemical characteristics of insect microsomes: NADPH oxidation by intact microsomes from the house fly, Musca domestica, Comp. Biochem. Physiol., 37:301-310.

Fontan, A. and E. Zerba, 1984. Integumental esteratic activity in Triatoma infestans and its contribution to the degradation of organophosphorous insecticides, Comp. Biochem. Physiol., 79C:183-188.

Fox, L. R. and P. A. Morrow, 1981. Specialization: species property or local phenomenon, Science, 211:887-893.

Fox, P. M. and J. S. Massare, 1976. Aspects of juvenile hormone metabolism in Periplaneta americana (L.), Comp. Biochem. Physiol., 53B:195-200.

Friedlander, A. and S. Navarro, 1984. The glutathione status of Ephestia cautella (Walker) pupae exposed to carbon dioxide. Comp. Biochem. Physiol., 79C:217-218.

Fujimoto, Y., M. Morisaki and N. Ikekawa, 1985. Enzymatic dealkylation of phytosterols in insects, Meth. Entymol., 111:346-352.

Garfinkel, D., 1958. Studies on pig liver microsomes. I. Enzymic and pigment composition of different microsomal fractions, Arch. Biochem. Biophys., 77:493-509.

Gatehouse, A. M. R., K. A. Fenton and J. H. Anstee, 1985. Carbohydrase and esterase activity in the gut of larval Callosobruchus maculatus, Experientia, 41:1202-1205.

Georghiou, G. P. and N. Pasteur, 1978. Electrophoretic esterase patterns in insecticide-resistant and susceptible mosquitoes, J. Econ. Entomol., 71:201-205.

Gessner, T. and M. Acara, 1968. Metabolism of thiols: S-glucosylation, J. Biol. Chem., 243:3142-3147.

Gil, D. L. and C. F. Wilkinson, 1977. Structure-activity relationships of 1,2,3-benzothiadiazole insecticide synergists as inhibitors of microsomal oxidation, Pestic. Biochem. Physiol., 7:183-193.

Gil, D. L., H. A. Rose, R. S. H. Yang, R. G. Young and C. F. Wilkinson, 1974. Enzyme induction by phenobarbital in the Madagascar cockroach, Gromphadorhina portentosa, Comp. Biochem. Physiol., 47B:657-662.

Gillette, J. R., A. H. Conney, G. J. Cosmides, R. W. Estabrook, J. R. Fouts and G. J. Mannering, eds., 1969. "Microsomes and drug oxidations", Academic Press, New York.

Glass, E. H. and P. J. Chapman, 1952. The redbanded leaf roller and its control, N. Y. State Agric. Exp. Sta. Bull. No. 755.

Goldman, P., 1982. Role of the intestinal microflora, in: "Metabolic Basis of Detoxication", W. B. Jakoby, J. R. Bend and J. Caldwell, eds., pp. 323-338, Academic Press, New York.

Gordon, H. T., 1961. Nutritional factors in insect resistance to chemicals, Annu. Rev. Entomol., 6:27-54.

Gorrod, J. W., 1973. Differentiation of various types of biological oxidation of nitrogen in organic compounds, Chem. Biol. Interact., 7:289-303.

Gould, F., 1984. Mixed-function oxidases and herbivore polyphagy: the devil's advocate position, Ecol. Entomol., 9:29-34.

Gould, F. and E. Hodgson, 1980. Mixed function oxidase and glutathione transferase activity in last instar Heliothis virescens larvae, Pestic. Biochem. Physiol., 13:34-40.

Greenwood, D. R. and H. H. Rees, 1984. Ecdysone 20-monooxygenase in the desert locust, Schistocerca gregaria, Biochem. J., 223:837-847.

Gunderson, C. A., L. B. Brattsten and J. T. Fleming, 1986. Microsomal oxidase and glutathione transferase as factors influencing the effects of pulegone in southern and fall armyworm larvae, Pestic. Biochem. Physiol., 26:238-249.

Haller, H. L., E. R. McGovran, L. D. Goodhue, and W. N. Sullivan, 1942. The synergistic action of sesamin with pyrethrum insecticides, J. Org. Chem., 7:183-185.

Halpin, R. A., K. P. Vyas, S. B. El-Naggar and D. M. Jerina, 1984. Metabolism and hepatotoxicity of the naturally occurring benzo(b)pyran precocene I, Chem. Biol. Interact., 48:297-315.

Hammock, B. D. and L. S. Hasagawa, 1983. Differential substrate selectivity of murine hepatic cytosolic and microsomal epoxide hydrolases, Biochem. Pharmacol., 32:1155-1164.

Hammock, B. D. and G. B. Quistad, 1976. The degradative metabolism of juvenoids by insects, in: "The Juvenile Hormones", L. I. Gilbert, ed., pp. 374-393, Plenum Publ. Corp., New York.

Hammock, B. D. and G. B. Quistad, 1981. Metabolism and mode of action of juvenile hormones, juvenoids, and other insect growth regulators, in: "Progress in Pesticide Biochemistry", D. H. Hutson and T. R. Roberts, eds, Vol. 1, pp. 1-83, John Wiley and Sons, New York.

Hammock, B. D., T. C. Sparks and S. M. Mumby, 1977. Selective inhibition of JH esterase from cockroach hemolymph, Pestic. Biochem. Physiol., 7:517-530.

Hammock, B. D., S. S. Gill, S. M. Mumby and K. Ota, 1980. Comparison of epoxide hydrolases in the soluble and microsomal fractions of mammalian liver, in: "Molecular Basis of Environmental Toxicity", R. S. Bhatnagar, ed., pp. 229-272, Ann Arbor Science Publ., Ann Arbor.

Hansen, L. G. and E. Hodgson, 1971. Biochemical characteristics of insect microsomes; N- and O-demethylation, Biochem. Pharmacol., 20:1569-1578.

Hayaoka, T. and W. C. Dauterman, 1983. The effect of phenobarbital induction on glutathione conjugation of diazinon in susceptible and resistant houseflies, Pestic. Biochem. Physiol., 19:344-349.

Heirweigh, K. P. M., J. A. T. P. Meuwissen and J. Fevery, 1971. Enzymic formation of β-D-monoglucuronide, β-D-monoglucoside and mixtures of β-D-monoxyloside and B-D-dixyloside of bilirubin by microsomal preparations from rat liver, Biochem. J., 125:28-29.

Hemingway, J., 1985. Malathion carboxylesterase enzymes in Anopheles arabiensis from Sudan, Pestic. Biochem. Physiol., 23:309-313.

Heymann, E., 1980. Carboxylesterases and amidases, in: "Enzymatic Basis of Detoxication", W. B. Jakoby, ed., Vol. 2, pp. 291-323, Academic Press, New York.

Hiltz, H. and F. Lipmann, 1955. The enzymatic activation of sulfate, Proc. Nat. Acad. Sci., USA, 41:880-890.

Himwich, W. A. and J. P. Saunders, 1948. Enzymatic conversion of cyanide to thiocyanate, Am. J. Physiol., 153:348-354.

Hipps, P. P. and D. R. Nelson, 1974. Esterases from the midgut and gastric cecum of the American cockroach, Periplaneta americana. Isolation and characterization, Biochem. Biophys. Acta., 341:421-436.

Hodgson, E., 1979. Comparative aspects of the distribution of cytochrome P-450-dependent monooxygenase systems; an overview, Drug Metab. Rev., 10:15-33.

Hodgson, E., 1983. Production of pesticide metabolites by oxidative reactions, J. Toxicol. Clin. Toxicol., 19:609-621.

Hodgson, E., 1985. Microsomal monooxygenases, in: "Comprehensive Insect Physiology, Biochemistry, and Pharmacology", G. A. Kerkut and L. I. Gilbert, Eds., Vol 11, pp. 206-321, Pergamon Press, New York.

Hodgson, E. and J. E. Casida, 1961. Metabolism of N, N-dialkylcarbamates and related compounds by rat liver, Biochem. Pharmacol., 8:179-191.

Hodgson, E. and A. P. Kulkarni, 1979. Characterization of cytochrome P-450 in studies of insecticide resistance, in: "Pest Resistance to Pesticides", G. P. Georghiou and T. Saito, eds., pp. 207-228, Plenum Publ. Corp., New York.

Hollingworth, R. M., 1976. The biochemical and physiological basis of selective toxicity, in: "Insecticide Biochemistry and Physiology", C. F. Wilkinson, ed., pp. 431-506. Plenum Publ. Corp., New York.

Hopf, H. S., 1954. Studies in the mode of action of insecticides. II. Inhibition of the acetylesterases of the locust nerve cord by some organic phosphoric esters, Ann. Appl. Biol., 41:248-260.

Hosel, W. and E. E. Conn, 1982. The aglycone specificity of plant β-glycosidases, Trends Biochem. Sci., 7:219-221.

Hughes, B. P. and D. A. Raftos, 1985. Genetics of an esterase associated with resistance to organophosphorous insecticides in the sheep blowfly, Lucilia cuprina (Wiedemann) (Diptera: Calliphoridae), Bull. Entomol. Res., 75:535-545.

Hwang-Hsu, K., G. Reddy, A. K. Kumaran, W. E. Bollenbacher and L. I. Gilbert, 1979. Correlations between juvenile hormone esterase activity, ecdysone titre, and cellular reprogramming in Galleria mellonella, J. Insect Physiol., 25:105-111.

International Union of Biochemistry (IUB), 1984. "Enzyme Nomenclature", Academic Press, New York.

Isaac, R. E., K. K. Phua and H. H. Rees, 1982. 3'-Phosphoadenosine-5'-phosphosulfate synthesis and involvement of sulphotransferase reactions in the insect, Spodoptera littoralis, Biochem. J., 204:127-133.

Ishaaya, I. and J. E. Casida, 1980. Properties and toxicological significance of esterases hydrolyzing permethrin and cypermethrin in Trichoplusia ni larval gut and integument, Pestic. Biochem. Physiol., 14:178-184.

Jakoby, W. B., ed., 1980. "Enzymatic Basis of Detoxication", Vol. 2, Academic Press, New York.

Jakoby, W. B. and W. H. Habig, 1980. Glutathione transferases, in: "Enzymatic Basis of Detoxication", W. B. Jakoby, ed., Vol. 2, pp. 63-94, Academic Press, New York.

Jakoby, W. B., J. R. Bend and J. Caldwell, eds., 1982. "Metabolic Basis of Detoxication, Metabolism of Functional Groups", Academic Press, New York.

Jameson, G. W., J. R. MacFarlane and T. W. Hogan, 1976. Esterases in relation to embryonic development in the field cricket Teleogryllus commodus, Insect Biochem., 6:59-63.

Jansen, M., A. J. Baars and D. D. Breimer, 1984. Cytosolic glutathione S-transferases in Drosophila melanogaster, Biochem. Pharmacol., 33:3655-3659.

Jao, L. T. and J. E. Casida, 1974. Insect pyrethroid-hydrolyzing esterases, Pestic. Biochem. Physiol., 4:465-472.

Jones, D., J. Parsons and M. Rothschild, 1962. Release of hydrocyanic acid from crushed tissues of all stages in the life-cylce of species of the Zygaenidae (Lepidoptera), Nature, 193:52-53.

Junge, W., 1984. Carboxylesterase, in: "Methods in Enzymatic Analysis" H. U. Bergmeyer, ed., 3rd edition, Vol. 4, pp. 2-8, Verlag Chemie, Miami.

Junge, W. and H. Klees, 1984. Arylesterase, in: ibid., pp. 8-14.

Kalyanaraman, B., 1982. Detection of toxic free radicals in biology and medicine, Rev. Biochem. Toxicol., 4:74-139.

Kaneko, A., Y. Yoshida, K. Enomoto, T. Kaku, K. Hirata and T. Onoe, 1979. Induction of a microsomal butyrylesterase in rat liver by phenobarbital treatment, Biochem. Biophys. Acta, 582:185-195.

Kao, L. R., N. Motoyama and W. C. Dauterman, 1985. The purification and characterization of esterases from insecticide-resistant and susceptible house flies, Pestic. Biochem. Physiol., 23:228-239.

Kapin, M. A. and S. Ahmad, 1980. Esterases in larval tissues of gypsy moth, Lymantria dispar (L.): optimum assay conditions, quantification and characterization. Insect Biochem., 10:331-337.

Karlson, P. and H. Ammon, 1963. Biogenesis and fate of the acetyl group of N-acetyl dopamine, Z. Physiol. Chem., 330:161-168.

Kato, R., K. Iwasaki and H. Noguchi, 1978. Reduction of tertiary amine N-oxides by cytochrome P-450, Mol. Pharmacol., 14:654-664.

Keevil, T. and H. S. Mason, 1978. Molecular oxygen in biological oxidations, an overview, Meth. Enzymol., 52:3-40.

Kent, P. W. and P. C. J. Brunet, 1959. The occurrence of protocathechuic acid and its 4-O-β-D-glucoside in Blatta and Periplaneta, Tetrahedron, 7:252-256.

Klages, G. and H. Emmerich, 1979. Juvenile hormone metabolism and juvenile hormone esterase titer in hemolymph and peripheral tissues of Drosophila hydrei, J. Comp. Physiol., 132:319-325.

Klingenberg, M., 1958. Pigments of rat liver microsomes, Arch. Biochem. Biophys., 75:376-386.

Koren, B., A. Yawetz and A. S. Perry, 1984. Biochemical properties characterizing the development of tolerance to malathion in Ceratitis capitata Wiedemann (Diptera: Tephritidae), J. Econ. Entomol., 77:864-867.

Kramer, S. J., M. Wieten and C. A. D. deKort, 1977. Metabolism of juvenile hormone in Colorado potato beetle, Leptinotarsa decemlineata, Insect Biochem., 7:231-236.

Krieger, R. I. and C. F. Wilkinson, 1969. Microsomal mixed-function oxidases in insects. 1. Localization and properties of an enzyme system effecting aldrin epoxidation in larvae of the southern armyworm (Prodenia eridania), Biochem. Pharmacol., 18:1403-1415.

Krieger, R. I. and C. F. Wilkinson, 1970. An endogenous inhibitor of microsomal mixed-function oxidases in homogenates of the southern armyworm (Prodenia eridania), Biochem. J., 116:781-789.

Krieger, R. I. and C. F. Wilkinson, 1971. The metabolism of 6,7-dihydroisodrin by microsomes and southern armyworm larvae, Pestic. Biochem. Physiol., 1:92-100.

Krieger, R. I., P. P. Feeny and C. F. Wilkinson, 1971. Detoxification enzymes in the guts of caterpillars: an evolutionary answer to plant defenses, Science, 172:579-581.

Kuhr, R. J., 1970. Metabolism of carbamate insecticide chemicals in plants and insects, J. Agric. Food Chem., 18:1023-1030.

Kupfer, D. and L. L. Bruggeman, 1966. Determination of enzymic demethylation of p-chloro N-methylaniline. Assay of aniline and p-chloroaniline, Anal. Biochem., 17:502-512.

Levine, W. G., 1982. Glutathione, lipid peroxidation and regulation of cytochrome P-450 activity, Life Sci., 31:779-784.

Lewis, S. E., C. F. Wilkinson and J. W. Ray, 1967. The relationship between microsomal epoxidation and lipid peroxidation in house flies and pig liver and the inhibitory effect of derivatives of 1,3-benzodioxole (methylenedioxyphenyl), Biochem. Pharmacol., 16:1195-1210.

Lichtenstein, E. P. and J. E. Casida, 1963. Myristicin, an insecticide and synergist occurring naturally in the edible parts of parsnip. J. Agric. Food Chem., 11:410-415.

Long, K. Y. and L. B. Brattsten, 1982. Is rhodanese important in the detoxification of cyanide in southern armyworm (Spodoptera eridania Cramer) larvae? Insect Biochem., 12:367-375.

Lord, K. A. and C. Potter, 1951. Studies on the mechanism of insecticidal action of organophosphorous compounds with particular reference to their anti-esterase activity, Ann. Appl. Biol., 38:495-507.

Lu, A. Y. H. and S. B. West, 1978. Reconstituted mammalian mixed function oxidases: requirements, specificities, and other properties, Pharmac. Ther., A2:337-358.

Lu, A. Y. H., W. Levin, M. Vore, A. H. Conney, D. R. Thakker, G. Holder and D. M. Jerina, 1976. Metabolism of benzo(a)pyrene by purified liver microsomal cytochrome P-448 and epoxide hydrase, in: "Carcinogenesis", R. I. Freudenthal and P. W. Jones, eds., Vol. 1, pp. 115-126, Raven Press, New York.

Maa, W. C. J. and L. C. Terriere, 1983. Age-dependent variation in enzymatic and electrophoretic properties of flesh fly (Sarcophaga bullata) and blow fly (Phormia regina) carboxylesterases, Comp. Biochem. Physiol., 74C:451-460.

MacGibbon, D. B. and R. M. Allison, 1971. An electrophoretic separation of cabbage aphid and plant glucosinolases, N. Z. J. Sci., 14:134-140.

Mackness, M. I., C. H. Walker, D. G. Rowlands and N. R. Price, 1983. Esterase activity in homogenates of three strains of the rust red flour beetle Tribolium castaneum (Herbst), Comp. Biochem. Physiol., 74C:65-68.

Marangos, A. and R. Hill, 1974. The hydrolysis and absorption of thioglucosides of rapeseed meal, Proc. Nutr. Soc., 33:90A.

Martin, M. M., 1983. Cellulose digestion in insects, Comp. Biochem. Physiol., 75A:313-324.

Martin, W. R. and J. W. Foster, 1955. Production of trans-L-epoxysuccinic acid by fungi and its microbial conversion to meso-tartaric acid, J. Bacteriol., 70:405-414.

Marty, M. A. and R. I. Krieger, 1984. Metabolism of uscharidin, a milkweed cardenolide, by tissue homogenates of monarch butterfly larvae, Danaus plexippus, L., J. Chem. Ecol., 10:945-956.

Marty, M. A., S. J. Gee and R. I. Krieger, 1982. Monooxygenase activities of fatbody and gut homogenates of monarch butterfly larvae, Danaus plexippus, fed four cardenolide-containing milkweeds, Asclepias spp., J. Chem. Ecol., 8:797-805.

Matsumura, F. and A. W. A. Brown, 1963. Studies on carboxyesterase in malathion-resistant Culex tarsalis, J. Econ. Entomol., 56:381-388.

Matsumura, F. and K. Sakai, 1968. Degradation of insecticides by esterases of the American cockroach, J. Econ. Entomol., 61:598-605.

Mayer, R. T., A. C. Chen and J. R. DeLoach, 1983. Characterization of mannosyl transferases during the pupal instars of Stomoxys calcitrans (L), Arch. Insect Biochem. Physiol., 3:1-15.

Mayer, R. T., J. A. Svoboda, and G. F. Weirich, 1978. Ecdysone 20-hydroxylase in midgut mitochondria of Manduca sexta (L.), Hoppe-Seyler's Z. Physiol. Chem., 359:1247-1257.

Mayer, R. T., J. W. Jermyn, M. D. Burke and R. A. Prough, 1977. Methoxyresorufin as a substrate for the fluorometric assay of insect microsomal, O-dealkylases, Pestic. Biochem. Physiol., 7:349-354.

Mehendale, H. M. and H. W. Dorough, 1972. In vitro glucosylation of 1-naphthol by insects, J. Insect Physiol., 18:981-987.

Menguelle, J., S. Fuzeau-Braesch and C. Papin, 1985. The influence of glutathione on the resistance to lindane of the migratory locust, Locusta migratoria cinerascens, Comp. Biochem. Physiol., 80C:401-405.

Menzie, C. M., 1969. Metabolism of pesticides, United States Department of the Interior, Bureau of Sport fisheries and Wildlife, Special Scientific report: Wildlife No. 127, Washington.

Mercot, H., 1985. A molecular approach to the role of historicity in evolution. 1. Experimental design with enzyme polymorphism in Drosophila melanogaster, Evolution, 39:819-830.

Metcalf, R. L., 1967. Mode of action of insecticide synergists, Annu. Rev. Entomol., 12:229-256.

Metcalf, R. L., R. B. March and M. G. Maxon, 1955. Substrate preferences of insect cholinesterases, Ann. Entomol. Soc. Am., 48:222-228.

Metcalf, R. L., M. F. Osman and T. R. Fukuto, 1967. Metabolism of C^{14}-labeled carbamate insecticides to $C^{14}O_2$ in the house fly, J. Econ. Entomol., 60:445-450.

Metcalf, R. L., M. G. Maxon, T. R. Fukuto and R. B. March, 1956. Aromatic esterase in insects, Ann. Entomol. Soc. Am., 49:274-279.

Miranda, C. L., P. R. Cheeke and D. R. Buhler, 1980. Effects of pyrrolizidine alkaloids from tansy ragwort (Senecio jacobaea) on hepatic drug-metabolizing enzymes in male rats, Biochem. Pharmacol., 29:2645-2649.

Miyakado, M., I. Nakayama, N. Ohno and H. Yoshioka, 1983. Structure, chemistry and actions of the Piperaceae amides: new insecticidal constituents isolated from the pepper plant, in: "Natural Products for Innovative Pest Management", D. L. Whitehead and W. S. Bowers, eds, pp. 369-382, Pergamon Press, New York.

Miyata, T., 1983. Detection and monitoring methods for resistance in arthropods based on biochemical characteristics, in: "Pest Resistance to Pesticides", G. P. Georghiou and T. Saito, eds., pp. 99-116, Plenum Publ. Corp., New York.

Moldenke, A. F., D. R. Vincent, D. E. Farnsworth and L. C. Terriere, 1984. Cytochrome P-450 in insects: 4. Reconsitution of cytochrome P-450 dependent monooxygenase activity in the house fly, Pestic. Biochem. Physiol., 21:358-367.

Morgenstern, R., J. W. DePierre and H. Jornvall, 1985. Microsomal glutathione transferase; primary structure, J. Biol. Chem., 260:13976-13983.

Morris, C. E., 1983. Uptake and metabolism of nicotine by the CNS of a nicotine-resistant insect, the tobacco hornworm (Manduca sexta), J. Insect Physiol., 29:807-817.

Morton, R. A. and B. C. Holwerda, 1985. The oxidative metabolism of malathion and malaoxon in resistant and susceptible strains of Drosophila melanogaster, Pestic. Biochem. Physiol., 24:19-31.

Motoyama, N., and W. C. Dauterman, 1980. Glutathion S-transferases: their role in the metabolism of organophosphorous insecticides, Rev. Biochem. Toxicol., 2:49-69.

Motoyama, N., T. Hayaoka, K. Nomura and W. C. Dauterman, 1980. Multiple factors for organophosphorus resistance in housefly, Musca domestica L., J. Pestic. Sci., 5:393-402.

Motoyama, N., L. R. Kao, P. T. Lin and W. C. Dauterman, 1984. Dual role of esterases in insecticide resistance in the green rice leafhopper, Pestic. Biochem. Physiol., 21:139-147.

Moxon, L. N., R. S. Holmes and P. A. Parsons, 1982. Comparative studies of aldehyde oxidase, alcohol dehydrogenase and aldehyde resource utilization among Australian Drosophila species, Comp. Biochem. Physiol., 71:387-395.

Mullin, C. A., 1985. Detoxification enzyme relationships in arthropods of differing feeding strategies, in: "Bioregulators for Pest Control", P. A. Hedin, ed., pp. 267-278, Symp. Ser. No. 276, Amer. Chem. Soc., Washington.

Mullin, C. A. and B. A. Croft, 1984. Trans-epoxide hydrolase: a key indicator enzyme for herbivory in arthropods, Experientia, 40:176-178.

Mullin, C. A. and B. A. Croft, 1985. An update on development of selective pesticides favoring arthropod natural enemies, in: "Biological Control in Agricultural Integrated Pest Management Systems", M. A. Hoy and D. C. Herzog, eds., pp. 123-150, Academic Press, New York.

Mullin, C. A. and B. D. Hammock, 1982. Chalcone oxides; potent selective inhibitors of cytosolic epoxide hydrolase, Arch. Biochem. Biophys., 216:423-439.

Mullin, C. A. and C. F. Wilkinson, 1980. Insect epoxide hydrolases: properties of a purified enzyme from the southern armyworm (Spodoptera eridania), Pestic. Biochem. Physiol., 14:192-207.

Mullin, C. A., F. Matsumura and B. A. Croft, 1984. Epoxide forming and degrading enzymes in the spider mite Tetranychus urticae, Comp. Biochem. Physiol., 79C:85-92.

Mullin, C. A., B. A. Croft, K. Strickler, F. Matsumura and J. R. Miller, 1982. Detoxification enzyme differences between a herbivorous and predatory mite, Science, 217:1270-1272.

Mumby, S. M. and B. D. Hammock, 1979. Substrate selectivity and stereochemistry of enzymatic epoxide hydration in the soluble portion of the mouse liver, Pestic. Biochem. Physiol., 11:275-284.

Nakatsugawa, T. and P. A. Dahm, 1962. Activation of guthion by tissue preparations from the American cockroach, J. Econ. Entomol., 55:594-599.

Nakatsugawa, T. and P. A. Dahm, 1967. Microsomal metabolism of parathion, Biochem. Pharmacol., 16:25-38.

Nakatsugawa, T. and M. A. Morelli, 1976. Microsomal oxidation and insecticide metabolism, in: "Insecticide Biochemistry and Physiology", C. F. Wilkinson, ed., pp. 61-114, Plenum Publ. Corp., New York.

Nebert, D. W. and N. M. Jensen, 1979. The Ah locus: genetic regulation of the metabolism of carcinogens, drugs and other environmental chemicals by cytochrome P-450-mediated monooxygenases. Crit. Rev. Biochem., 6:401-437.

Nelson, J. D. and F. Matsumura, 1973. Dieldrin (HEOD) metabolism in cockroaches and house flies, Arch. Environ. Contam. Toxicol., 1:224-244.

Newman, A. A., 1962. The occurrence, genesis and chemistry of the phenolic methylenedioxy ring in nature, Chem. Prod., 25:161-166.

Ngah, W. Z. U. and J. N. Smith, 1983. Acidic conjugate of phenols in insects: glucoside phosphate and glucoside sulphate derivatives, Xenobiotica, 13:383-389.

Oginsky, E. L., A. E. Stein, and M. A. Greer, 1965. Myrosinase activity in bacteria as demonstrated by the conversion of progoitrin to goitrin, Proc. Soc. Exp. Biol. Med., 119:360-364.

Okey, A. B., G. P. Bondy, M. E. Mason, G. F. Kahl, H. J. Eisen, T. M. Guenther and D. W. Nebert, 1979. Regulatory gene product of the Ah locus; characterization of the cytosolic inducer-receptor complex and evidence for its nuclear translocation, J. Biol. Chem., 254:11636-11648.

Omura, T. and R. Sato, 1964. The carbon monoxide-binding pigment of liver microsomes. J. Biol. Chem., 239:2370-2378 and 2379-2385.

Oosterbaan, R. A. and H. S. Jansz, 1965. Cholinesterases, esterases and lipases, in: "Comprehensive Biochemistry", M. Florkin and E. H. Stolz, eds., Vol. 16, pp. 1-54, Elsevier Publ. Co., New York.

Oppenoorth, F. J. and K. van Asperen, 1960. Allelic genes in the house fly producing modified enzymes that cause organophosphate resistance, Science, 132:298-299.

Oppenoorth, F. J., H. R. Smissaert, W. Welling, L. T. J. van der Pas and K. T. Hitman, 1977. Insensitive acetylcholinesterase, high glutathione S-transferase, and hydrolytic activity as resistance factors in a tetrachlorvinphos-resistant strain of house fly, Pestic. Biochem. Physiol., 7:34-47.

Orrenius, S., M. Berggren, P. Moldeus and R. I. Krieger, 1971. Mechanism of inhibition of microsomal mixed-function oxidation by the gut contents inhibitor of the southern armyworm (Prodenia eridania), Biochem. J., 124:427-430.

Ortiz de Montellano, P. R. and M. A. Correia, 1983. Suicidal destruction of cytochrome P-450 during oxidative drug metabolism, Annu. Rev. Pharmacol. Toxicol., 23:481-503.

Ottea, J. A. and F. W. Plapp, Jr., 1981. Induction of glutathione S-aryl transferase by phenobarbital in the house fly, Pestic. Biochem. Physiol., 15:10-13.

Ozaki, K. and H. Koike, 1965. Naphthyl acetate esterase in the green rice leafhopper, Nephotettix cincticeps Uhler, with special reference to the resistant colony to the organophosphorous insecticides, Jap. J. Appl. Ent. Zool., 9:53-59.

Palade, G. E. and P. Siekevitz, 1956. Liver microsomes. An integrated morphological and biochemical study, J. Biophys. Biochem. Cytol., 2:171-200.

Pasteels, J. M., M. Rowell-Rahier, J. C. Braekman and A. Du Pont, 1983. Salicin from host plant as precursor of salicylaldehyde in defensive secretion of Chrysomeline larvae, Physiol. Entomol., 8:307-314.

Pasteur, N., G. P. Georghiou and A. Iseki, 1984. Variation in organophosphate resistance and esterase activity in Culex quinquefasciatus Say from California, Genet. Sel. Evol., 16:271-284.

Pasteur, N., A. Iseki and G. P. Georghiou, 1981. Genetic and biochemical studies of the highly active esterases A' and B associated with organophosphate resistance in mosquitoes of the Cules pipiens complex, Biochem. Genet., 19:909–919.

Paulson, G. D., J. Caldwell, D. H. Hutson and J. J. Mann, eds., 1986. "Xenobiotic Conjugation Chemistry", Symp. Ser. No. 299, American Chemical Society, Washington.

Perry, A. S. and A. J. Buckner, 1970. Studies on microsomal cytochrome P-450 in resistant and susceptible house flies, Life Sci., 9:335–350.

Plapp, F. W., Jr., 1984. The genetic basis of insecticide resistance in the house fly: evidence that a single locus plays a major role in metabolic resistance to insecticides, Pestic. Biochem. Physiol., 22:194–201.

Plapp, F. W., Jr. and D. L. Bull, 1978. Toxicity and selectivity of some insecticides to Chrysopa carnea, a predator of the tobacco budworm. Env. Entomol., 7:431–434.

Plapp, F. W., Jr. and S. B. Vinson, 1977. Comparative toxicities of some insecticides to the tobacco budworm and its ichneumonid parasite, Campoletis sonorensis, Env. Entomol., 6:381–384.

Poulsen, L. L., 1981. Organic sulfur substrates for the microsomal flavin-containing monooxygenase, Rev. Biochem. Toxicol., 3:33–49.

Pratt, G. E., 1975. Inhibition of juvenile hormone carboxyesterase of locust haemolymph by organophosphates in vitro, Insect Biochem., 5:595–607.

Prestwich, G. D., M. Angelastro, A. De Palma and M. A. Perino, 1985. Fucosterol epoxide lyase of insects: synthesis of labeled substrates and development of a partition assay, Anal. Biochem., 151:315–326.

Raffa, K. F. and T. M. Priester, 1985. Synergists as research tools and in agriculture, J. Agric. Entomol., 2:27–45.

Raftell, M., K. Berzins and F. Blomberg, 1977. Immunochemical studies on a phenobarbital-inducible esterase in rat liver microsomes, Arch. Biochem. Biophys., 181:534–541.

Rafter, J. J., J. Bakke, G. Larsen, B. Gustafsson and J. A. Gustafsson, 1983. Role of the intestinal microflora in the formation of sulfur-containing conjugates of xenobiotics, Rev. Biochem. Toxicol., 5:387–408.

Respicio, N. C., 1975. Toxicological and biochemical transformation capabilities in the gypsy moth, Porthetria dispar (Linn.), larvae, Ph.D. thesis, Rutgers University, New Brunswick, New Jersey.

Riskallah, M. R., 1983. Esterases and resistance to synthetic pyrethroids in the Egyptian cotton leafworm, Pestic. Biochem. Physiol., 19:184–189.

Robbins, W. E., J. N. Kaplanis, J. A. Svoboda and M. J. Thompson, 1971. Steroid metabolism in insects, Annu. Rev. Entomol., 16:53–72.

Robinson, D., 1956. The fluorometric determination of β-glucosidase: its occurrence in the tissues of animals, including insects, Biochem. J., 63:39–44.

Roe, R. M., A. M. Hammond, Jr. and T. C. Sparks, 1983. Characterization of the plasma juvenile hormone esterase in synchronous last stadium female larvae of the sugar cane borer, Diatraea saccharalis (F.), Insect Biochem., 13:163-170.

Rose, H. A., 1985. The relationship between feeding specialization and host plants to aldrin epoxidase activities of midgut homogenates in larval Lepidoptera, Ecol. Entomol., 10:455-467.

Rose, H. A. and R. G. Young, 1973. Nitroreductases in the Madagascar cockroach, Gromphadorhina portentosa, Pestic. Biochem. Physiol., 3:243-252.

Rosenthal, G. A. and D. H. Janzen, 1985. Ammonia utilization by the bruchid beetle, Caryedes brasiliensis (Bruchidae), J. Chem. Ecol., 11:539-544.

Rosenthal, G. A., C. Hughes and D. H. Janzen, 1982. L-Canavanine, a dietary nitrogen source for the seed predator Caryedes brasilienis (Bruchidae), Science, 217:353-355.

Rosenthal, G. A., D. H. Janzen and D. L. Dahlman, 1976. Degradation and detoxification of canavanine by a specialized seed predator, Science, 196:658-660.

Rowland, I. R., 1986. Reduction by the gut microflora of animals and man, Biochem. Pharmacol., 35:27-32.

Ryan, D., A. Y. H. Lu, S. West and W. Levin, 1975. Multiple forms of cytochrome P-450 in phenobarbital and 3-methylcholanthrene-treated rats. Separation and spectral properties. J. Biol. Chem., 250:2157-2163.

Ryan, D. E., P. E. Thomas, L. M. Reik and W. Levin, 1982. Purification, characterization and regulation of five rat hepatic microsomal cytochrome P-450 isozymes, Xenobiotica, 12:727-744.

Sanchez-Bernal, M. C., J. Martin-Barrientos and J. A. Cabezas, 1984. Effect of tobramycin and gentamicin on the activity of some glycosidases in rat serum and urine, Comp. Biochem. Physiol., 79C:401-405.

Scheline, R. R., 1978. "Mammalian Metabolism of Plant Xenobiotics", Academic Press, New York.

Schenkman, J. B., H. Remmer and R. W. Estabrook, 1967. Spectral studies of drug interaction with hepatic microsomal cytochrome, Molec. Pharmacol., 3:113-123.

Shono, T., K. Ohsawa and J. E. Casida, 1979. Metabolism of trans- and cis-permethrin, trans- and cis-cypermethrin, and decamethrin by microsomal enzymes, J. Agric. Food Chem., 27:316-325.

Shono, T., T. Unai and J. E. Casida, 1978. Metabolism of permethrin isomers in American cockroach adults, house fly adults, and cabbage looper larvae, Pestic. Biochem. Physiol., 9:96-106.

Shyamala, M. B., 1964. Detoxication of benzoate by glycine conjugation in the silkworm, Bombyx mori L., J. Insect Physiol., 10:385-391.

Slade, M. and C. F. Wilkinson, 1974. Degradation and conjugation of cecropia juvenile hormone by the southern armyworm (Prodenia eridania), Comp. Biochem. Physiol., 49B:99-103.

Slade, M., H. K. Hetnarski and C. F. Wilksinson, 1976. Epoxide hydrolase activity and its relationship to development in the southern armyworm, Prodenia eridania, J. Insect Physiol., 22:619-622.

Slade, M., G. T. Brooks, H. K. Hetnarski and C. F. Wilkinson, 1975. Inhibition of the enzymatic hydration of the epoxide HEOM in insect, Pestic. Biochem. Physiol., 5:35-46.

Sladek, N. E. and G. J. Mannering, 1966. Evidence for a new P-450 hemoprotein in hepatic microsomes from methylcholanthrene treated rats, Biochem. Biophys. Res. Commun., 24:668-674.

Slama, K. and V. Jarolim, 1980. Fluorimetric method for the determination of juvenoid esterase activity in insects, Insect Biochem., 10:73-80.

Smith, G. J. and G. Litwack, 1980. Roles of ligandin and the glutathione S-transferases in binding steroid metabolites, carcinogens and other compounds, Rev. Biochem. Toxicol., 2:1-48.

Smith, J. N., 1964. Comparative biochemistry of detoxification, in: "Comparative Biochemistry", M. Florkin and H. S. Mason, eds., Vol. 6, 403-448. Academic Press, New York.

Smith, J. N., 1968. The comparative metabolism of xenobiotics, Adv. Comp. Physiol. Biochem., 3:173-232.

Smith, J. N. and H. B. Turbert, 1961. Enzymic glucoside synthesis in locusts, Nature, 189:600.

Sparks, T. C. and B. D. Hammock, 1979. A comparison of the induced and naturally occurring juvenile hormone esterases from last instar Trichoplusia ni, Insect Biochem., 9:411-421.

Sparks, T. C. and R. L. Rose, 1983. Inhibition and substrate specificity of the haemolymph juvenile hormone esterase of the cabbage looper, Trichoplusia ni (Hubner), Insect Biochem., 13:633-640.

Tanada, Y., R. Hess and E. M. Omi, 1980. Localization of esterase activity in the larval midgut of the armyworm (Pseudaletia unipuncta), Insect Biochem., 10:125-128.

Tate, L. B., S. S. Nakat and E. Hodgson, 1982. Comparison of detoxication activity in midgut and fatbody during fifth instar development of the tobacco hornworm, Manduca sexta, Comp. Biochem. Physiol., 72C:75-81.

Teas, H. J., 1967. Cycasin synthesis in Seirarctia echo (Lepidoptera) larvae fed methylazoxymethanol, Biochem. Biophys. Res. Comm., 26:686-690.

Terriere, L. C. and S. J. Yu, 1973. Insect juvenile hormone: induction of detoxifying enzymes in the house fly and detoxication by house fly enzymes, Pestic. Biochem. Physiol., 3:96-107.

Terriere, L. C. and S. J. Yu, 1974. The induction of detoxifying enzymes in insects, J. Agr. Food Chem., 22:366-373.

Terriere, L. C. and S. J. Yu, 1976. Microsomal oxidases in the flesh fly

(Sarcophaga bullata Parker) and the black blow fly [Phormia regina (Meigen)], Pestic. Biochem. Physiol., 6:223-228.

Terriere, L. C. and S. J. Yu, 1977. Juvenile hormone analogs: in vitro metabolism in relation to biological activity in blow flies and flesh flies, Pestic. Biochem. Physiol., 7:161-168.

Terriere, L. C., R. B. Boose and W. T. Roubal, 1961. The metabolism of naphthalene and 1-naphthol by house flies and rats, Biochem. J., 79:620-623.

Thongsinthusak, T., and R. I. Krieger, 1974. Inhibitory and inductive effects of piperonyl butoxide on dihydroisodrin hydroxylation in vivo and in vitro in black cutworm (Agrotis ypsilon) larvae, Life Sci., 14:2131-2141.

Townsend, M. G. and J. R. Busvine, 1969. The mechanism of malathion-resistance in the blowfly Chrysomya putoria, Entomol. Exp. Appl., 12:243-267.

Trammell, D. J., 1982. In vitro metabolism of (+)-pulegone and (-)-carvone by southern armyworm (Spodoptera eridania) microsomes, M. S. Thesis, Georgia Institute of Technology, Atlanta, Georgia.

Trivelloni, J. C., 1964. Estudio sobre la formacion de β-glucosidos en la langosta (Schistocerca cancellata), Enzymologia, 26:329-339.

Tsukamoto, M. and J. E. Casida, 1967. Metabolism of methylcarbamate insecticides by the NADPH requiring enzyme system from house flies, Nature, 213:49-51.

Turunen, S. and G. M. Chippendale, 1977. Ventricular esterases: comparison of their distribution within the larval midgut of four species of lepidoptera, Ann. Entomol. Soc. Am., 70:146-149.

Tynes, R. E. and E. Hodgson, 1985. Magnitude of involvement of the mammalian flavin-containing monooxygenase in the microsomal oxidation of pesticides, J. Agr. Food Chem., 33:471-479.

Ullrich, V., 1977. The mechanism of cytochrome P-450-catalyzed drug oxidations, in: "Drug Action at the Molecular Level", G. C. Roberts, ed., pp. 201-212, University Park Press, Baltimore.

Usui, K., J. I. Fukami and T. Shishido, 1977. Insect glutathione S-transferase: separation of transferases from fatbodies of American cockroaches active on organophosphorous triesters, Pestic. Biochem. Physiol., 7:249-260.

van Asperen, K., 1962. A study of house fly esterases by means of a sensitive colorimetric method, J. Insect Physiol., 8:401-416.

Vickery, M. L. and B. Vickery, 1981. "Secondary plant metabolism", University Park Press, Baltimore.

Villani, F., G. B. White, C. F. Curtis and S. J. Miles, 1983. Inheritance and activity of some esterases associated with organophosphate resistance in mosquitoes of the complex Culex pipiens L. (Diptera: Culicidae), Bull. Entomol. Res., 73:153-170.

Vogt, R. G., L. M. Riddiford and G. D. Prestwich, 1985. Kinetic properties of a sex pheromone-degrading enzyme: the sensillar esterase of Antheraea polyphemus, Proc. Natl. Acad. Sci.: USA, 82:8827-8831.

Volkova, R. I. and E. V. Titova, 1983. Multiple molecular forms of esterases from spring grain aphids: inhibitor identification and stereospecificity, Biokhimiya, 48:1634-1642.

von Wartburg, J. P. and B. Wermuth, 1980. Aldehyde reductase, in: "Enzymatic Basis of Detoxication", W. B. Jackoby, ed., Vol. 1, pp. 249-260, Academic Press, New York.

Walker, C. H. and M. I. Mackness, 1983. Esterases: problems of identification of classification, Biochem. Pharmacol., 32:3265-3269.

Weirich, G. F., J. A. Svoboda and M. J. Thompson, 1985. Ecdysone 20-monooxygenase in mitochondria and microsomes of Manduca sexta (L.) midgut: is the dual localization real? Arch. Insect Biochem. Physiol., 2:385-396.

Wells, D. S., G. C. Rock and W. C. Dauterman, 1983. Studies on the mechanisms responsible for variable toxicity of azinphosmethyl to various larval instars of the tufted apple budmoth, Platynota idaeusalis, Pestic. Biochem. Physiol., 20:238-245.

Wermuth, B., 1981. Purification and properties of an NADPH-dependent carbonyl reductase from human brain, J. Biol. Chem., 256:1206-1213.

Westlake, D. W. S., 1963. Microbiological degradation of quercitrin, Can. Microbiol., 9:211-220.

Westley, J., 1973. Rhodanese, Adv. Enzymol., 39:327-368.

Westley, J., 1981. Cyanide and sulfane sulfur, in: "Cyanide in Biology", B. Vennesland, E. E. Conn, C. J. Knowles, J. Westley and F. Wissing, eds., pp. 61-76. Academic Press, New York.

Whitmore, D., Jr., E. Whitmore and L. I. Gilbert, 1972. Juvenile hormone induction of esterases: a mechanism for the regulation of juvenile hormone titer, Proc. Nat. Acad. Sci., U.S.A., 69:1592-1595.

Whitten, C. J. and D. L. Bull, 1978. Metabolism and absorption of methyl parathion by tobacco budworms resistant or susceptible to organophosphorous insecticides, Pestic. Biochem. Physiol., 9:196-202.

Wickramasinghe, R. H., and C. A. Villee, 1975. Early role during chemical evolution for cytochrome P-450 in oxygen detoxification, Nature, 256:509-511.

Wiggleworth, V. B., 1958. The distribution of esterase in the nervous system and other tissues of the insect Rhodnius prolixus, Quart. J. Micros. Sci., 99:441-450.

Wilkinson, C. F., 1980. The metabolism of xenobiotics: a study in biochemical evolution, in: "The Scientific Basis of Toxicity Assessment", H. Witschi, ed., pp. 251-267, Elsevier, New York.

Wilkinson, C. F., 1986. Xenobiotic conjugation in insects, in: "Xenobiotic Conjugation Chemistry", G. D. Paulson, J. Caldwell, D. H. Hutson and J. J. Menn, eds., pp. 48-61, Symp. Ser. No. 299, Amer. Chem. Soc., Washington.

Wilkinson, C. F. and L. B. Brattsten, 1972. Microsomal drug metabolizing enzymes in insects, Drug Metab. Rev., 1:153-227.

Wilkinson, C. F. and L. J. Hicks, 1969. Microsomal metabolism of the 1,3-benzodioxole ring and its possible significance in synergistic action, J. Agric. Food Chem., 17:829-836.

Wilkinson, C. F., K. Hetnarski and L. J. Hicks, 1974a. Substituted imidazoles as inhibitors of microsomal oxidation and insecticide synergists, Pestic. Biochem. Physiol., 4:299-312.

Wilkinson, C. F., K. Hetnarski, G. P. Cantwell and F. J. Di Carlo, 1974b. Structure-activity relationships in the effects of 1-alkylimidazoles on microsomal oxidation in vitro and in vivo. Biochem. Pharmacol., 23:2377-2386.

Williams, R. T., 1974. Interspecies variation in the metabolism of xenobiotics, Biochem. Soc. Trans., 2:359-377.

Williams, R. T. and P. Millburn, 1975. Detoxification mechanisms, the biochemistry of foreign compounds, in: "Physiological and Pharmacological Biochemistry", H. K. F. Blaschko, ed., Ser. 1, Vol. 12, pp. 211-226, University Park Press, Baltimore.

Wing, K. D., M. Rudnicka, G. Jones and B. D. Hammock, 1984. Juvenile hormone esterases of Lepdioptera II. Isoelectric points and binding affinities of hemolymph juvenile hormone esterase and binding protein activities, J. Comp. Physiol., 154B:213-223.

Wislocki, P. G., G. T. Miwa and A. Y. H. Lu, 1980. Reactions catalyzed by the cytochrome P-450 system, in: "Enzymatic Basis of Detoxication", W. B. Jacoby, ed., Vol. 1, pp. 136-182, Academic Press, New York.

Wongkrobat, A. and D. L. Dahlman, 1976. Larval Manduca sexta hemolymph carboxylesterase activity during chronic exposure to insecticide-containing diets, J. Econ. Entomol., 69:237-240.

Wray, V., R. H. Davis and A. Nahrstedt, 1983. Biosynthesis of cyanogenic glucosides in butterflies and moths: incorporation of valine and isoleucine into linamarin and lotaustralin by Zygaena and Heliconius species (Lepidoptera), Z. Naturforsch., 38C:583-588.

Yang, R. H. S., 1976. Enzymatic conjugation and insecticide metabolism, in: "Insecticide Biochemistry and Physiology", C. F. Wilkinson, ed., pp. 177-225, Plenum Publ. Corp., New York.

Yang, R. S. H. and C. F. Wilkinson, 1973. Sulfotransferases and phosphotransferases in insects, Comp. Biochem. Physiol., 46B:717-726.

Yang, R. S. H., J. G. Pellicia and C. F. Wilkinson, 1973. Age-dependent aryl-sulfatase and sulfotransferase activities in the southern armyworm: a possible insect endocrine regulatory mechanism? Biochem. J., 136:817-820.

Yasutomi, K., 1971. Studies on diazinon-resistance and esterase activity in Cules tritaeniorhynchus I., Jap. J. Sanit. Zool., 22:9-13.

Yawetz, A. and B. Koren, 1984. Purification and properties of the Mediterranean fruit fly, Ceratitis capitata W. glutathione S-transferase, Insect Biochem., 14:663-670.

Yu, S. J., 1982. Host plant induction of glutathione S-transferase in the fall armyworm, Pestic. Biochem. Physiol., 18:101-106.

Yu, S. J., 1983. Age variation in insecticide suceptibility and detoxification capacity of fall armyworm (Lepidoptera: Noctuidae) larvae, J. Econ. Entomol., 76:219-222.

Yu, S. J., 1984. Interactions of allelochemicals with detoxication enzymes of insecticide-susceptible and resistant fall armyworms, Pestic. Biochem. Physiol., 22:60-68.

Yu, S. J., 1985. Microsomal sulfoxidation of phorate in the fall armyworm, Spodoptera frugiperda (J. E. Smith), Pestic. Biochem. Physiol., 23:273-281.

Yu, S. J. and E. L. Hsu, 1985. Induction of hydrolases by allelochemicals and host plants in fall armyworm (Lepidoptera: Noctuidae) larvae, Environ. Entomol., 14:512-515.

Yu, S. J., and L. C. Terriere, 1971. Hormonal modification of microsomal oxidase activity in the house fly, Life Sci., 10:1173-1185.

Yu, S. J. and L. C. Terriere, 1978. Juvenile hormone epoxide hydrase in house flies and blow flies, Insect Biochem., 8:349-352.

Yu, S. J. and L. C. Terriere, 1979. Cytochrome P-450 in insects. 1. Differences in the forms present in insecticide resistant and susceptible house flies, Pestic. Biochem. Physiol., 12:239-248.

Yu, S. J., F. A. Robinson and J. L. Nation, 1984. Detoxication capacity in the honey bee, Apis mellifera L., Pestic. Biochem. Physiol., 22:360-368.

CONSEQUENCES OF INDUCTION OF FOREIGN COMPOUND-METABOLIZING ENZYMES IN INSECTS

S. J. Yu

Department of Entomology and Nematology
University of Florida
Gainesville, FL 32611

1. INTRODUCTION

Detoxifying enzymes such as microsomal cytochrome P-450-dependent monooxygenases (polysubstrate monooxygenases, PSMOs), glutathione transferases, esterases and epoxide hydrolases can be induced by various chemicals in animals. Enzyme induction is a commonly occurring phenomenon representing an effective mechanism of adaptation to external conditions. Most of the induction work has been done with the PSMO system in mammals because of its important role in drug metabolism. The first report of induction of the PSMO enzymes in insects was that of Morello (1964) who found that nymphs of Triatoma infestans Klug (Hemiptera: Reduviidae) were less susceptible to DDT after pretreatment with 3-methylcholanthrene and produced more polar metabolites from DDT. Since then, induction has been demonstrated in at least 16 other species of insects (Terriere and Yu, 1974; Terriere, 1983, 1984). The PSMOs can be induced by a variety of chemicals with diverse structures including insecticides such as DDT and cyclodienes (Walker and Terriere, 1970; Plapp and Casida, 1970; Yu and Terriere, 1972; Khan and Matsumura, 1972); insect hormones and growth regulators such as ecdysone and juvenile hormone (Yu and Terriere, 1971, 1975; Terriere and Yu, 1976), organic solvents such as pentamethylbenzene (Brattsten and Wilkinson, 1973; Brattsten and Wilkinson, 1977), drugs such as phenobarbital and 3-methylcholanthrene (Agosin et al., 1969; Yu and Terriere, 1973; Gil et al., 1974; Brattsten et al., 1976; Yu and Ing, 1984), butylated hydroxytoluene and triphenyl phosphate (Perry et al., 1971), and allelochemicals such as terpenoids, indoles and flavonoids (Brattsten et al., 1977, 1984; Yu et al., 1979; Yu, 1982a; Yu, 1983).

It is now well established that the induction of PSMO activity involves synthesis of new enzyme, i.e., de novo protein synthesis, rather than activation of pre-existing enzyme or a block in the rate of degradation (Agosin, 1985). The exact mechanism of induction is not yet clear (see also Ahmad et al., Chapter 3 in this text). In mammals, according to Nebert et al. (1981), certain exogenous inducers enter the cell and bind to a semi-specific cytosolic receptor protein. The inducer-receptor complex is then transferred to the nucleus. The complex activates the appropriate structural genes which results in the production of the appropriate cytochrome P-450 and associated components. The maintenance of high levels of PSMOs is energetically expensive for the insect. Therefore, induction when needed would be an effective way for an insect to save energy (Brattsten, 1979a).

In recent years, it has become evident that insect cytochrome P-450, the terminal oxidase of PSMO system, exists in isoenzymic forms as found in mammals. As many as six forms of cytochrome P-450 have been isolated from the house fly, Musca domestica L. (Diptera: Muscidae) (Schonbrod and Terriere, 1975; Capdevila et al., 1975; Yu and Terriere, 1979), four forms from the flesh fly, Sarcophaga bullata Parker (Diptera: Sarcophagidae) and the black blow fly, Phormia regina (Meigen) (Diptera: Calliphoridae) (Terriere and Yu, 1979), and three forms from the pomace fly, Drosophila melanogaster Meigen (Diptera: Drosophilidae) (Naquira et al., 1980).

Because of the multiplicity of cytochrome P-450, selective induction of P-450 caused by different inducers has been observed. For example, phenobarbital, pentamethylbenzene, naphthalene and α-pinene induced cytochrome P-450 which exhibited lower absorption maxima (1-3 nm) of the reduced cytochrome P-450-CO difference spectrum compared to control insects (Stanton et al., 1978; Capdevila et al., 1973; Brattsten et al., 1980; Rose and Terriere, 1980). Other inducers such as β-naphthoflavone had no effect on the absorption maximum (Rose and Terriere, 1980; Stanton et al., 1978). Likewise, the induction of various PSMO activities is affected by the presence of multiple P-450 forms which show different but overlapping substrate specificities. For example, the order of the induction of the PSMO activities by phenobarbital in house flies was: heptachlor epoxidation > methoxyresorufin O-demethylation > aldrin epoxidation > 7-methoxy-4-methylcoumarin O-demethylation (Vincent et al., 1985a). In black blow flies, however, the order of induction by phenobarbital was: 7-methoxy-4-methylcoumarin O-demethylation > heptachlor epoxidation > aldrin epoxidation > methoxyresorufin O-demethylation (Vincent et al., 1985b). It is thus clear that each cytochrome P-450 isozyme may respond differently to an inducer resulting in differential induction of PSMO activities observed. Furthermore, cytochrome P-450s may differ among species.

Very recently, induction of other detoxifying enzymes, e.g., glutathione transferases (Yu, 1982b, 1983, 1984; Brattsten et al., 1984; Ottea and Plapp, 1984), esterases (Dowd et al., 1983; Yu and Hsu, 1985) and epoxide hydrolases (Yu and Hsu, 1985) has also been shown. I shall here summarize the present knowledge regarding the induction of detoxifying enzymes by plants and plant allelochemicals and discuss its demonstrated and potential consequences for the insects (see Ahmad et al., Chapter 3 in this text, for biochemical details of the enzymes and Brattsten, Chapter 6 in this text, for metabolism).

2. INDUCTION OF CYTOCHROME P-450-DEPENDENT MONOOXYGENASES

Although insect PSMOs were shown to be induced by various chemicals such as insecticides, drugs, insect hormones and organic solvents (Terriere, 1984), the first evidence that plant allelochemicals could induce this enzyme system was obtained by Brattsten et al. (1977). They found that dietary allelochemicals (e.g., (+)-α-pinene, sinigrin, trans-2-hexenal) induced midgut microsomal oxidase activity of southern armyworm, Spodoptera eridania (Cramer) (Lepidoptera: Noctuidae), larvae up to four-fold. The induction was a rapid process resulting in a significant increase in N-demethylation activity already 30 min after treatment. The larvae with induced enzymes were less susceptible to the toxic tobacco alkaloid nicotine, suggesting an adaptive advantage for the phenomenon.

Midgut PSMO activity in variegated cutworm, Peridroma saucia Hubner (Lepidoptera: Noctuidae) larvae was induced up to 45-fold by feeding on peppermint leaves compared with activity in larvae fed a meridic diet (Yu et al., 1979). The induction was apparently due to high concentrations of certain monoterpenes such as (-)-menthol, l-menthone, (+)-α-pinene and (-)-β-pinene in the peppermint leaves. These allelochemicals, when fed to cutworm larvae, all increased midgut activity of aldrin epoxidation up to 24-fold and the cytochrome P-450 content was increased six-fold (Yu et al., 1979). Induction by peppermint leaves in midgut and other unspecified tissues was also observed in larvae of the alfalfa looper, Autographa californica (Speyer) (Lepidoptera: Noctuidae) and the cabbage looper, Trichoplusia ni (Hubner) (Lepidoptera: Noctuidae) (Farnsworth et al., 1981).

In a similar study, Brattsten (1979b) showed several food plants to be inducers of midgut aldrin epoxidation and p-chloro N-methylaniline (PCMA) N-demethylation in southern armyworm larvae. Remarkably, two of the plants, parsley and coriander, simultaneously inhibited epoxidation and induced N-demethylation. Carrot foliage was the best inducer of both enzymes. The induction was caused by a mixture of monoterpenes such as α-pinene, β-pinene, limonene and terpinene present in carrot foliage (Brattsten et al., 1984).

PSMO induction by plants was also demonstrated in fall armyworm, Spodoptera frugiperda (J. E. Smith) (Lepidoptera: Noctuidae) larvae (Yu, 1982a). Of ten plants studied, soybean and millet leaves did not have any effect on aldrin epoxidation compared with the artificial diet. However, alfalfa, sorghum, peanuts, cabbage, cowpeas, cotton, Bermudagrass, and corn stimulated the epoxidation activity with corn being the strongest inducer. The induction was likely due to allelochemicals in the plants since various monoterpenes, e.g. (+)-α-pinene, (-)-α-pinene, (+)-limonene, (-)-menthol, and essential oil of peppermint, indoles, e.g. indole-3-carbinol and indole 3-acetonitrile, and flavone were all found to stimulate aldrin epoxidation (Yu, 1982a, 1983). Maximum induction was found with indole-3-carbinol which caused a 4.8-fold increase in epoxidation. However, other allelochemicals such as quercetin, gossypol, 2-phenylethyl isothiocyanate, stigmasterol, sitosterol, and β-carotene inhibited the activity (Yu, 1983).

In addition to aldrin epoxidation, plants and allelochemicals also induced other PSMO activities such as biphenyl hydroxylation, p-nitroanisole (PNA) O-demethylation, PCMA N-demethylation, phorate sulfoxidation, and parathion desulfuration in fall armyworm larvae. The induction of six PSMO reactions in this insect by five host plants is shown in Table 1. The patterns of induction by the host plants were similar with some exceptions. For example, corn was the best inducer for epoxidation and hydroxylation, yet cotton was the best inducer for O-demethylation, N-demethylation, desulfuration and sulfoxidation. In all instances, soybean leaves inhibited the activities to 19 - 65 percent below control levels; desulfuration was most inhibited. The results of these studies showing selective induction of PSMO activities by an inducer clearly indicate the multiplicity of cytochrome P-450 in southern and fall armyworm larvae. In this respect, it would be erroneous to use one type of enzyme activity to express general PSMO levels in the induction studies since each enzyme activity may respond differently to an inducer.

Our recent studies (Yu, 1983, 1985, 1986a, Yu and Ing, 1984) with allelochemicals as inducers also implicate the existence of isoenzymic forms of cytochrome P-450 in the fall armyworm midgut. Data in Table 2 show that the patterns of induction by the three allelochemicals were different for the four reactions assayed. The order of activities after induction for aldrin epoxidation was: indole-3-carbinol > flavone > indole-3-acetonitrile; for biphenyl hydroxylation: flavone > indole-3-carbinol > indole-3-acetonitrile; for parathion desulfuration: indole-3-acetonitrile >

Table 1. Effect of host plants on microsomal oxidase activities in fall armyworm larvae.

Host plant	Specific activity (pmol/min/mg protein)					
	Aldrin epoxidation[a]	Biphenyl hydroxylation[b]	PNA O-demethylation	PCMA N-demethylation	Parathion desulfuration[c]	Phorate sulfoxidation[d]
Control	159.5 ± 22.2	444.8 ± 23.47	55.7 ± 8.7	419.7 ± 27.2	137.2 ± 11.3	2.50 ± 0.13
Soybeans	140.8 ± 25.5	233.6 ± 7.64	44.9 ± 3.3	310.5 ± 38.5	47.5 ± 8.0	1.81 ± 0.10
Peanuts	251.2 ± 66.2	844.9 ± 16.59	134.8 ± 7.8	681.1 ± 47.3	89.4 ± 5.4	2.50 ± 0.04
Cowpeas	261.2 ± 67.6	494.3 ± 27.00	313.3 ± 5.8	692.7 ± 109.2	202.6 ± 17.8	2.83 ± 0.28
Cotton	342.1 ± 16.2	761.1 ± 25.59	359.4 ± 35.2	983.0 ± 149.7	339.6 ± 16.0	5.73 ± 0.28
Corn	628.9 ± 46.5	1156.6 ± 18.38	257.0 ± 12.9	769.2 ± 17.3	165.0 ± 17.5	3.46 ± 0.22

[a] Data from Yu (1982a).
[b] Data from Yu and Ing (1984).
[c] Data from Yu (1986a).
[d] Data from Yu (1985), nmol/min/mg protein

Table 2. Induction of microsomal oxidase activities by allelochemicals in fall armyworm larvae.

Allelochemical[a]	Specific activity (pmol/min/mg protein)			
	Aldrin epoxidation[b]	Biphenyl hydroxylation[c]	Parathion desulfuration[d]	Phorate sulfoxidation[e]
Control (meridic diet)	185.6 ± 20.2	416.4 ± 22.4	47.5 ± 8.0	2.41 ± 0.05
Indole-3-carbinol	890.9 ± 87.8	2671.2 ± 192.4	337.1 ± 30.1	12.19 ± 1.24
Indole-3-acetonitrile	666.8 ± 41.3	1614.6 ± 248.8	463.3 ± 27.4	11.83 ± 0.96
Flavone	809.2 ± 27.7	3041.8 ± 564.6	369.1 ± 24.8	14.73 ± 1.54

[a] Newly molted sixth-instar larvae were fed meridic diets containing the compounds (0.2%) for two days prior to enzyme assays.
[b] Data from Yu (1983).
[c] Data from Yu and Ing (1984).
[d] Data from Yu (1986a).
[e] Data from Yu (1985), nmol/min/mg protein.

flavone > indole-3-carbinol; and for phorate sulfoxidation: flavone > indole-3-carbinol > indole-3-acetonitrile. Furthermore, xanthotoxin increased cytochrome P-450 content and heptachlor epoxidation activity but inhibited aldrin epoxidation, biphenyl hydroxylation, and PCMA N-demethylation in fall armyworm larvae (Yu, 1984).

Induction was also observed in the semi-specialist (oligophagous) velvetbean caterpillar, Anticarsia gemmatalis Hubner (Lepidoptera: Noctuidae) (Christian and Yu, 1986). The order of midgut epoxidation activity after the larvae had eaten various plants was cowpeas > cotton > peanuts > soybeans > hairy indigo. The maximum induction was four-fold compared to activity in midguts from caterpillars fed meridic diet. In agreement with results for the fall armyworm, dietary allelochemicals such as indole-3-carbinol, indole-3-acetonitrile, and (-)-menthol were good inducers of this enzyme.

Levels of allelochemicals are known to be affected by the stage of growth in plants (Battaile and Loomis, 1961). This would explain why age and developmental stage of plants influence its inducing ability. Mature corn leaves were more potent than younger leaves in inducing PSMO activities in fall armyworm larvae (Yu, 1982a). Also, corn leaves were more potent as an inducer than silk, corn husks, or developing corn cobs (Yu, 1983).

In most cases, induction of various PSMO activities by allelochemicals was higher than that of cytochrome P-450 content (Yu, 1982a, 1983, 1985; Yu and Ing, 1984; Moldenke et al., 1983) and there was no positive correlation between enzyme activity and P-450 content after induction. For example, among the allelochemicals indole-3-carbinol, indole-3-acetonitrile, flavone, and β-naphthoflavone as inducers of microsomal phorate sulfoxidase activity, flavone was the best inducer of sulfoxidation activity, but was a weak inducer of P-450 content (Yu, 1985). These studies indicate that allelochemicals selectively induce certain forms of cytochrome P-450 in insects.

Gut microsomal N-demethylation of the highly polyphagous Japanese beetle, Popillia japonica Newman (Coleoptera: Scarabaeidae), was moderately affected by plants. The activity was highest in field collected beetles, intermediate in beetles laboratory reared on many different plants to simulate polyphagy, and lowest in beetles fed one single plant (Ahmad, 1983). This indicates greater involvement of the P-450 enzymes in populations with a generalist or polyphagous feeding habit and supports such conclusions by Krieger et al. (1971) and Brattsten et al. (1977). Mullin and Croft (1983) found that host plant-related changes of aldrin epoxidation in the two-spotted spider mite, Tetranychus urticae Koch (Acari: Tetranychidae) ranged from 0.4- to 1.5-fold when the mites were feeding on plants other than snap beans which served as the control. Thus, plant allelochemicals influence the PSMO activity in many lepidopterous, other herbivores and omnivorous insect species and herbivorous mites. The inducibility of each species could depend on intrinsic or exogenous factors such as inducer used, dosage, developmental stage, sex, genetic makeup, etc.

3. INDUCTION OF GLUTATHIONE TRANSFERASES

GSH transferases are induced by xenobiotics, e.g. barbiturates and pesticides in mammals (Kulkarni et al., 1980) and in insects. Ottea and Plapp (1981) reported that GSH transferase was induced nearly three-fold in house flies by dietary phenobarbital. Hayaoka and Dauterman (1982) also obtained induction in house flies by various dietary insecticides with chlorinated hydrocarbons being most active.

Plants and plant allelochemicals also induce GSH transferase activities (Yu, 1982b, 1984). Among several plants tested with the fall armyworm, the following were inducers and the order of induction was: parsnip > parsley > mustard > turnip > radish > cowpeas > collards > cabbage > Chinese cabbage > peanuts > cotton. Parsnip caused a 39-fold increase compared with activity in larvae fed artificial diet. However, plants such as soybeans, sorghum, millet, Bermudagrass, corn, potato, cucumber, carrot and broccoli had no effect on this enzyme. Time-course studies showed that the maximum induction of the transferase by cowpeas occurred two days after feeding began. The transferase exists in isoenzymic forms in the house fly (Clark et al., 1984). Ottea and Plapp (1984) reported that phenobarbital induced different forms of GSH transferase in house flies. However, kinetic studies of the enzyme from fall armyworm midguts revealed no qualitative differences between soybean- and cowpea-fed larvae, suggesting that the allelochemicals in these plants did not selectively induce different forms in this species (Yu, 1982b).

The identity of the GSH transferase inducer in parsnip leaves was determined by thin-layer chromatography, high-pressure liquid chromatography, gas chromatography, and mass spectrometry as xanthotoxin, a linear furanocoumarin (Yu, 1984).

Other allelochemicals such as indole-3-carbinol, indole-3-acetonitrile, flavone and sinigrin also induce the transferase in fall armyworm larvae (Yu, 1983). When indole-3-acetonitrile, indole-3-carbinol and flavone were used as inducers, the inducing patterns of GSH transferases was the same toward the substrates 1,2-dichloro-4-nitrobenzene, 1-chloro-2,4-dinitrobenzene and methyl iodide (Yu, 1984). These results again supported the notion that these transferases were not selectively induced by these compounds. The transferase was also inducible by allelochemicals in larvae of a carbaryl-resistant strain (Yu, 1984).

Of considerable interest was the observation that the monoterpenes (+)-α-pinene, (-)-β-pinene, (-)-menthol and essential oil of peppermint, which were inducers of the microsomal oxidases in fall armyworm larvae, had no effect on the transferase when included in the artificial diet (Yu, 1982b). In southern armyworm larvae, however, monoterpenes included in a semisynthetic diet moderately stimulated GSH transferase (Brattsten et al., 1984). Dietary coumarin induced the transferase up to seven-fold in southern armyworm larvae (Brattsten et al., 1984).

4. INDUCTION OF ESTERASES

Very little is known about plant-related modifications of esterase activity (see Ahmad et al., Chapter 3 in this text) in insects. Dowd et al. (1983a) reported that it was induced in cabbage looper larvae and reduced in soybean looper Pseudoplusia includens (Walker) (Lepidoptera: Noctuidae) larvae by leaf extracts from a resistant variety of soybeans. Large differences (0.4- to 2.4-fold) in esterase activity toward 1-naphthyl acetate (1-NA) were observed in two-spotted spider mites when they were fed plants other than snap beans (Mullin and Croft, 1983). According to these authors, umbellifers such as celery and carrot appeared to be good inducers of esterases in the mite.

The monoterpenes (+)-α-pinene, (-)-menthol and peppermint oil, and the sesquiterpene lactone santonin were all moderate inducers of the esterase, causing increases of 35 to 65 percent in activity, Table 3. The plant hormone analogs indole-3-acetonitrile and indole-3-carbinol, the

Table 3. Effect of allelochemicals on 1-naphthyl acetate esterase activity in fall armyworm larvae[a].

Allelochemical[b] (0.2% in diet)	Specific activity (μmol naphthol/min/mg protein)
Control (meridic diet)	0.89 ± 0.03 [c]
(+)-α-pinene	1.20 ± 0.03
(-)-menthol	1.34 ± 0.09
(+)-pulegone	0.99 ± 0.02
Peppermint oil	1.33 ± 0.01
Santonin	1.47 ± 0.06
Indole-3-acetonitrile	1.90 ± 0.05
Indole-3-carbinol	1.40 ± 0.02
Flavone	1.60 ± 0.06
Quercetin	0.89 ± 0.07
β-naphthoflavone	1.33 ± 0.04
sinigrin	0.95 ± 0.07
2-phenylethyl isothiocyanate (0.02%)	1.34 ± 0.02
Gossypol (0.1%)	1.04 ± 0.06
Colchicine (0.1%)	1.13 ± 0.11
Quinine	1.20 ± 0.02
Xanthotoxin (0.05%)	1.57 ± 0.04

[a] Data from Yu and Hsu (1985).
[b] Newly molted sixth-instar larvae were fed meridic diets containing the compounds for 2 days prior to enzyme assays.
[c] Mean ± SE of two experiments, each with duplicate determinations.

flavonoids flavone and β-naphthoflavone, and the alkaloid quinine also stimulated the esterase, resulting in increases of 35 to 114 percent. The furanocoumarin xanthotoxin is a good inducer of the esterase, causing an increase of 80 percent in activity. Although sinigrin (allylglucosinolate, did not show any inducing activity, the hydrolysis product of another glucosinolate, 2-phenylethyl isothiocyanate (mustard oil), significantly increased the enzyme activity. Plants such as celery, potato, and parsley were also active in inducing the esterase, whereas corn, peanuts, cotton, soybeans, cowpeas, carrot, sweet potato, peppermint, radish, turnip and tomato had no significant effect. It is apparent that allelochemicals were responsible for the host plant induction of hydrolase activity in this study.

Whitmore et al. (1972) found that injection of juvenile hormone (JH) I into pupae of _Hyalophora gloverii_ (Strecker) (Lepidoptera: Saturniidae) caused the production of JH esterases in the haemolymph. The stimulation can be blocked by protein synthesis inhibitors, suggesting _de novo_ synthesis of the carboxylesterase. Sparks and Hammock (1979) showed that JH I, JH II, and juvenoids induced JH esterase in cabbage looper larvae when they were topically treated with the inducers. Induced esterases in some cases were specific for JH and were unable to hydrolyze 1-NA.

5. INDUCTION OF EPOXIDE HYDROLASES

Epoxide hydrolase is inducible in insects. JH epoxide hydrolase was induced in house flies by the classical enzyme inducer sodium phenobarbital (Yu and Terriere, 1978a). Large differences in _trans_-epoxide hydrolase (0.5-1.5-fold) and _cis_-epoxide hydrolase (1-3.4-fold) activities were also observed in two-spotted spider mites when they were fed various host plants as compared to snap beans (Mullin and Croft, 1983). More recently, we have found that the styrene oxide epoxide hydrolase activity was increased by the allelochemicals indole-3-carbinol (by 42 percent) and essential oil of peppermint (by 34 percent), Table 4, in fall armyworm larvae (Yu and Hsu, 1985). The barbiturate phenobarbital also induced the hydrolase as much as 50 percent. However, none of the plants studied including corn, cotton, peanuts, soybeans, cowpeas, peppermint and sweet potato showed any inductive effect on the hydrolase activity.

6. CONSEQUENCES OF ENZYME INDUCTION

6.1. Toxicological implications

The significance of the microsomal cytochrome P-450-dependent monooxygenases, glutathione transferases, esterases and epoxide hydrolases in insecticide metabolism and detoxification as well as in insecticide resistance is well documented (Wilkinson, 1976; Georghiou and Saito, 1983). It is therefore logical to expect that an increase in these enzyme activities resulting from induction by plant allelochemicals would decrease the toxicity of an insecticide due to enhanced metabolism. Indeed, dietary α-pinene caused southern armyworm larvae to become more tolerant of a botanical insecticide, nicotine (Brattsten et al., 1977).

Enhanced tolerance to synthetic insecticides was also demonstrated in herbivorous insect pests when they were fed plants capable of inducing PSMO activities. Variegated cutworm larvae fed peppermint leaves were more tolerant of the insecticides carbaryl, acephate, methomyl, and malathion than larvae fed snap bean leaves (Yu et al., 1979; Berry et al., 1980). Increased tolerance for carbaryl and methomyl was also observed in larvae of the alfalfa looper and cabbage looper when they were fed peppermint plants instead of their favored host plants alfalfa and broccoli (Farnsworth et al., 1981).

Further studies of enzyme induction by plant allelochemicals revealed that corn leaves which are potent inducers of PSMO cause fall armyworm larvae to become less susceptible to the insecticides methomyl, acephate, methamidophos, diazinon, trichlorfon, monocrotophos, permethrin, and cypermethrin than soybean-fed larvae (Yu, 1982a).

Induction of glutathione transferase in fall armyworm larvae protects the larvae against organophosphorous (OP) insecticides (Yu, 1982b). Larvae fed cowpeas, a potent inducer of the transferase, were twice as tolerant of diazinon, methyl parathion, and methamidophos as those fed soybeans.

Induction of esterases and epoxide hydrolases would also affect the toxicity of insecticides. For example, host plant induction of 1-NA esterase would decrease the toxicity of certain insecticides containing an ester linkage such as organophosphates, pyrethroids, and some juvenile hormone analogs, and possibly carbamates. This would probably explain why soybean looper larvae that were fed foliage of the resistant soybean variety, ED73-371, were more susceptible to methyl parathion compared to larvae fed foliage of the susceptible variety, Bragg (Kea et al., 1978). Induction of epoxide hydrolases could possibly reduce the toxicity of epoxide-containing insecticides such as certain cyclodienes although cyclodiene epoxides are unusually stable.

Microsomal oxidations such as sulfoxidation and desulfuration of OP insecticides result in activation, i.e., the metabolites are more toxic than the parent compound. Figs. 1 and 2 show the toxicity of various OP

Table 4. Effect of allelochemicals and barbiturate on epoxide hydrolase activity in fall armyworm larvae[a].

Compound[b] (0.2% in diet)	Specific activity[c] (nmol styrene glycol/min/mg protein)
Control (meridic diet)	22.59 ± 0.21
Indole-3-acetonitrile	28.07 ± 0.21
Indole-3-carbinol	32.14 ± 0.59
Flavone	28.07 ± 1.41
Peppermint oil	30.17 ± 0.75
Phenobarbital	33.99 ± 1.57

[a] Data from Yu and Hsu (1985).
[b] Newly molted sixth-instar larvae were fed meridic diets containing the compounds for 2 days prior to enzyme assays.
[c] Mean ± SE of two experiments, each with duplicate measurements.

insecticides to fall armyworm larvae as influenced by host plant induction (Yu, 1985). In this study, parsley was used as inducer for microsomal sulfoxidation (Yu, 1986a), whereas cotton was used as inducer for microsomal desulfuration, Table 1). Feeding cotton leaves to the larvae enhanced the toxicity of parathion, malathion, and isofenphos as much as four-fold, Fig. 1. The same treatment caused a decrease in toxicity of diazinon and azinphos methyl. On the other hand, feeding parsley leaves to the larvae enhanced the toxicity of phorate, disulfoton, carbophenothion, demeton, oxydemeton methyl and fenthion more than three-fold, Fig. 2. In most cases, mortality occurred earlier in the induced larvae than in the controls.

The enhanced toxicity was apparently due to an increased activity of microsomal desulfuration and sulfoxidation caused by cotton or parsley, respectively. Parathion, malathion, azinphos methyl, diazinon and isofenphos are phosphorothionate or phosphorothiolothionate compounds containing the P=S moiety. These proinsecticides are known to be activated by microsomal desulfuration to become more potent acetylcholinesterase inhibitors (Nakatsugawa and Dahm, 1965; Matsumura, 1975; Yang et al., 1971; Motoyama and Dauterman, 1972). It is not known why enhanced desulfuration caused by cotton did not increase the toxicity of diazinon and azinphos methyl to the larvae. One possible explanation could be that cotton also induced GSH transferases (Yu, 1982b) resulting in an increased degradation of diazinon and diazoxon (Yang et al., 1971), and azinphos methyl and its oxygen analog (Motoyama and Dauterman, 1972). The possibility of induction by cotton of PSMO activity that oxidatively hydrolyzes these two insecticides and their oxygen analogs (Yang et al., 1971; Motoyama and Dauterman, 1972) can not be ruled out, however. In any case, the detoxification rate of diazinon and azinphos methyl exceeded the activation rate, resulting in the protective effect observed.

On the other hand, phorate, disulfoton, carbophenothion, demeton, fenthion and oxydemeton methyl are all thioether-containing OP compounds. These proinsecticides are known to be activated by microsomal sulfoxidation to become more potent acetylcholinesterase inhibitors (Yu, 1985; Bull, 1965; Menzie, 1969; March et al., 1955; Stone, 1969). It should be mentioned that with the exceptions of oxydemeton methyl and the thiol isomer of demeton, these insecticides also contain the P=S group. Therefore in addition to sulfoxidation, they can also undergo desulfuration even though the desulfuration of thioether-containing compounds might be

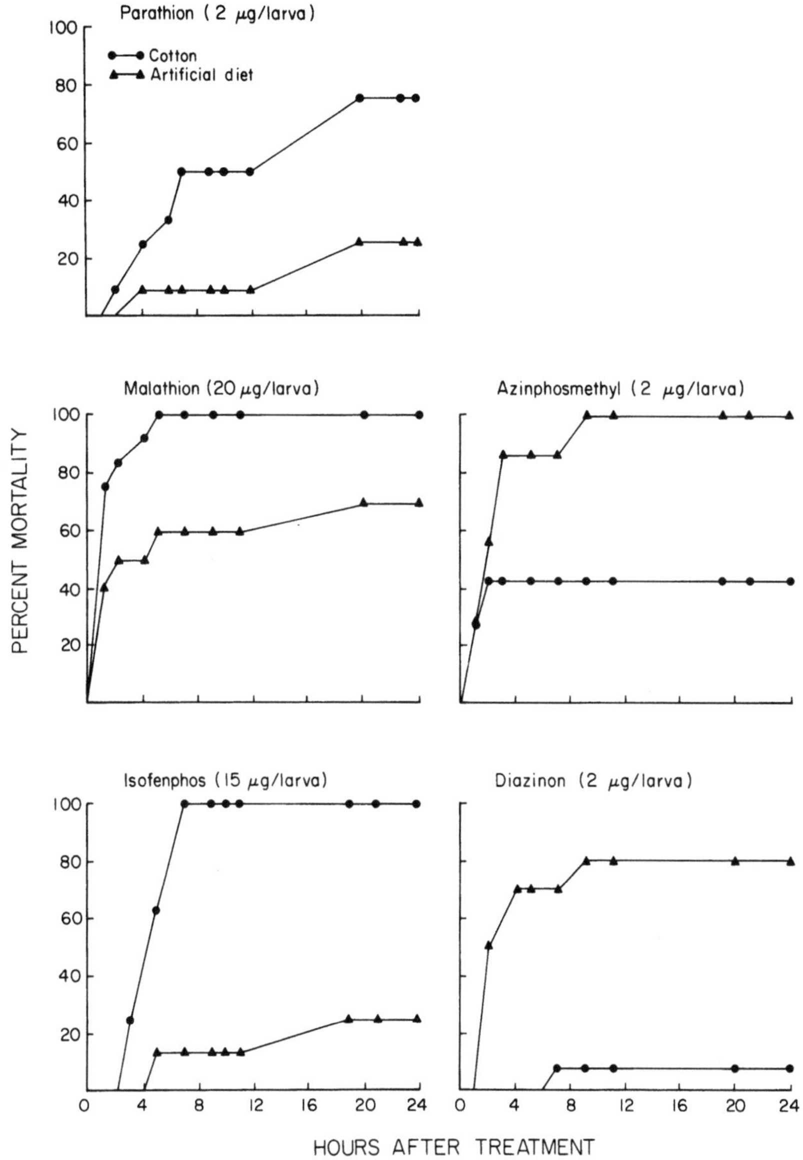

Fig. 1. Effect of cotton on the toxicity of various phosphorothionate insecticides to fall armyworm larvae. Newly molted sixth-instar larvae were fed cotton leaves or artificial diet for two days prior to insecticide treatments. Body weights in g per larva (Mean ± SE): cotton-fed, 0.28 ± 0.007; artificial diet-fed, 0.30 ± 0.01 (Yu, 1986a; reprinted with permission).

rather insignificant in the fall armyworm larvae (Yu, 1985).

Table 5 summarizes the results obtained from the time-course analysis of internal insecticide residues in the control and parsley-fed fall armyworm larvae after topical treatment with phorate. Contrary to the *in vivo* results, the induced larvae retained less phorate and its oxidative metabolites up to eight hours after treatment. However, two hours after treatment the induced larvae had more phorate sulfoxide and phorate sulfone than the controls. This suggests that induction of sulfoxidation activity resulted in an overall increase in the rate of phorate metabolism.

These results, in conjunction with those reviewed earlier, clearly demonstrate that various important detoxification enzymes can be induced by plants and the resulting induction provides protection for herbivorous insects against different types of commonly used insecticides in laboratory experiments. If field studies corroborate these observations, we may utilize the enzyme activation approach to improve insecticide efficacy in the field and hence reduce insecticide application rates, since organophosphate activation enzyme activity (desulfuration and sulfoxidation) can be induced.

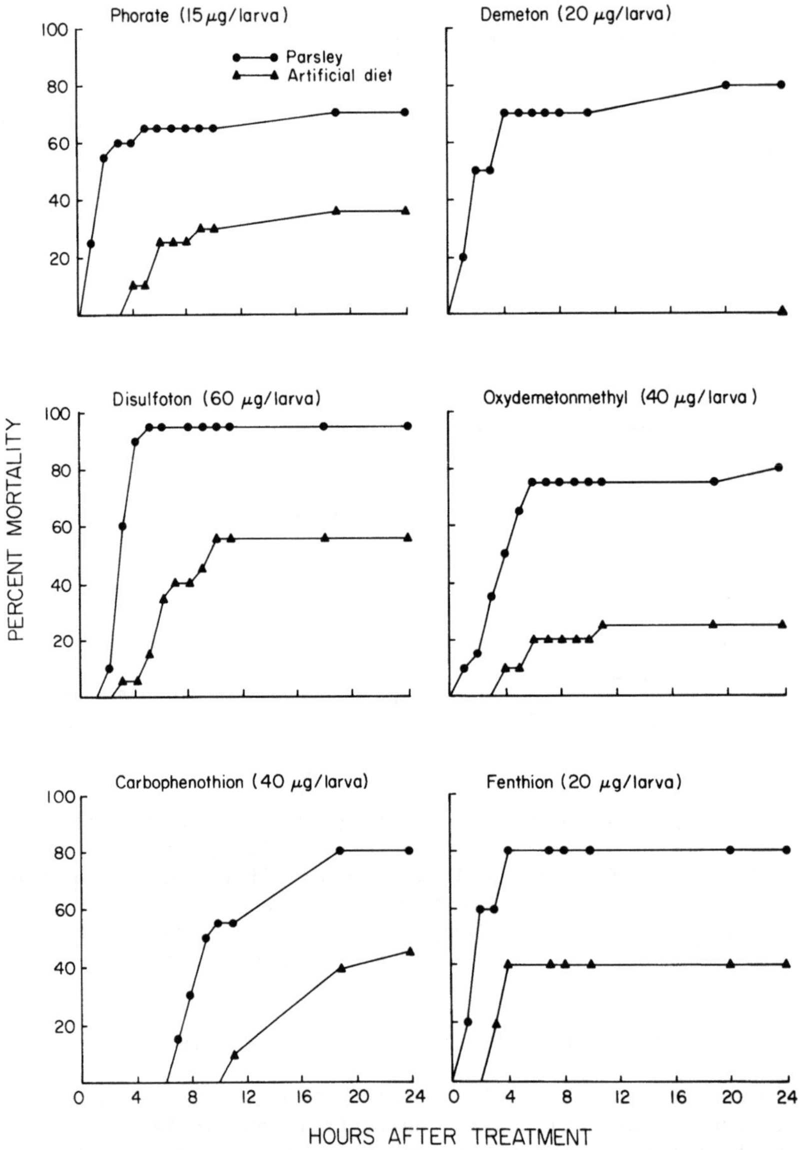

Fig. 2. Effect of parsley on the toxicity of various thioether-containing organophosphorus insecticides to fall armyworm larvae. Newly molted sixth instar larvae were fed parsley leaves or artificial diet for two days prior to insecticide treatments. Body weights in g per larva (mean + SE): Parsley-fed, 0.27 + 0.02; artificial diet-fed, 0.29 ± 0.01 (Yu, 1986a; reprinted with permission).

Table 5. Effect of microsomal enzyme induction on the in vivo metabolism of phorate in fall armyworm larvae.[a]

Enzyme inducer[b]	Time after treatment (hr)	Internal phorate and metabolites (μg /larva)[c]: Phorate	Phorate sulfoxide	Phorate sulfone
Control	2	8.18 ± 0.82[d]	0.83 ± 0.04	0.29 ± 0.02
	4	5.84 ± 1.04	0.76 ± 0.04	0.31 ± 0.01
	6	3.73 ± 0.53	0.77 ± 0.05	0.39 ± 0.03
	8	3.36 ± 0.03	0.71 ± 0.01	0.37 ± 0.01
Parsley	2	2.47 ± 0.28	0.90 ± 0.04	0.42 ± 0.01
	4	0.54 ± 0.15	0.47 ± 0.08	0.22 ± 0.05
	6	0	0	0
	8	0	0	0

[a] Data from Yu (1986a).
[b] Newly molted sixth-instar larvae were fed parsley leaves or artificial diet (control) for two days prior to enzyme assays.
[c] Larvae (after induction) were topically treated with phorate at 10 μg/larva in 1 μl acetone.
[d] Mean ± SE of two experiments, each assayed in duplicate.

6.2. Ecological implications

The PSMO system has a wide substrate specificity (Ahmad et al., Chapter 3 in this text) and, therefore, we would expect it to metabolize plant allelochemicals. Krieger et al. (1971) showed that aldrin epoxidation activity in lepidopterans was significantly higher in polyphagous larvae than in oligophagous ones which, in turn, had significantly higher activity than monophagous larvae. This suggests that microsomal oxidases play an important role in the adaptation to host plants in phytophagous insects. However, unlike pesticides, very little is known about the microsomal metabolism of allelochemicals in insects (Dowd et al., 1983b; Brattsten, Chapter 6 in this text). Most of the work has been done in mammals because of the toxicological and pharmacological importance of many plant allelochemicals in people and domestic animals. If the PSMO system indeed plays an important role in the detoxification of allelochemicals in insects, the question then arises as to what consequences induction would have. A PSMO system from the bark beetle, <u>Dendroctonus terebrans</u> (Olivier) (Coleoptera: Scolytidae), oxidizes α-pinene after induction by α-pinene (White et al., 1979). Pulegone is hydroxylated to 9-hydroxypulegone and 10-hydroxypulegone by microsomes from southern armyworm larvae treated with α-pinene (Brattsten, 1983). Pyrethrum oxidation as measured by NADPH oxidation is induced in microsomes from southern armyworm larvae treated with α-pinene, β-pinene, (+)-limonene, (-)-limonene, α-terpinene or γ-terpinene (Brattsten et al., 1984). In order to learn more about the consequences, we have recently studied the metabolism of allelochemicals by midgut microsomes of fall armyworm larvae and the effect of induction. Ten allelochemicals were studied, all of which were toxic to first instar larvae when they were fed meridic diets containing 0.1 percent of the allelochemicals. By measuring disappearance of substrate (allelochemical) in microsomal incubations fortified with NADPH (Terriere and Yu, 1973), we found that all were metabolized, Table 6. The metabolism can be inhibited by piperonyl butoxide or carbon monoxide, indicating the involvement of cytochrome P-450. The oxidative metabolism

Table 6. Oxidative metabolism of plant toxins by fall armyworm microsomes.[a]

Enzyme substrate	Allelochemical metabolized (nmol/hr/mg protein)[b] Control	Indole-3-carbinol-fed	Peppermint oil-fed
Xanthotoxin	6.33 ± 0.35	24.65 ± 1.17	-
Indole-3-acetonitrile	2.27 ± 0.28	4.59 ± 0.62	-
Myristicin	2.74 ± 0.10	-	8.25 ± 1.6
Safrole	3.93 ± 1.06	8.75 ± 0.08	-
Flavone	1.41 ± 0.56	5.75 ± 0.72	-
Rotenone	31.35 ± 1.65	48.97 ± 5.57	-
Nicotine	0.86 ± 0.16	2.33 ± 0.25	-
Pulegone	10.35 ± 1.59	-	39.71 ± 5.2
trans-Anethole	14.39 ± 0.98	-	33.93 ± 3.1
Estragole	34.34 ± 2.55	-	93.72 ± 15.2

[a] Yu (1986b).

[b] Midgut microsomes were prepared from two-day-old sixth instar larvae. Mean ± SE of two experiments, each with duplicate measurements.

can be enhanced by feeding the larvae allelochemicals such as indole-3-carbinol and essential oil of peppermint, both of which are inducers of PMSO activities in the fall armyworm (Yu, 1982a, 1985; Yu and Ing, 1984). The induction ranges from 1.5- to 4-fold depending on the allelochemical used as substrate, Table 6.

Further studies of allelochemical metabolism were made by measuring allelochemical-dependent NADPH oxidation by midgut microsomes (Yu, 1986b). Data in Table 7 show that fall armyworm microsomes generally metabolize monoterpenes more favorably than other types of terpenes indicating a preference for these compounds. In addition to terpenoids, other plant toxins such as alkaloids, indoles, glucosinolates, coumarins, a cardenolide and a phenylpropene were also metabolized by the PSMO system, Table 8. In all instances, the oxidative metabolism of these allelochemicals can be induced by dietary indole-3-carbinol ranging from 1.3- to 9.5-fold in this insect. It is interesting to note that in the case of triterpenes, the tetraterpene, certain coumarins and the cardenolide digitoxin, metabolic activity can only be observed after induction. As to plant allelochemical metabolism, evidence accumulated so far suggests that enzyme induction caused by plant chemicals would help herbivorous insects detoxify plant toxins in their diets. As a result, they would be able to expand their host plant range if enzyme inducers happen to co-exist with plant toxins in the same plants, or inducing plants and toxin-containing plants are all available in the same area. As shown earlier, PSMO enzymes are capable of detoxifying various allelochemicals but the involvement of glutathione transferases in detoxification of plant toxins is not yet clear. Since GSH transferases are able to attack, α,β-unsaturated carbonyl compounds that are commonly distributed in the plants, efforts should be made to learn if these allelochemicals are metabolized by the transferase. If so, the consequence of induction should be studied. The induction of esterase activity by allelochemicals may have some implications with respect to the metabolism of certain plant toxins containing an ester linkage such as pyrrolizidine alkaloids, pyrethrins and certain cucurbitacins. In the case of epoxide hydrolases, the induction of these enzymes would enhance epoxide hydration following microsomal epoxidation of various olefinic

Table 7. Oxidative metabolism of terpenoids by fall armyworm microsomes measured by NADPH oxidation[a].

Substrate (50 μM)	NADPH oxidation (nmol/min/mg protein)[b]	
	Control	Indole-3-carbinol-fed
Monoterpenes		
(+)-α-pinene	3.18 ± 0.31	16.56 ± 0.93
(-)-α-pinene	3.90 ± 0.21	21.66 ± 0.20
(-)-β-pinene	2.67 ± 0.21	15.94 ± 0.54
(+) limonene	3.69 ± 0.41	27.86 ± 0.90
(-) menthol	5.33 ± 0.41	24.08 ± 1.38
l-menthone	5.54 ± 0.62	32.18 ± 3.43
β-myrcene	4.72 ± 0.21	28.22 ± 0.54
(+) pulegone	3.69 ± 0.41	22.11 ± 0.54
d-carvone	2.25 ± 0.22	13.90 ± 1.27
(+)-camphor	1.84 ± 0.20	12.04 ± 0.42
(+)-camphene	2.86 ± 0.83	17.86 ± 0.75
(-)-camphene	3.52 ± 0.26	11.28 ± 0.43
Geraniol	2.05 ± 0.41	16.37 ± 0.37
Sesquiterpenes		
Nerolidol	3.80 ± 0.11	35.95 ± 1.86
Farnesol	3.70 ± 0.42	34.96 ± 2.51
Santonin	0	0
Diterpenes		
Phytol	3.70 ± 0.42	29.07 ± 0.98
Gibberellic acid	0	0
Triterpenes		
Stigmasterol	0	1.97 ± 0.19
Sitosterol	0	4.15 ± 0.54
Squalene	0	1.12 ± 0.38
Cholesterol	0.62 ± 0.21	3.35 ± 0.37
Ergosterol	0	3.51 ± 0.39
Tetraterpene		
β-carotene	0	1.97 ± 0.19

[a] Yu (1986b).
[b] Newly-molted sixth instar larvae were fed a meridic diet containing 0.2% of the compound. Mean ± SE of two experiments, each with duplicate measurements. Allelochemical-dependent NADPH oxidation was measured as follows. The 4.6-ml mixture which contained 0.5-1 mg microsomal protein, 0.1 M sodium phospate buffer, pH 7.5, was first incubated for 3 min at 30°C and then 0.1 mM NADPH in 0.4 ml of the same buffer was added and mixed. After the addition of 50 M substrate (allelochemical) in 5 μl of methyl Cellosolve to 2.5 ml of the mixture, the rate of NADPH oxidation was recorded at 340 nm against the same reation mixture in the absence of substrate in a Beckman model 5260 uv/vis spectrophotometer.

allelochemicals such as furanocoumarins, phenylpropenes and precocenes.

In addition to the exogenous compounds, the metabolism of endogenous chemicals may also be affected by enzyme induction in insects. For example, the insect hormones ecdysone and juvenile hormone are known to be metabolized by cytochrome P-450-dependent monooxygenases, esterase and

Table 8. Oxidative metabolism of allelochemicals by fall armyworm microsomes measured by NADPH oxidation[a].

Substrate (50 μM)	NADPH oxidation (nmol/min/mg protein)[b] Control	Indole-3-carbinol-fed
Alkaloids		
Atropine	0.39 ± 0.04	1.57 ± 0.17
Strychnine	0.73 ± 0.15	3.13 ± 0.35
Caffeine	0.47 ± 0.12	1.22 ± 0.17
Indoles		
Indole-3-carbinol	0.46 ± 0.11	1.80 ± 0.20
Indole-3-acetaldehyde	1.92 ± 0.18	4.20 ± 0.20
Glucosinolates		
Sinigrin	0.30 ± 0.06	0.70 ± 0.10
2-phenylethyl isothiocyanate	0.55 ± 0.03	5.00 ± 0.70
Coumarins		
Coumarin	0	0.86 ± 0.14
Umbelliferone	0	0
Scopoletin	0.99 ± 0.14	1.29 ± 0.15
Cardenolide		
Digitoxin	0	0.71 ± 0.15
Phenylpropene		
Eugenol	1.02 ± 0.13	4.80 ± 0.80

[a] Yu (1986b).
[b] Newly molted sixth-instar larvae were fed a meridic diet containing 0.2% of the compound for two days prior to enzyme assays. Mean ± SE of two experiments, each with duplicate measurements.

epoxide hydrolase in insects (Yu and Terriere, 1978b; Slade and Wilkinson, 1973, 1974; Hammock, 1975; Terriere and Yu, 1976; Feyereisen and Durst, 1978). No information is currently available as to whether plant chemicals induce hormone-specific cytochrome P-450, esterase and epoxide hydrolase activities. However, evidence that phenobarbital induction of PSMO activity disrupted metamorphosis and reproduction of house flies (Yu and Terriere, 1974) suggests that induction of these detoxifying enzymes by allelochemicals is likely to influence the biosynthesis and inactivation of ecdysones and juvenile hormones in insects. If so, an insect that feeds on an inducing plant may have to make a necessary adjustment to cope with the higher rate of its own hormone metabolism. Thus, the effect of allelochemical-mediated induction of PSMO, esterase, and epoxide hydrolase activities on the hormone regulation of lepidopterous insects warrants active investigation.

7. SUMMARY

Detoxification enzymes including microsomal cytochrome P-450-dependent monooxygenases, glutathione transferases, esterases and epoxide hydrolases are induced in herbivorous insects by many plants and their allelochemicals. The inducing patterns of various PSMO activities including epoxidation, hydroxylation, N-demethylation, O-demethylation, desulfuration, and sulfoxidation were different in the fall armyworm, used as a model insect, probably due to the multiplicity of cytochrome P-450. Xanthotoxin, the GSH transferase inducer in parsnip leaves, also induces the PSMO system, increasing cytochrome P-450 content and heptachlor epoxidation but inhibiting aldrin epoxidation, biphenyl hydroxylation, and PCMA N-demethylation. However, selective induction of various GSH transferases by allelochemicals was not observed in this insect.

Induction of PSMO in fall armyworm larvae feeding corn leaves increases their tolerance to commonly used insecticides such as organophosphates, carbamates and pyrethroids. Likewise, induction of GSH transferase by cowpea leaves resulted in increasing insecticide tolerance. Induction of PSMO by plant allelochemicals in the larvae enhances the oxidative metabolism of plant toxins such as terpenoids, flavonoids, indoles, coumarins, alkaloids, a cardenolide, glucosinolates and phenylpropenes. Modification of detoxifying enzymes by allelochemicals results in changing insecticide susceptibility and has a strong potential for influencing host plant selections in herbivorous insects.

8. ACKNOWLEDGEMENT

The original research reported here was supported in part by USDA (Competitive Research Grants Office) Grant No. 82-CRCR-1-1091. I thank Drs. S. H. Kerr and F. Slansky for reviewing the manuscript and Mrs. Glinda Burnett for typing the manuscript. Florida Agricultural Experiment Station Publication No. 6283.

9. REFERENCES

Agosin, M., 1985. Role of microsomal oxidations in insecticide degradation, in: "Comprehensive Insect Physiology, Biochemistry and Pharmacology", Vol. 12, G. A. Kerkut and L. I. Gilbert, eds., pp. 647-712, Pergamon Press, New York.

Agosin, M., N. Scaramelli, L. Gil and M. E. Letelier, 1969. Some properties of the microsomal system metabolizing DDT in Triatoma infestans, Comp. Biochem. Physiol., 29:785-793.

Ahmad, S., 1983. Mixed function oxidase activity in a generalist herbivore in relation to its biology, food plants, and feeding history, Ecology, 64:235-243.

Battaile, J. and W. B. Loomis, 1961. Biosynthesis of terpenes. II. The site and sequence of terpene formation in peppermint, Biochim. Biophys. Acta., 51:545-552.

Berry, R. E., S. J. Yu, and L. C. Terriere, 1980. Influence of host plants on insecticide metabolism and management of variegated cutworm, J. Econ. Entomol., 73:771-774.

Brattsten, L. B., 1979a. Biochemical defense mechanisms in herbivores against plant allelochemicals, in: "Herbivores: Their Interaction with Secondary Plant Metabolites", G. A. Rosenthal and D. H. Jansen, eds., pp. 199-270, Academic Press, New York.

Brattsten, L. B., 1979b. Ecological significance of mixed-function oxidations, Drug Metab. Rev., 10:35-58.

Brattsten, L. B., 1983. Cytochrome P-450 involvement in the interactions between plant terpenes and insect herbivores, in: "Plant Resistance to Insects", P. A. Hedin, ed., pp. 173-195, Symp., Ser. No. 208, Amer. Chem. Soc., Washington.

Brattsten, L. B., C. K Evans, S. Bonetti and L. H. Zalkow, 1984. Induction by carrot allelochemicals of insecticide-metabolizing enzymes in the southern armyworm (Spodoptera eridania), Comp. Biochem. Physiol., 77C:29-37.

Brattsten, L. B., S. L. Price and C. A. Gunderson, 1980. Microsomal oxidases in midgut and fatbody tissues of a broadly herbivorous insect larvae, Spodoptera eridania Cramer (Noctuidae), Comp. Biochem. Physiol., 66C:231-237.

Brattsten, L. B. and C. F. Wilkinson, 1973. Induction of microsomal enzymes in the southern armyworm (Prodenia eridania), Pestic. Biochem. Physiol., 3:393-407.

Brattsten, L. B. and C. F. Wilkinson, 1977. Insecticide solvents: Interference with insecticidal action, Science, 196:1211-1213.

Brattsten, L. B., C. F. Wilkinson, and T. Eisner, 1977. Herbivore-plant interactions: Mixed-function oxidases and secondary plant substances, Science, 196:1349-1352.

Brattsten, L. B., C. F. Wilkinson and M. M. Root, 1976. Microsomal hydroxylation of aniline in the southern armyworm, Spodoptera eridania, Insect Biochem., 6:615-620.

Bull, D. L., 1965. Metabolism of di-syston by insects, isolated cotton leaves, and rats, J. Econ. Entomol., 58:249-254.

Capdevila, J., N. Ahmad and M. Agosin, 1975. Soluble cytochrome P-450 from house fly microsomes. Partial purification and characterization of two hemoprotein forms, J. Biol. Chem., 250:1048-1060.

Capdevila, J., A. Morello, A. S. Perry and M. Agosin, 1973. Effect of phenobarbital and naphthalene on some of the components of the electron transport system and the hydroxylating activity of house fly microsomes, Biochem., 12:1445-1451.

Christian, M. F. and S. J. Yu, 1986. Cytochrome P-450-dependent monooxygenases in the velvetbean caterpillar, Anticarsia gemmatalis Hubner, Comp. Biochem. Physiol., 83C:23-27.

Clark, A. G., N. A. Shamaan, W. C. Dauterman and T. Hayaoka, 1984. Characterization of multiple glutathione transferases from the house fly, Musca domestica (L.), Pestic. Biochem. Physiol., 22:51-59.

Dowd, P. F., C. M. Smith and T. C. Sparks, 1983a. Influence of soybean leaf extracts on ester cleavage in cabbage and soybean loopers (Lepidoptera: Noctuidae), J. Econ. Entomol., 76:700-703.

Dowd, P. F., C. M. Smith and T. C. Sparks, 1983b. Detoxication of plant toxins by insects, Insect Biochem., 13:453-468.

Farnsworth, D. E., R. E. Berry, S. J. Yu and L. C. Terriere, 1981. Aldrin epoxidase activity and cytochrome P-450 content of microsomes prepared from alfalfa and cabbage looper larvae fed various plant diets, Pestic. Biochem. Physiol., 15:158-165.

Feyereisen, R. and F. Durst, 1978. Ecdysone biosynthesis: A microsomal cytochrome P-450-linked ecdysone 20-monooxygenase from tissues of the African migratory locust, Eur. J. Biochem., 88:37-47.

Georghiou, G. P. and T. Saito, eds., 1983. "Pest Resistance to Pesticides", Plenum Publ. Corp., New York, 809 pp.

Gil, D. L., H. A. Rose, R. S. H. Yang, R. G. Young and C. F. Wilkinson, 1974. Enzyme induction by phenobarbital in the Madagascar cockroach, Gromphadorhina portentosa, Comp. Biochem. Physiol., 47B:657-662.

Hammock, B. D., 1975. NADPH dependent epoxidation of methyl farnesoate to juvenile hormone in the cockroach, Blaberus giganteus, L., Life Sci., 17:323-328.

Hayaoka, T. and W. C. Dauterman, 1982. Induction of glutathione S-transferase by phenobarbital and pesticides in various house fly strains and its effect on toxicity, Pestic. Biochem. Physiol., 17:113-119.

Kea, W. C., S. G. Turnipseed and G. R. Carner, 1978. Influence of resistant soybeans on the susceptibility of lepidopterous pests to insecticides, J. Econ. Entomol., 71:58-60.

Khan, M. A. Q. and F. Matsumura, 1972. Induction of mixed-function oxidase and protein synthesis by DDT and dieldrin in German and American cockroaches, Pestic. Biochem. Physiol., 2:236-243.

Krieger, R. I., P. P. Feeny and C. F. Wilkinson, 1971. Detoxication enzymes in the guts of caterpillars: An evolutionary answer to plant defenses? Science, 172:579-581.

Kulkarni, A. P., D. L. Fabacher and E. Hodgson, 1980. Pesticides as inducers of hepatic drug metabolizing enzymes. II. Glutathione S-transferases, Gen. Pharmacol., 11:437-441.

March, R. B., R. L. Metcalf, T. R. Fukuto and M. G. Maxon, 1955. Metabolism of systox in the white mouse and American cockroach, J. Econ. Entomol., 48:355-363.

Matsumura, F., 1975. "Toxicology of Insecticides", Plenum Publ. Corp., New York, 503 pp.

Menzie, C. M., 1969. "Metabolism of Insecticides", U.S. Dept. of the Interior Fish and Wildlife Service, Special Scientific Report: Wildlife No. 127, Washington, DC.

Moldenke, A. F., R. E. Berry and L. C. Terriere, 1983. Cytochrome P-450 in insects. 5. Monoterpene induction of cytochrome P-450 and associated monooxygenase activities in the larva of the variegated cutworm Peridroma saucia (Hubner), Comp. Biochem. Physiol., 74C:365-371.

Morello, A., 1964. Role of DDT-hydroxylation in resistance, Nature, 203:785-786.

Motoyama, N. and W. C. Dauterman, 1972. In vitro metabolism of azinphosmethyl in susceptible and resistant house flies, Pestic. Biochem. Physiol., 2:113-122.

Mullin, C. A., and B. A. Croft, 1983. Host-related alterations of detoxication enzymes in Tetranychus urticae (Acari: Tetranychidae), Environ. Entomol., 12:1278-1281.

Nakatsugawa, T. and P. A. Dahm, 1965. Parathion activation enzymes in the fat body microsomes of the American cockroach, J. Econ. Entomol., 58:500-509.

Naquira, C., R. A. White, Jr. and M. Agosin, 1980. Multiple forms of Drosophila cytochrome P-450, in: "Biochemistry, Biophysics and Regulation of Cytochrome P-450", J. A. Gustafsson, J. Carlstedt-Duke, A. Mode and J. Rafter, eds., pp. 105-108, Elsevier, New York.

Nebert, D. W., H. J. Eisen, M. Negishi, M. A. Lang and L. M. Hjilmeland, 1981. Genetic mechanisms controlling the induction of polysubstrate monooxygenase (P-450) activities, Annu. Rev. Pharmacol. Toxicol., 21:431-62.

Ottea, J. A. and F. W. Plapp, Jr., 1981. Induction of glutathione S-transferase by phenobarbital in the house fly, Pestic. Biochem. Physiol., 15:10-13.

Ottea, J. A. and F. W. Plapp, Jr., 1984. Glutathione S-transferase in the house fly: Biochemical and genetic changes associated with induction and insecticide resistance, Pestic. Biochem. Physiol., 22:203-208.

Perry, A. S., W. E. Dale and A. J. Buckner, 1971. Induction and repression of microsomal mixed-function oxidases and cytochrome P-450 in resistant and susceptible house flies, Pestic. Biochem. Physiol., 1:131-142.

Plapp, F. W., Jr. and J. Casida, 1970. Induction of DDT and dieldrin of insecticide metabolism by house fly enzymes, J. Econ. Entomol., 63:1091-1092.

Rose, H. A. and L. C. Terriere, 1980. Microsomal oxidase activity of three blow fly species and its induction by phenobarbital and - naphthoflavone, Pestic. Biochem. Physiol., 14:275-281.

Schonbrod, R. D. and L. C. Terriere, 1975. The solubilization and separation of two forms of microsomal cytochrome P-450 from the house fly, Musca domestica L., Biochem. Biophys. Res. Commun., 64:829-835.

Slade, M. and C. F. Wilkinson, 1973. Juvenile hormone analogs: A possible case of mistaken identity? Science, 181:672-674.

Slade, M. and C. F. Wilkinson, 1974. Degradation and conjugation of cecropia juvenile hormone by the southern armyworm (Prodenia eridania), Comp. Biochem. Physiol., 49B:99-103.

Sparks, T. C. and B. D. Hammock, 1979. Induction and regulation of juvenile hormone esterases during the last larval instar of the cabbage looper, Trichoplusia ni, J. Insect Physiol., 25:551-560.

Stanton, R. H., F. W. Plapp, Jr., R. A. White and M. Agosin, 1978. Induction of multiple cytochrome P-450 species in house fly microsomes. SDS-gel electrophoresis studies, Comp. Biochem. Physiol., 61B:297-305.

Stone, B. F., 1969. Metabolism of fenthion by the southern house mosquito, J. Econ. Entomol., 62:977-981.

Terriere, L. C., 1983. Enzyme induction, gene amplification and insect resistance to insecticides, in: "Pest Resistance to Pesticides", G. P. Georghiou and T. Saito, eds., pp. 265-297, Plenum Publ. Corp., New York.

Terriere, L. C., 1984. Induction of detoxication enzymes in insects, Annu. Rev. Entomol., 29:71-88.

Terriere, L. C. and S. J. Yu, 1973. Insect juvenile hormones: Induction of detoxifying enzymes in the house fly and detoxication by house fly enzymes, Pestic. Biochem. Physiol., 3:96-107.

Terriere, L. C. and S. J. Yu, 1974. The induction of detoxifying enzymes in insects, J. Agric. Food Chem., 22:366-373.

Terriere, L. C. and S. J. Yu, 1976. Interaction between microsomal enzymes of the house fly and the moulting hormones and some of their analogs, Insect Biochem., 6:109-114.

Terriere, L. C. and S. J. Yu, 1979. Cytochrome P-450 in insects. 2. Multiple forms in the flesh fly (*Sarcophaga bullata* Parker), and the blow fly (*Phormia regina* (Meigen)), Pestic. Biochem. Physiol., 12:249-256.

Vincent, D. R., A. F. Moldenke, D. E. Farnsworth and L. C. Terriere, 1985a. Cytochrome P-450 in insects. 6. Age dependency and phenobarbital induction of cytochrome P-450, P-450 reductase, and monooxygenase activities in susceptible and resistant strains of *Musca domestica*, Pestic. Biochem. Physiol., 23:171-181.

Vincent, D. R., A. F. Moldenke, D. E. Farnsworth and L. C. Terriere, 1985b. Cytochrome P-450 in insects. 7. Age dependency and phenobarbital induction of cytochrome P-450, P-450 reductase, and monooxygenase activities in black blow fly (*Phormia regina* (Meigen)), Pestic. Biochem. Physiol., 23:182-189.

Walker, C. R. and L. C. Terriere, 1970. Induction of microsomal oxidases by dieldrin in *Musca domestica*, Ent. Exp. Appl., 13:260-274.

White, R. A., Jr., R. T. Franklin and M. Agosin, 1979. Conversion of α-pinene to α-pinene oxide by rat liver and the bark beetle *Dendroctonus terebrans* microsomal fractions, Pestic. Biochem. Physiol., 10:233-242.

Whitmore, D., Jr., E. Whitmore, and L. I. Gilbert, 1972. Juvenile hormone induction of esterases: A mechanism for the regulation of juvenile hormone titer, Proc. Nat. Acad. Sci., 69:1592-1595.

Wilkinson, C. F., ed., 1976. "Insecticide Biochemistry and Physiology", Plenum Publ. Corp., New York, 768 pp.

Yang, R. S. H., E. Hodgson and W. C. Dauterman, 1971. Metabolism *in vitro* of diazinon and diazoxon in susceptible and resistant house flies, J. Agric. Food Chem., 19:14-19.

Yu, S. J., 1982a. Induction of microsomal oxidases by host plants in the fall armyworm, *Spodoptera frugiperda* (J. E. Smith), Pestic. Biochem. Physiol., 17:59-67.

Yu, S. J., 1982b. Host plant induction of glutathione *S*-transferase in the fall armyworm, Pestic. Biochem. Physiol., 18:101-106.

Yu, S. J., 1983. Induction of detoxifying enzymes by allelochemicals and host plants in the fall armyworm, Pestic. Biochem. Physiol., 19:330-336.

Yu, S. J., 1984. Interactions of allelochemicals with detoxication enzymes of insecticide-susceptible and resistant fall armyworms, Pestic. Biochem. Physiol., 22:60-68.

Yu, S. J., 1985. Microsomal sulfoxidation of phorate in the fall armyworm, Spodoptera frugiperda (J. E. Smith), Pestic. Biochem. Physiol., 23:273-281.

Yu, S. J., 1986a. Host plant induction of microsomal monooxygenases in relation to organophosphate activation in fall armyworm larvae, Florida Entomol., 69:579-587.

Yu, S. J., 1986b. Microsomal oxidation of allelochemicals in a generalist (Spodoptera frugiperda) and semi-specialist (Anticarsia gemmatalis) insect, J. Chem. Ecol., In Press.

Yu, S. J., R. E. Berry, and L. C. Terriere, 1979. Host plant stimulation of detoxifying enzymes in a phytophagous insect, Pestic. Biochem. Physiol., 12:280-284.

Yu, S. J. and E. L. Hsu, 1985. Induction of hydrolases by allelochemicals and host plants in fall armyworm (Lepidoptera: Noctuidae) larvae, Environ. Entomol., 14:512-515.

Yu, S. J. and R. T. Ing, 1984. Microsomal biphenyl hydroxylase of fall armyworm larvae and its induction by allelochemicals and host plants, Comp. Biochem. Physiol., 78C:145-152.

Yu, S. J. and L. C. Terriere, 1971. Hormonal modification of microsomal oxidase activity in the house fly, Life Sci., 10:1173-1185.

Yu, S. J. and L. C. Terriere, 1972. Enzyme induction in the house fly: The specificity of the cyclodiene insecticides, Pestic. Biochem. Physiol., 2:184-190.

Yu, S. J. and L. C. Terriere, 1973. Phenobarbital induction of detoxifying enzymes in resistant and susceptible house flies, Pestic. Biochem. Physiol., 3:141-148.

Yu, S. J. and L. C. Terriere, 1974. A possible role for microsomal oxidases in metamorphosis and reproduction in the house fly, J. Insect Physiol., 20:1901-1912.

Yu, S. J. and L. C. Terriere, 1975. Activities of hormone metabolizing enzymes in house flies treated with some substituted urea growth regulators, Life Sci., 17:619-626.

Yu, S. J. and L. C. Terriere, 1978a. Juvenile hormone epoxide hydrase in house flies, flesh flies and blow flies, Insect Biochem., 8:349-352.

Yu, S. J. and L. C. Terriere, 1978b. Metabolism of juvenile hormone I by microsomal oxidase, esterase, and epoxide hydrase of Musca domestica and some comparisons with Phormia regina and Sarcophaga bullata, Pestic. Biochem. Physiol., 9:237-246.

Yu, S. J. and L. C. Terriere, 1979. Cytochrome P-450 in insects. 1. Differences in the forms present in insecticide resistant and suceptible house flies, Pestic. Biochem. Physiol., 12:239-248.

ADAPTIVE DIVERGENCE OF CHEWING AND SUCKING

ARTHROPODS TO PLANT ALLELOCHEMICALS

Christopher A. Mullin

The Pennsylvania State University
Department of Entomology
Pesticide Research Laboratory and Graduate Study Center
University Park, PA 16802

1. INTRODUCTION

Plant-feeding arthropods, in the course of evolutionary time, have adapted differentially to plant hosts and thereby achieved optimized use of available resources. Divergent patterns of niche orientation have lead to guilds of herbivores that may specialize on a single common plant host. Plants, in turn, have responded by developing chemical and morphological defenses to discourage potential consumers, particularly where a low apparency (sensu Feeny, 1976) is not sufficient to assure the plant's survival. These arthropod-plant coadaptations have resulted in herbivores that represent very distinct feeding modes, and, expectedly, contrast in their capabilities for eluding plant defenses. Leaf-chewing lepidopterans, coleopterans, and orthopterans should have very different exposures to plant allomones compared to phloem-sucking homopterans and hemipterans. The possible differences in metabolic mechanisms within chewing and sucking arthropods in response to their phytochemical niches are discussed in this review.

While visual cues and morphological features are important for finding and accepting a host (Bell and Carde, 1984; Stipanovic, 1983), much evidence suggests that chemical factors dominate the orientation to and consumption of a plant by an insect (Rosenthal and Janzen, 1979). Suckers selectively feed on high energy portions of the plant such as phloem and seed, whereas chewers consume external or entire plant structures, sometimes seemingly indiscriminately. Plant defensive chemicals are thought to be allocated mostly to specialized organelles or tissues of external structures, and to occur only at low concentrations in phloem (McKey, 1979; Waller and Nowacki, 1978), perhaps due to lack of sufficient solubility. However, physicochemical factors controlling phloem loading and translocation of chemicals within plants are poorly understood (Giaquinta, 1985; Crisp and Larson, 1983). In any case, chewing herbivores presumably consume higher concentrations of plant toxicants than phloem-sucking species. Thus, metabolic adaptations to toxic chemicals should be better developed in chewing relative to sucking herbivores. It follows that phloem-feeders may be more susceptible to increased allocations of toxic chemicals into plant tissues essential to their finding and accepting phloem.

Knowledge about mechanisms by which herbivores adapt to defensive chemicals within plant hosts would aid development of resistant plant cultivars and synthetic chemical control agents for pests, since adaptations to synthetic pesticides or toxic dietary chemicals are biochemically and physiologically related (Brattsten, 1979a; Yu, 1983; Mullin, 1985). Arthropods in dissimilar feeding niches tend to have widely different susceptibilities to pesticides. For example, chewing herbivores such as lepidopterans and coleopterans are often less susceptible to pesticides than phloem-sucking aphids (Hollingworth, 1976; Saito, 1969; Weiden, 1971). These selectivities may be explained, in part, by preadaptations to toxic dietary chemicals.

2. PLANT RESISTANCE TO CHEWING AND SUCKING INSECTS

Successful integration of host plant resistance into pest management strategies most often occurs when multiresistant cultivars are used. While multiresistance often involves a primary insect pest and a major disease, occasionally a cultivar resistant to both chewing and sucking insects is bred. This approach can be highly effective since resistance to primary pests, which are usually of the chewing type, often leads to a plant's increased susceptibility to secondary sucking pests which then become the primary pests (Harris, 1979; Hedin et al., 1977; Kennedy, 1978; Maxwell and Jennings, 1980). Plant breeding programs devoted to control of a chewing-sucking complex of insects may be hampered by the complicated polygenic inheritance to be expected in multiresistant cultivars (Gould, 1983). Nevertheless, resistance to both chewing and sucking insects can often be explained by a common chemical factor.

Chemical factors are more important than morphological factors in explaining plant coresistances to chewing and sucking insects. Potato resistance to both the Colorado potato beetle, Leptinotarsa decemlineata (Say) (Coleoptera: Chrysomelidae), and the potato leafhopper, Empoasca fabae (Harris) (Homoptera: Cicadellidae), is positively correlated with leaf glycoalkaloid content (Tingey, 1984), presumably as a glandular trichome exudate. In tomatoes, resistance to lepidopterous larvae and aphids is associated with increasing trichome exudates of 2-tridecanone (Williams et al., 1980). Terpenoids such as gossypol and heliocides in subepidermal pigment glands have been implicated as resistant factors in cotton for a host of chewing and sucking insects (Harris, 1979; Maxwell and Jennings, 1980). The high correlation between susceptibility of corn hybrids to corn-leaf aphid, Rhopalosiphum maidis (Fitch) (Homoptera: Aphididae), and subsequent infestation by first generation European corn borer, Ostrinia nubilalis (Hubner) (Lepidoptera: Pyralidae), recognized by Huber and Stringfield (1940), was later associated with decreasing foliar content of 2,4-dihydroxy-7-methoxy-1,4-benzoxazin-3-one (DIMBOA) for both the lepidopterous larva (Beck, 1965; Klun et al., 1967) and for this and other grain-feeding aphids (Long et al., 1977; Argandona, 1983). While constituent chemical factors within glandular trichomes (Stipanovic, 1983) are often responsible for these resistances, more recent studies suggest that induced resistances, in response to previous insect herbivory or other stress factors, may be important chemical barriers to both chewing and phloem-feeding insects (Campbell et al., 1984; Hart et al., 1983; Schultz and Baldwin, 1982). However, isoflavonoid phytoalexins such as coumestrol are feeding deterrents to legume chewers (Kogan and Paxton, 1983) but not necessarily to aphids (Loper, 1968).

3. SUSCEPTIBILITY DIFFERENCES TO TOXICANTS

Large species differences in susceptibility to pesticides or dietary toxicants is a persuasive indicator of dissimilar metabolic adaptations to toxic chemicals. However, in comparing bioassay data, many factors including route of entry, insecticide solvents, penetration enhancers, temperature, insect phenology, anesthesia, diet, test chamber design, and so forth, may influence mortality and sublethal responses. More accurate comparisons are made if test animals of similar size, life stage, and cuticular bulk can be bioassayed concurrently by the same procedure. For example, parallel study of soft-bodied insects, e.g. aphids and lepidopterous larvae, or hard-shelled ones such as membracids and chrysomelids on the same host plant can provide clues to differences in basic mechanisms in the metabolic basis for phloem- and foliar- feeding species. Species that differ in susceptibility to a topically applied toxicant by two orders or greater of magnitude usually do so through biochemical and physiological (enzymes, receptors) rather than physical (penetration, sequestration) mechanisms. Large ($> 10^3$-fold) within species variability in chemical susceptibility may be observed due to differing preexposure of populations to the chemical.

It is prudent to have a good knowledge of the chemical history of both the test organism and the environment from which it was originally collected. For example, comparison of a sucking insect collected from an intensively sprayed agricultural field and an inbred laboratory population of a chewing species with no previous history of chemical selection would be inappropriate.

Four very important physicochemical properties of toxicants largely dictate their mobility in an animal, and hence the success of comparative bioassays. The first, the water solubility of the chemical, is determined by the nature of substituents such as polar or nonpolar, and ionic or nonionic groups. Water solubility will determine the propensity of the chemical to be transported in a water phase like hemolymph, or to be excreted. Secondly, the partition coefficient of a chemical, which is its solubility in an organic solvent relative to water, is an index of the affinity of the chemical for lipid-rich sinks such as membranes and fat tissues. Lipophilic (fat-loving) compounds are usually the most toxic ones since target sites are often membrane associated. Ease of membrane penetration is determined mostly by the hydrophilic to lipophilic balance of the chemical and, thirdly, a steric factor dictated by the size, shape and molecular weight of a chemical. Fourthly, knowing the vapor pressure of a chemical will help predict if it has sufficient volatility to enter an animal by inhalation as well as by ingestion or contact.

3.1. Susceptibility to synthetic toxicants

Many studies have demonstrated that conventional insecticides including organochlorines, organophosphates, carbamates, and synthetic pyrethroids do not always achieve equivalent control of chewing and sucking insects in the field. Indeed, piercing-sucking insects often predominate in major crops such as cotton and apples when effective chemical control of the major chewing insect is attained (Croft and Whalon, 1982; Ware, 1980). While the higher reproductive rate and shorter generation time of sucking insects such as aphids compared to chewing species such as lepidopterous larvae may explain some of these field observations, selective toxicity is often indicated. Busvine (1942) and others have long recognized the importance of screening promising insecticide candidates on multiple insect species rather than solely resorting to a "prototype" species most often represented by a fly or cockroach.

Generally, chewing herbivores are less susceptible than sucking ones to equivalent doses of insecticides. For example, DDT is 23 times more lethal to the milkweed bug, *Oncopeltus fasciatus* (Dallas) (Hemiptera:

Lygaeidae), than to the differential grasshopper, Melanoplus differentialis (Thomas) (Orthoptera: Acrididae); dimethoate is four-fold more toxic to O. fasciatus than to the Colorado potato beetle; and carbaryl is over 1000 times more toxic to the tarnished plant bug, Lygus lineolaris (P. deBeauvois) (Hemiptera: Lygaeidae), than to the American cockroach, Periplaneta americana (L.) (Orthoptera: Blattidae), by topical application, Table 1. The bean aphid, Aphis fabae Scopoli (Homoptera: Aphididae), and buckthorn aphid, A. nasturtii Kaltenbach (Ibid.), were up to 160 times more susceptible than either the southern armyworm, Spodoptera eridania (Cramer) (Lepidoptera: Noctuidae), or Mexican bean beetle, Epilachna varivestis Mulsant (Coleoptera: Coccinellidae), to a wide variety of carbamates and carbamoyloximes (Weiden, 1971; D'Silva et al., 1985). Dieldrin and parathion were much more toxic to the cabbage aphid, Brevicoryne brassicae (L.) (Homoptera: Aphididae), than to the cowpea curculio, Chalcodermus aeneus Boheman (Coleoptera: Curculionidae) Wolfenbarger and Holscher, 1967). The mustard aphid, Lipaphis erysimi (Kaltenbach) (Homoptera: Aphididae), was much more susceptible to various organophosphates and carbaryl than a chrysomelid beetle or two species of weevils; yet, this trend was not apparent with the two organochlorine insecticides DDT and endosulfan (Kumar and Lal, 1966). Winteringham (1969) has tabulated data that also indicate a tendency among the organochlorinated insecticides, e.g. cyclodienes, DDT and lindane, to be more or equitoxic to chewing relative to sucking herbivores. These latter studies, however, have been conducted using somewhat nonequivalent contact lethality assessments where the test species were fed different host plants or where the mobility of the arthropod would bias the degree of exposure.

This tendency towards susceptibility is not limited to conventional insecticides, but extends into newer classes of synthetic morphogenic agents. A number of juvenile hormone analogues including hydroprene and kinoprene are more effective on a wide variety of plant feeding homopterans and hemipterans than on chewing lepidopterans and coleopterans (Henrick, 1982; Hollingworth, 1976). Hypocholesterolemic (antilipemic) analogues of clofibrate were also more active on O. fasciatus than on the mealworm, Tenebrio molitor L. (Coleoptera: Tenebrionidae) (Hammock et al., 1978).

3.2. Susceptibility to naturally-derived toxicants

Pyrethrins, the insecticidal principles of Chrysanthemum spp., and their synthetic analogues that lack the α-cyano moiety on the alcohol are less toxic to O. fasciatus, L. lineolaris, and a number of aphid species than to some chewing herbivores (Busvine, 1942; Jao and Casida, 1974a), although the reverse tendency was found when α-cyanophenoxybenzyl pyrethroids were screened on the aphid, A. fabae, relative to S. eridania and E. varivestis (Ayad and Wheeler, 1984). Distinct selectivities between chewing and sucking pests for the recently isolated avermectin insecticides from Streptomyces avermitilis are equivocal (Putter et al., 1981). However, sucking insects such as O. fasciatus are clearly more susceptible to the antijuvenile hormone action of plant chromenes (precocenes) from Ageratum spp. than holometabolous chewing herbivores (Bowers, 1982; Pratt, 1983).

Based on synthetic diet bioassays, many polyhydroxylated flavonoid and related phenolics are strongly deterrent to aphids including Schizaphis graminum (Rondani) (Homoptera: Aphididae) and the green peach aphid, Myzus persicae (Sulz.) (Ibid.), where the aglycones are usually more active than the corresponding glycosides (Todd et al., 1971; Dreyer and Jones, 1981; Dreyer et al., 1981; Schoonhoven and Dersken-Koppers, 1976). Methylation of the phenolic group usually decreased deterrency (Dreyer and Jones, 1981; Todd et al., 1971). Aphids are more susceptible

Table 1. Variation in insecticidal activity among different species.

Species	Topical LD_{50} relative to housefly = 1		
	DDT[a]	Dimethoate[a]	Carbaryl[b]
American cockroach	1	13	0.2
Honey bee	11	---	0.003
Milkweed bug	41	387	0.036
Tarnished plant bug	---	---	0.0002
Differential grasshopper	940	---	---
Colorado potato beetle	---	1453	---

[a] Data from R. D. O'Brien (1967)
[b] Data from L. B. Brattsten and R. L. Metcalf (1970; 1973)

to the antifeedant action of flavonoids than the corn earworm, Heliothis zea Boddie (Lepidoptera: Noctuidae), Table 2. Moreover, the pattern of flavonoid toxicity to H. zea differs from that to aphids; adjacent (ortho) hydroxyl groups are necessary for anti-growth activity in H. zea (Elliger et al., 1980), but not in aphids (Dreyer and Jones, 1981). Consistently, morin (lacking o-hydroxy groups) was strongly deterrent to aphids yet inert to H. zea, Table 2. Extensive work on the chemical basis of resistance of Sorghum bicolor to insect pests has revealed that cyanogenesis is more important than phenolic compounds in reducing the feeding by chewing grasshoppers and lepidopterans, while the reverse tendency is apparent for phloem feeders including a planthopper and an aphid (Woodhead et al., 1980; Fish, 1980). Feeding bioassays demonstrated that the polyhydroxylated flavan from sorghum, procyanidin, is at least twice more deterrent than dhurrin, the major cyanogenic glucoside of sorghum, to the aphid S. graminum (Dreyer et al., 1981).

Another major group of plant secondary compounds associated with phloem-feeders is the alkaloids. Many of these weak bases are phloem-mobile, including Papaveraceae alkaloids and ricinine (Waller and Nowacki, 1978). The apparently phloem-mobile quinolizidine alkaloids inhibit the feeding of a generalist aphid, Acrythosiphon pisum (Harris) (Homoptera: Aphididae), yet stimulate the feeding by aphid specialists on lupines (Smith, 1966; Wink et al. 1982). Many alkaloids are strong feeding deterrents to M. persicae in artificial diet bioassays (Schoonhoven and Derksen-Koppers, 1976; Junde and Lidao, 1984). Although M. persicae avoids the xylem-mobile nicotine by imbibing phloem (Guthrie et al., 1962), aphid resistance in Nicotiana sp. readily correlates with elevated alkaloid content of trichome exudates (Thurston et al., 1966; Gibson and Plumb, 1977). Aphids, leafhoppers, whiteflies and other homopterans have long been recognized as more susceptible than chewing herbivores to the inhalation toxicity of nicotine (Richardson and Casanges, 1942).

4. THE APHID AS A MODEL PHYTOPHAGE OF PHLOEM CHEMICALS

Prior to embarking on a discussion of mechanisms that arthropod herbivores of differing ecologies may use to deal with toxic phytochemicals, it may be necessary first to understand the chemical classes that would impact more on sucking relative to chewing herbivores. Chewers tend to be somewhat indiscriminate, and often consume all of the above ground parts

Table 2. Feeding deterrency and growth inhibition of flavonoids to three insect species.

	ED (μmole/g diet)		
	Feeding deterrency[a]		Anti-growth[b]
Flavonoid	*S. graminum*	*M. persicae*	*H. zea*
Eriodictyol	0.7	0.7	6.2
Luteolin	1.0	-	5.4
Morin	1.3	1.3	> 10.0
Quercetin	2.6	1.0	3.5
Dihydroquercetin	2.0	2.0	3.5
Rutin	0.33	> 16.0	4.0
Quercitrin	1.3	1.3	4.5
Naringenin	5.5	9.2	> 10.0
Naringin	> 17.0	> 17.0	> 10.0
Neohesperidin	> 10.0	> 16.0	> 10.0

[a] After 24 hours; from Dreyer and Jones (1981).
[b] Growth inhibition after 12 days; Elliger et al. (1980).

of a plant where toxins are largely contained in non-plasmatic cell compartments such as vacuoles (Matile, 1984; McKey, 1979). Hence, all of the numerous allelochemical groups including flavonoids and other phenolics, alkaloids, terpenoids, glucosinolates, cyanogenic glycosides, saponins, etc., would be ingested. Suckers, in contrast, feed selectively on plant tissues and often avoid the vacuolated parenchyma of plants.

4.1. Phloem toxicology

The chemistry of phloem is important to phloem-feeders since phloem is their primary source of nutrition. Much evidence indicates that considerable proton gradients exist between phloem and other plant tissues. Phloem is alkaline (average pH of 8), whereas xylem (pH 5), mesophyll (pH 6), and other external plant tissues are usually slightly acidic (Fife and Frampton, 1936; Giaquinta, 1983; Smith and Raven, 1979; Ziegler, 1975). A proton gradient of this magnitude (1.5 to 3 pH units) could drive solutes across the phloem membrane. In particular, ionizable compounds that are mostly neutral at pH 5 yet ionized at pH 8 should selectively pass through the plasmalemma membranes into the phloem. For example, a weak acid with a pKa of 7 would be only one percent dissociated at pH 5, yet 90 percent dissociated at pH 8. It is not surprising that weak acids such as carboxylic acids are often translocated by phloem (Crisp and Larson, 1983; Ziegler, 1975). This "ion trap" mechanism has been proposed to drive the passive absorption of acidic herbicides from the apoplast into the more alkaline phloem (Crisp and Larson, 1983; Jacob and Neumann, 1983; Giaquinta, 1985).

Much of our knowledge concerning physicochemical features directing phloem mobility comes from developmental work on systemic pesticides. Early studies with organophosphate insecticides (Geary, 1953) indicated the importance of moderate lipophilicity (i.e. some water solubility) for systemic effect. Structure-activity studies with herbicides have clearly shown that a dissociable acid group is optimal for phloem transport (Crisp and Larson, 1983; Ashton and Crafts, 1981; Chamberlain et al., 1984). Highly phloem-mobile herbicides include aliphatic carboxylic acids such as 2,4-D, dalapon, and glyphosate, aromatic carboxylic acids such as amiben and picloram, and other weak acids such as maleic hydrazide. Most of these compounds exhibit pKa values or pI values (polyprotic acids) of 3 to 6.

Phenols are another major group of weak organic acids. Phenolic compounds are second only to carbohydrates in abundance in plants (Levin, 1971; Harborne and Mabry, 1982), and comprise many simple phenols, phenylpropanoids, flavonoids, coumarins, stilbenes, and quinones. Although phenol has a pKa of 9.9, many of these substituted phenolics are extensively conjugated and rendered considerably more acidic (pKa 7-9) by electron withdrawing groups (Rappoport, 1967; Kennedy et al., 1984). Fragmentary evidence, mostly by indirect aphid stylet (see later) and honeydew analysis, indicates that phenolics with and without carboxylic acid groups, including flavonoids, are translocated in phloem (Hussain et al., 1974; Macleod and Pridham, 1965; Ziegler, 1975). Polar flavonoids such as catechins and proanthocyanidins are readily found in the phloem of woody species (Schmid and Feucht, 1981; Hemingway et al., 1981). Schultz (1969) did not observe mobility of a nonpolar flavonoid biochanin A, although the more polar naringenin and liquiritigenin were apparently translocatable in clover.

The presence of an ionizable group and/or relatively high water solubility are not always essential for phloem mobility. Even lipophilic terpenoids such as phytosterols and gibberellins are phloem mobile in small amounts, and detectable in the excretory honeydew of aphids (Dixon, 1975; Forrest and Knights, 1972).

Flavonoids and related phenolic compounds are the most often cited class of plant allelochemicals associated with both plant resistance to and host acceptance by homopterans. Flavonoids are found universally in higher plants with about 3000 compounds characterized. They are extremely abundant with concentrations in plant tissue averaging five percent of dry weight, and ranging from 0.3 to 30 percent of dry weight depending on the plant part (Harborne and Mabry, 1982). These chemicals act as strong feeding deterrents to many insects, Table 2, but stimulatory effects are sometimes observed also. Various rice flavonoids were probing but not sucking stimulants for the brown planthopper, *Nilaparvata lugens* (Stal.) (Homoptera: Delphacidae) (Sogawa, 1976). The dihydrochalcone glucoside phlorizin from *Malus* was unusual in stimulating or having no effect on feeding by the apple-specialist *Aphis pomi* DeGeer (Homoptera: Aphididae), although non-specialized aphids such as *M. persicae* and *Acrythosiphon pisum* were strongly deterred by this chemical (Klingauf, 1971; Montgomery and Arn, 1974). Strong inverse correlations between total phenolic, including flavonoid, concentrations and the suitability of a leaf or tree for galling aphids have also been noted (Zucker, 1982). Moreover, aphid feeding strongly elicits increased concentrations of the isoflavonoid coumestrol in the herbage of *Medicago sativa* (Loper, 1968, Kain and Biggs, 1980).

These studies indicate a defensive role for flavonoids and related phenolics against homopteran herbivory. However, failures in early attempts to verify phloem mobility of these allelochemicals by indirect methods such as aphid whole body and honeydew analyses (Edwards, 1965; Montgomery and Arn, 1974), and grafting experiments (Schultz, 1969) have led some to suggest that flavonoids are not translocatable at concentrations high enough to influence aphid-plant interactions (cf. McClure, 1975; Dreyer et al., 1981). Many of these studies lacked sensitive methods to detect the compounds and/or did not consider that many flavonoids and related phenolics are readily oxidizable either in air or by phenol oxidases, and perhaps would not survive transport across grafts or through an aphid body. Other investigations (Macleod and Pridham, 1975; Hussain et al., 1974; Ziegler, 1975; see above) have indicated that at least some flavonoids are phloem-mobile.

Relatively high concentrations of allelochemicals are usually required to elicit feeding deterrency, lethality, or other biological effects in aphids, whereas phloem concentrations of these chemicals are thought to be lower. However, sampling phloem from plants is difficult (Ziegler, 1975) and is often accomplished by severing the stylets of feeding aphids and then collecting the sap exudate. A sufficient volume of phloem to detect low levels of plant allelochemicals is rarely available, and the dynamics of phloem-loading of these chemicals is largely unknown. Thus, the relevance of phloem-derived allelochemicals as a defensive strategy against phytophagous homopterans remains debatable.

Aphids consume high volumes of phloem sap and are wasteful feeders (Kennedy and Fosbrooke, 1972, Dixon, 1975, 1985) with low efficiencies for conversion of ingested food, compared to chewing insects. Low levels of allelochemicals in the sap could impact on the fitness of this insect group. In addition, aphids must contend with the phytochemical defenses they inadvertently encounter while reaching the sieve elements. For example, flavonoids are common components of trichome exudates (Stipanovic, 1983), and stylet penetration to the phloem proceeds through many flavonoid-rich plant tissues including the cuticle (Baker et al., 1982) and vacuolated mesophyll (McKey, 1979; Harborne and Mabry, 1982).

Many studies have indicated the importance of flavonoids and other phenolics in limiting herbivory by chewing arthropods (Harborne, 1979; Swain, 1979; Waiss et al., 1977). Both acute and subacute effects including growth inhibition, feeding deterrency, and reduced fecundity have been noted in lepidopterans (Shaver and Lukefahr, 1969; Elliger et al. 1980), coleopterans (Sutherland et al., 1980), and mites (Larson and Berry, 1984). Correspondingly, increased levels of specific flavonoids correlate with host plant resistance to several lepidopteran and dipteran insects (Hedin et al., 1983; Pathak and Dale, 1983; Chiang and Norris, 1984). However, as in sucking insects (Sogawa, 1976; see above), flavonoids may sometimes stimulate feeding and growth in chewing insects (McFarlane and Distler, 1982; Nielsen et al., 1979).

4.2. Chemical ecology of aphids

Familiarity with the chemicals a herbivore group is exposed to provides clues to metabolic strategies that may characterize a peculiar feeding niche. Aphids elaborate a variety of functional chemicals that are largely limited to Aphididae and related phloem-feeding homopterans, Fig. 1. Included among these are unusual pigments called aphins that comprise up to two percent of the live weight of the aphid. These polycyclic phenolic quinones (C_{30} polyketides) occur mainly as β-glucosides (Brown, 1975; Banks and Cameron, 1973; Todd, 1963). Most aphid cornicles readily exude defensive secretions that contain rare triglycerides and the alarm pheromone trans-β-farnesene (Nault et al., 1976). Up to one-third of the weight of aphids can consist of triglycerides containing exclusively the short chain fatty acid myristate, and the highly unusual sorbic acid (Brown, 1975; Fallon and Shimizu, 1977). The exceptional content of myristate in aphids, compared to other insects (Fast, 1970; Thompson, 1973), is due to an unusual acylthioesterase that terminates fatty acid synthesis after 14 carbons (Ryan et al., 1982). This is most remarkable in light of the excess carbon content of aphid food. Aphid triglycerides are one of the few natural sources of the fungistatic sorbic acid in

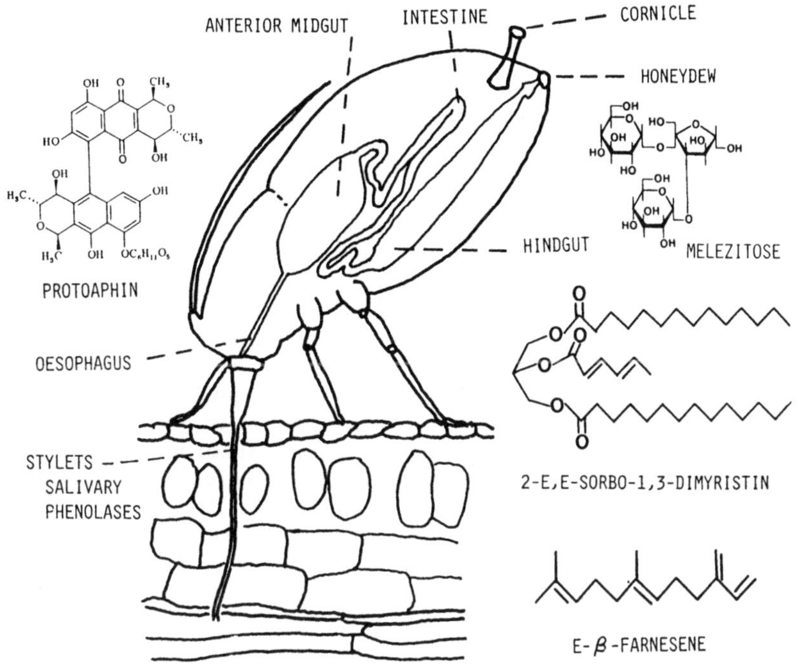

Fig. 1. Chemical ecology of an aphid as a model phloem herbivore.

either the plant or animal kingdom (Luck, 1971). The incorporation of sorbate-containing triglycerides into defensive secretions may be a mechanism to void this fungistat, and thus preserve microbial symbionts, or a means of keeping the honeydew-laden environment of the aphid sterile. The highly unusual branched-trisaccaride melezitose (3^f-glucosylsucrose) is a frequent major component of aphid honeydew (Brown, 1975).

Aphids synthesize pigmentary phenolic glycosides, trisaccharides, triglycerides, and a sesquiterpenoid alarm pheromone that are either unique to this animal group or peculiar to phloem-imbibing insects. This may reflect metabolic adaptations not only to excess carbon but also to phenolic defensive chemicals within the phloem. Their distinctive chemical ecology and the nonbehavioral strategies they may use to avoid the deleterious effects of plant allelochemicals will now be discussed by way of contrast to chewing herbivores.

5. ADAPTIVE DETOXIFICATION

Investigations on chewing insect herbivores are reviewed by Ahmad et al. (Chapter 3 in this text). Possible mechanisms that sucking insects use to deal with plant allelochemicals include sequestration in nonessential tissues such as fat, efficient excretion of the imbibed chemical, and metabolic detoxification. Aphids may use microbial symbionts rather than solely endogenous enzymes to detoxify compounds such as flavonoids and other phenolic compounds. However, one of the earliest biological activities recognized for this group of compounds is their antimicrobial activity (cf. Harborne and Mabry, 1982; Mabry and Ulubelen, 1980), and thus the role of microbes in detoxification may be limited.

5.1. Sequestration

Sequestration of toxicants into less essential tissues may be an energetically favorable mechanism for an aphid to avoid low concentrations of dietary toxicants, since it has a short generation time and high fecundity. If only low concentrations accumulated, aphid populations would suffer less than longer-lived species, if stored toxicants were suddenly mobilized out of fat deposits at a time of physiological stress. Aphids can readily sequester plant toxins (Duffey, 1080). For example, the aposematic oleander aphid, Aphis nerii Fonscholombe (Homoptera: Aphididae), sequesters cardiac glycosides (Rothschild et al., 1970; Botha et al., 1977), A. pisum stores phenolics (Dixon, 1975), Aphis cytisorum Hartig (Homoptera: Aphididae) sequesters quinolizidine alkaloids (Wink et al., 1982), and many species store carotenoids (Brown, 1975; Czeczuga, 1976). Chewing herbivorous insects are able to incorporate dietary phenolics into their cuticles during sclerotization apparently to conserve tyrosine (Bernays et al., 1983; Bernays and Woodhead, 1982), although this might also be an adaptation to sequester toxicants in general. Aphids are soft-bodied, and lack extensive areas of hardened cuticle, including their crop (Goodchild, 1966), and thus would probably not incorporate phenolics into cuticle as a primary mechanism of phenolic detoxification. However, aphids do exhibit distinctive metabolic capabilities including the synthesis of polycyclic phenolic quinones as pigments (Brown, 1975; Banks and Cameron, 1973; Todd, 1963). The biosynthesis of these acetate-malonate (polyketide) derived pigments is poorly understood, and endosymbionts might be involved (Brown, 1975). Dietary phenolics may be detoxified in aphids by oxidative coupling to form pigments; this idea needs experimental testing.

Work with herbivores of Asclepias spp. and Nerium oleander do not indicate major differences between chewing (Seiber et al., 1984; Brower et al., 1984) and sucking insects (Rothschild, 1972; Duffey, 1980) in cardenolide sequestration. Nevertheless, qualitative differences have been noted. For example, Rothschild et al. (1970) could not find the major foliar cardenolide of Nerium oleander in the phloem-feeder A. nerii, but a minor epoxycardenolide, adynerin, was readily apparent. This cardenolide was not present in an oleander-feeding lygaeid (Rothschild, 1972). These results probably reflect differences in the cardenolide chemistry of the phloem, leaf, and seed tissues.

Flavonoid profiles are sufficiently diversified to constitute a taxonomic fingerprint that can be used to distinguish apparently morphologically identical plant species (Classen et al., 1982). To investigate the dynamics between dietary consumption of flavonoids and their resultant metabolic fate in insects, we have developed a two-dimensional thin-layer chromatographic (tlc) method to obtain profiles of flavonoid and phenolic aglycones from the insect or plant (Mullin and Reeves, unpubl.). Briefly, 30 mg of fresh tissue is homogenized and hydrolyzed in 2 N hydrochloric acid (0.5 ml) for 30 min at 100°C, filtered through 1 μm regenerated cellulose, and then extracted first by petroleum ether followed by ethyl acetate. A concentrate of the ethyl acetate extract in ethanol is spotted on silica gel 60 plates and chromatographed with four solvent systems as shown in Fig. 2. In each dimension, the first solvent system (A or C) is developed to four cm followed by development to eight cm with system B or D. Spot patterns are analyzed by mobilities, UV quench at 254 nm, fluorescence at 366 nm, and interaction with various flavonoid spray reagents including diphenylboric acid - ethanolamine and ferric chloride - potassium ferricyanide (Markham, 1982; Harborne and Mabry, 1982).

Results using this method indicate higher overlap of aphid flavonoid and phenolic profiles (seven spots) than of the western corn rootworm extract (four spots) with corn leaf compounds, Fig. 3. This indicates that sucking herbivores are at least equally exposed to phytochemicals as chewing herbivores, although concentrations may differ. Profiles for plants (10 species) and aphids (three species) are readily distinguishable. Host-related differences in phenolic profiles in the potato aphid,

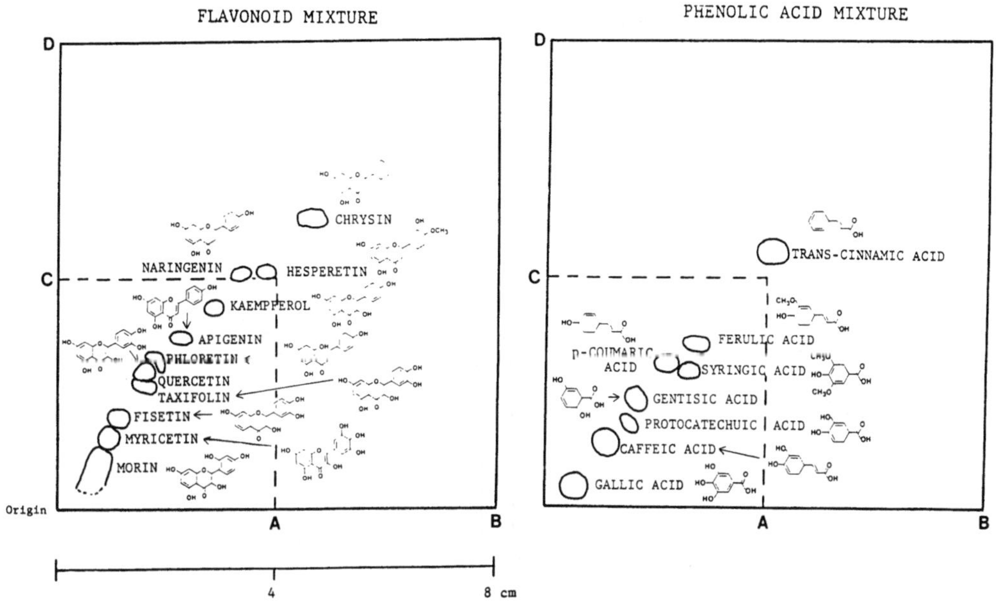

Fig. 2. Two-dimensional silica gel thin-layer chromatography of reference flavonoid and phenolic compounds; A = toluene, ethyl acetate, formic acid (60:40:2.5), B = Chloroform, ethyl acetate, methanol (90:10:1), C = toluene, ethyl acetate, acetic acid (60:40:2.5), D = toluene, chloroform, ethyl acetate, methanol (65:15:15:5).

Macrosiphum euphorbiae (Thomas) (Homoptera: Aphididae), as well as different profiles between this generalist and the specialist (A. nerii) aphid on milkweed, Fig. 4, suggest that these species adapt differently to dietary phytochemicals. The plant species tested were readily accepted by these primarily phloem-feeding species, and it is unlikely that much feeding was occurring outside of the phloem tissue, although this possibility exists in other aphid-plant associations (Raven, 1983; van Emden et al., 1969; van Emden, 1972; Lowe, 1967). Indeed, the specialist species A. nerii is primarily a phloem-feeder, and is known to sequester defensive phytochemicals such as cardenolides (Botha et al., 1977). These results strongly indicate that aphids are imbibing and sequestering phloem-derived phenolics.

5.2. Excretion

Aphids are unlikely to have active excretory mechanisms for dietary toxicants. All aphids lack Malpighian tubules, and most, except the larger species, do not possess a filter chamber as a means of transferring excess water directly from the anterior to the posterior of the gut (Auclair, 1963; Goodchild, 1966). Copious amounts of honeydew are excreted due to the large volumes of carbon-rich and nitrogen-poor phloem imbibed to satisfy nutrient needs. Much of the osmoregulation to maintain a hypotonic blood that is not diluted by excess dietary water with a resulting loss of useful solutes is performed passively in the midgut. While aphids can excrete phenolic compounds via their honeydew (cf. Dixon, 1975), efficient excretion of these compounds has been noted only in chewing herbivores such as lepidopterous larvae (Isman and Duffey, 1983). Interestingly, glucose transport across Malpighian tubules in chewing insects is strongly inhibited by the flavonoids phlorizin and phloretin (Rafaeli-Bernstein and Mordue, 1979); perhaps aphids are less sensitive to this inhibition since the tubules are absent.

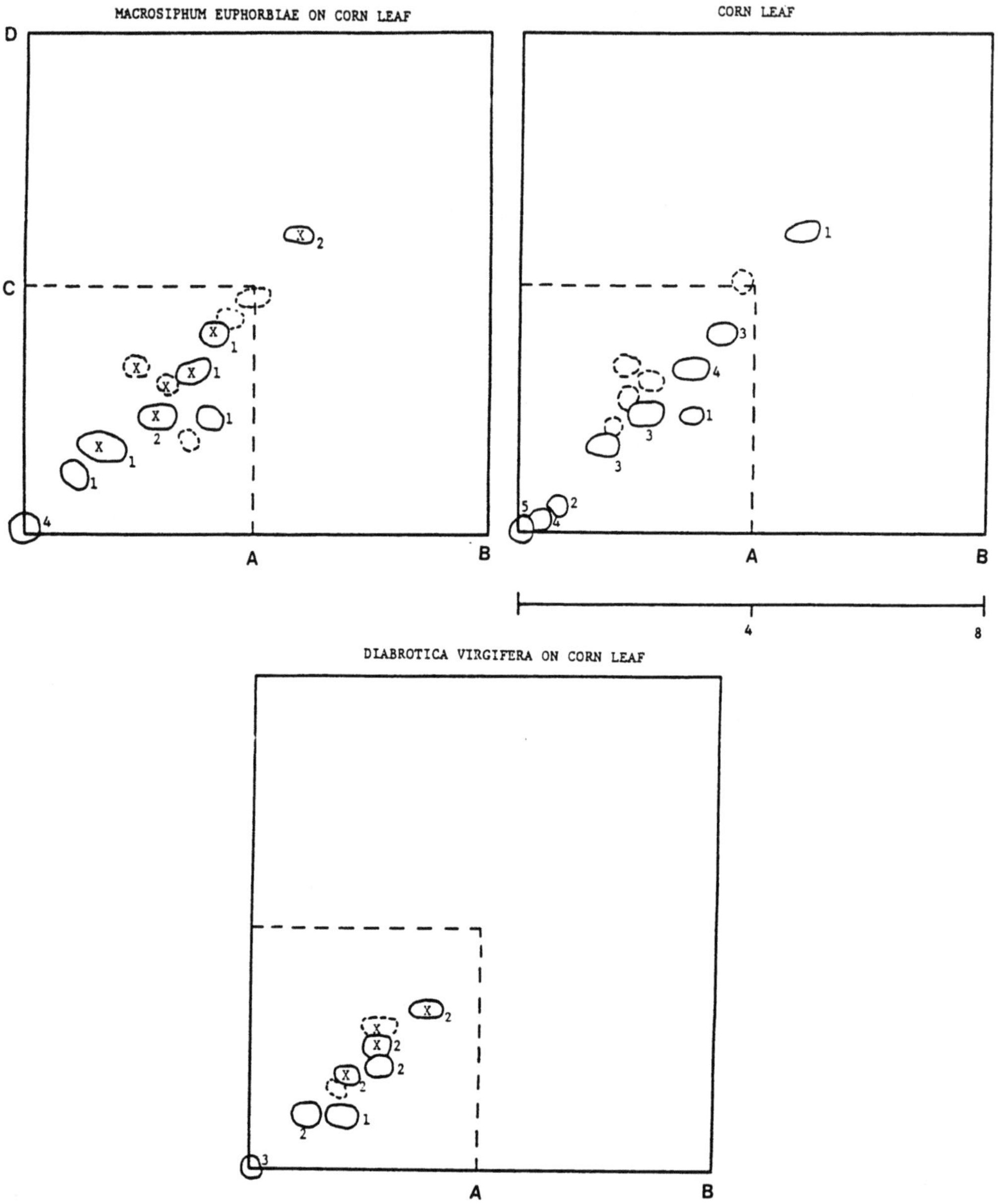

Fig. 3. Flavonoid and phenolic profiles in corn leaf, potato aphid (*M. euphorbiae*), and the western corn rootworm (*D. virgifera*). Increasing UV quench at 254 nm (1 to 5) or fluorescence at 366 nm (broken lines) shown on chromatogram; X = apparent cochromatography between insect and host plant. Solvent abbreviations as in Fig. 2.

Cornicles might also serve as an excretory organ in aphids. However, to date, only defensive secretions including unusual triglycerides and the alarm pheromone trans-β-farnesene (Nault et al., 1976) have been found. Nevertheless, the copious triglyceride secretions from cornicles might be an adaptation to divert excess acetate resulting from the carbohydrate-rich content of aphid food.

Efficient excretion of alkaloids has been observed in tobacco-feeding insects (Self et al., 1964). Ingested nicotine is excreted ultra-rapidly

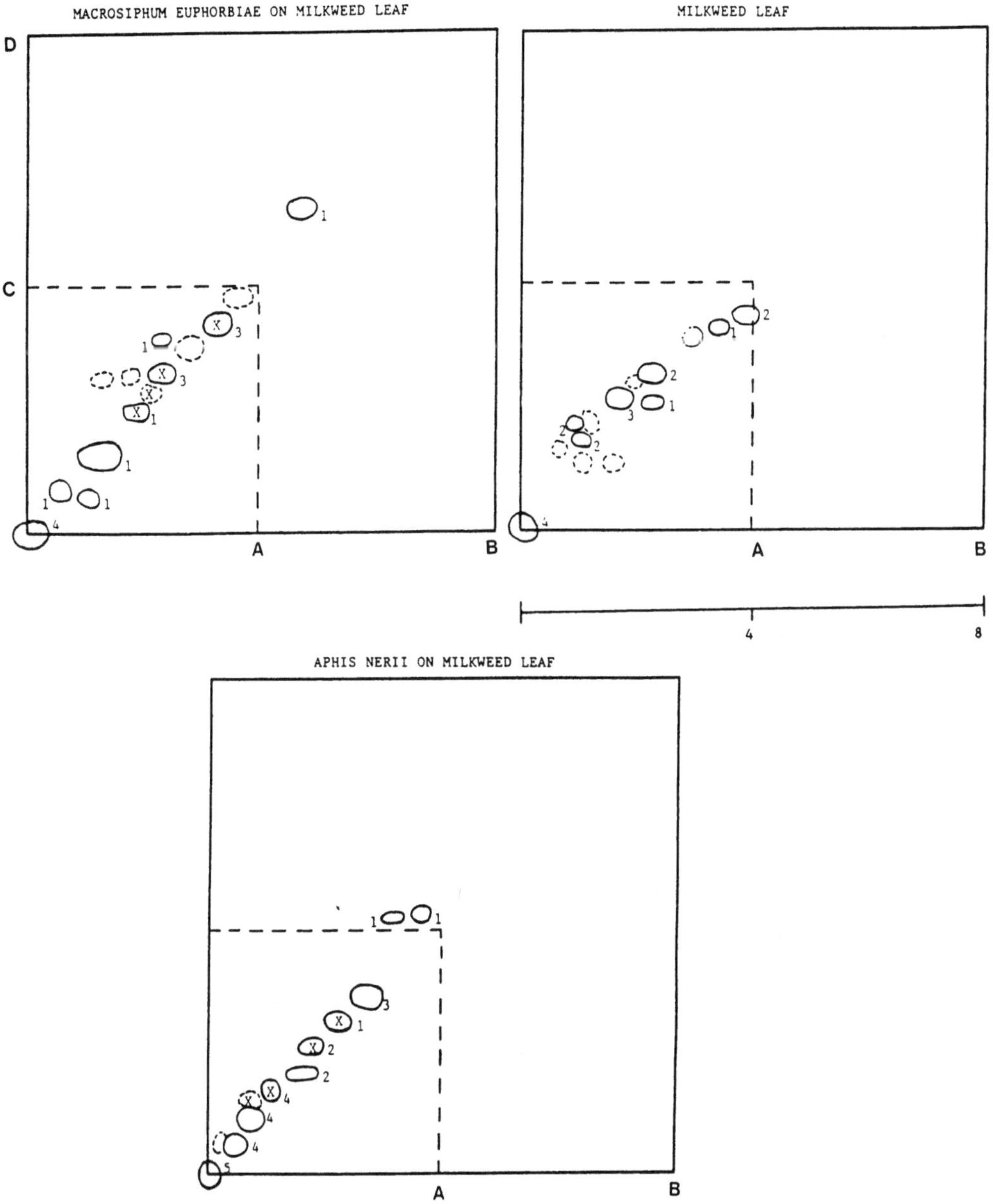

Fig. 4. Flavonoid and phenolic profiles in milkweed leaf, the generalist (M. euphorbiae), and specialist (A. nerii) aphids that had fed on milkweed. See legend for Fig. 3. Solvent abbreviations as in Fig. 2.

in the tobacco hornworm, Manduca sexta (L.) (Lepidoptera: Sphingidae), and two other chewing herbivores (e.g., 90 percent in fours hours), whereas M. persicae avoids the xylem-translocated nicotine in tobacco by selectively feeding on phloem.

5.3. Metabolism of plant allelochemicals

Much of what is known about arthropod detoxification of plant allelochemicals concerns chewing herbivores (Ahmad et al., Chapter 3 in this text; Brattsten, 1979b; Dowd et al., 1983). Thus, an attempt will be made here to concentrate on work that relates to sucking insects.

The metabolic fate of the flavonoids in aphids, as in other insects, is largely unknown. Preliminary studies by Dreyer and Jones (1981) with two flavonoid glycosides fed to S. graminum via artificial diet did not result in detectable phenolics in the honeydew, indicating that degradation or sequestration had occurred. In contrast, only small amounts of the ingested glycosides chlorogenic acid and rutin were degraded in Heliothis zea, and the majority was excreted unchanged (Isman and Duffey, 1983). Oxidative and hydrolytic routes are available in cockroaches and houseflies for metabolism of the isoflavonoid rotenone (Fukami et al., 1967; cf. Dowd et al., 1983). Lepidoptera can readily sequester flavonoids in external structures such as wings (Harborne, 1979).

Detoxification of flavonoids might even precede their ingestion into aphids and other sucking insects. Polyphenol oxidases (phenolases, tyrosinases) and pectinases (McAllan and Adams, 1961; Campbell and Dreyer, 1985) are characteristic of the saliva from phloem-sucking insects, and are secreted in high concentrations along with the stylet-sheath precursors (Miles, 1972). The oxidase could be involved in hardening of the sheath, converting free tryptophan to the plant hormone indoleacetic acid, or in the detoxification of dietary phenolics (Miles, 1978). Several flavonoids have been detected in aphid saliva (Miles, 1972; Schaller, 1968) and, presumably, they are plant derived. It may be interesting to study if phenolics are incorporated into the stylet-sheath as an alternative detoxification strategy. Chewing herbivores, in contrast, generally macerate plant tissues thereby releasing plastid phenol oxidases that may neutralize flavonoids released from vacuoles during ingestion. Hence, salivary oxidases are not as essential to a chewing herbivore as to a phloem-feeder.

Metabolism has a crucial role on the allatocidal activity of precocenes. Low rates of detoxification in peripheral tissues allows oxidative bioactivation to a cytotoxic epoxide in the corpora allata; this appears to determine species sensitivity to these plant chromenes (Bergot et al., 1980; Bowers, 1982; Pratt, 1983). Many of the insects most sensitive to precocenes are sucking species, and not holometabolous chewing herbivores.

Unusual differences in steroid metabolism have been associated with feeding guilds. Phytophagous hemipterans including O. fasciatus and Dysdercus cingulatus (F.) (Hemiptera: Pyrrhocoridae) are unable to dealkylate the C24 position of phytosterols readily to produce the dietary requirement of cholesterol for C-27 ecdysteroid biosynthesis. To compensate, they use makisterone A, a C-28 ecdysteroid, as molting hormone (Svoboda et al., 1984). Chewing phytophages easily dealkylate dietary phytosterols via reactions catalyzed by cytochrome P-450 to satisfy steroid needs.

Recently, the aphid, S. graminum was found capable of dealkylating sitosterol to cholesterol. The ability of aphids to sequester and utilize phytosterols suggests that sterol synthesis via symbionts is not obligatory (Campbell and Nes, 1983). Obvious differences in the metabolism of the cardenolides, another steroid group, by chewing and sucking herbivores are not indicated at present (cf. Duffey, 1980; Dowd et al., 1983; Marty and Krieger, 1984).

5.4. Metabolism of pesticides and other synthetic toxicants

The role of arthropod detoxification enzymes in pesticide metabolism is known largely for chewing herbivores and saprophagous insects (Wilkinson and Brattsten, 1972; Dauterman, 1983; Hodgson, 1983; Kulkarni and Hodgson, 1984). Direct measurement of cytochrome P-450 activities in phloem-sucking insects is lacking. However, in vivo bioassays with synergists indicate that aphids and some heteropterans have low levels of the activities (Brattsten and Metcalf, 1970; Al-Rajhi, 1982; Osman and Brindley, 1981). Nevertheless, hydrolytic pathways within piercing-sucking herbivores appear well developed. Myzus persicae can develop resistance to organophosphate, carbamate, and pyrethroid insecticides through increased amounts of carboxylesterase (Sawicki et al., 1978; Devonshire and Moores, 1982; Wachendorff and Klingauf, 1978). Esterases for trans-pyrethroids are also highly active in preparations from O. fasciatus (Jao and Casida, 1974b), and may explain, in part, the general higher tolerance of these organisms over chewing herbivores for pyrethroids (Casida et al., 1983; Soderlund et al., 1983b). A low level of epoxide hydrolase, another hydrolytic enzyme, has also been noted in M. persicae (Mullin and Croft, 1984). Moreover, Brevicoryne brassicae, an aphid specialist on glucosinolate-containing host plants, has high glucosinolase activity, presumably, for hydrolyzing these toxins (MacGibbon and Allison, 1971). Demethylation of two organophosphonates, probably by a carboxylesterase with phosphatase activity, was reported to occur faster in sucking insects than in chewing insects (Saito, 1969). Glutathione conjugation with arylhalides, on the other hand, is very low in the cotton stainer, Dysdercus sp., compared to some chewing herbivores (Cohen et al., 1964), although glucose conjugation in Dysdercus and some aphid species is better developed (Smith, 1955; 1968). In contrast, hydrolysis of glucose conjugates catalyzed by β-glucosidase is particularly active in piercing-sucking insects, and is probably responsible for the selective activation of glucosidic juvenogens in Dysdercus and A. pisum (Slama and Romanuk, 1976; Slama et al., 1978; Robinson, 1956).

We examined in vitro enzyme activities in preparations from two broadly polyphagous aphids, M. euphorbiae and M. persicae, and one specialist, A. nerii. Aphids were collected onto 243 μm controlled pore nylon mesh filters and then homogenized with a Teflon pestle in a smooth glass tube in 0.15 M potassium phosphate and 50 mM sucrose buffer, pH 7.4, at 15 percent (w/v). Four detoxification enzyme activities were measured in the post-mitochondrial supernatant fraction from late nymph and fundatrix homogenates by using analytical methods established for other arthropod whole body preparations (Mullin and Croft, 1983; Mullin et al., 1984). The cytochrome P-450-dependent epoxidation of aldrin to dieldrin was followed by electron-capture gas-liquid chromatography; epoxide hydration of trans-β-ethylstyrene oxide and cis-stilbene oxide by radiometric isooctane partitioning; and hydrolysis of 1-naphthyl acetate by diazo coupling of 1-naphthol with Fast Blue B.

Aphids had low levels of both NADPH-dependent aldrin epoxidation (E.C.1.14.14.1) and trans-epoxide hydrolase (E.C.3.3.2.3) activity, Table 3. This is the first in vitro measurement of a cytochrome P-450 catalyzed activity in aphids; epoxide hydrolase has been measured previously (Mullin and Croft, 1984). Recently, the in vivo conversion of aldrin to dieldrin was detected in M. persicae (R. Wadleigh, pers. commun., 1985). Levels of epoxidase and trans-epoxide hydrolase were on the average 20 and 80 times lower, respectively, in these aphids than in chewing herbivores that often cohabit the same host plant (Mullin, 1985). These results support the view that phloem-feeders have less enzyme reserves than chewing herbivores for dealing with plant allelochemicals, since concentrations of defensive chemicals in phloem are probably lower than in external tissues. Nevertheless, comparison of the highly polyphagous M. euphorbiae and M. persicae with the oleander aphid, a specialist on hosts from Asclepiadaceae and Apocynaceae, indicates that the 19 times higher epoxidation rate and five to 14 times higher epoxide hydrolase activity in the generalists correlate with increasing encounter with phytochemicals, as estimated by host plant range, Table 3. The association of both aldrin epoxidation (Krieger et al., 1971; Brattsten, 1979a; Yu, 1983) and trans-epoxide hydrolase

Table 3. Detoxification enzymes in phloem-feeding insects.

Species	Enzyme activity (pmol/min, mg protein)					
	Aldrin	Epoxide hydrolase			Esterase	Plant Families
	Epoxidation	trans	cis	trans/cis	(X 10^{-3})	Consumed
Green Peach Aphid	1.36[a]	177[a]	817[a]	0.22[a]	433[a]	72
Potato Aphid	2.01[a]	220[a]	1032[a]	0.21[a]	171[b]	30
Oleander Aphid	0.09[b]	38[b]	66[b]	0.57[a]	226[b]	2

Species with same letter are not significantly different at $p < 0.05$. Host plant range based mainly on Patch (1938).

activity (Mullin and Croft, 1984) with herbivory in chewing insects has been previously demonstrated.

The maintenance of a tightly coupled cytochrome P-450-epoxide hydrolase complex may have important consequences for arthropod herbivores (Mullin et al., 1984). Cytochrome P-450 can produce harmful epoxides from olefins (Brooks, 1977; Casida, 1983). Epoxide hydrolase detoxifies the epoxide through the hydrolytic addition of water. Organisms often invest a great deal of nutrient resources into an olefin-to-diol pathway, for detoxification or defensive purposes. Indeed, plants assemble cuticular layers containing copolymers of epoxidized and polyhydroxylated fatty acids using the same enzyme pathways, Fig. 5, presumably to protect against invading pathogens and insect grazing (Kolattukudy, 1980). These protective epidermal layers may contain up to 60 percent of monomeric epoxy fatty acids and considerable quantities of phenolic acids (Kolattukudy, 1976; Holloway and Deas, 1973). Such barriers would have greater consequences for chewing than sucking herbivores.

Epoxides and their olefinic precursors, many of which are toxic per se, are common constituents of plants (Cross, 1960; Dean, 1963; Schoental, 1976). These phytochemicals often exhibit trans-geometry, or are higher substituted epoxides and olefins, while animals preferably biosynthesize cis-olefins (cf. Harwood, 1980; Luckner, 1984; Mullin, 1985). Due to the multiplicity of epoxide hydrolases available to insects (Brooks, 1977; Hammock and Quistad, 1981), plant feeders may have isozymes more selective for trans- and higher substituted epoxides than for cis-epoxides. This was verified recently by comparison of herbivores and carnivores (Mullin et al., 1982; Mullin and Croft, 1984). Chewing herbivores (20 species) had on average 21 times higher trans-selective epoxide hydrolase than carnivorous arthropods (seven species). Besides fatty acid epoxides, terpenoid epoxides are also widely occurring in the plant kingdom (Brattsten, 1983; Seaman, 1982) and readily consumed by chewing herbivores. The terpenoid epoxide pulegone-1,2-oxide was recently isolated as the insecticidal principle of *Lippia stoechadifolia* (Grundy and Still, 1985). Neither of these phytochemical groups would be readily translocated in phloem because of their high lipophilicity. This may, in part, explain the low trans-epoxide hydrolase characteristic of phloem-sucking insects. This enzyme is also used by insects to degrade the sesquiterpenoid juvenile hormones (JH); preliminary studies do not indicate major differences in JH epoxide degradation between chewing and sucking

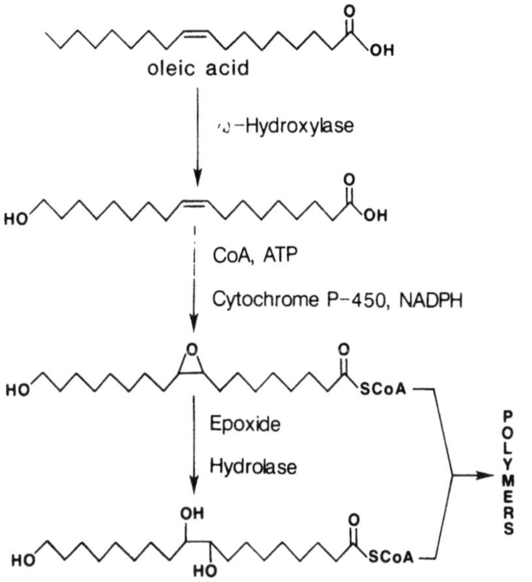

Fig. 5. Biosynthetic pathway for cutin. Based on Kolattukudy (1980).

herbivores or in entomophagous arthropods (Hammock and Quistad, 1981; Ajami and Riddiford, 1973).

Chewing herbivores which probably consume more olefinic and epoxide phytochemicals than phloem-suckers would benefit more from a well-developed cytochrome P-450 - epoxide hydrolase complex. Evidence for this was obtained recently in our laboratory, Table 4. The western, Diabrotica virgifera L. (Coleoptera: Chrysomelidae), and northern, D. barberi Smith and Lawrence (ibid.), corn rootworms were compared with the phloem-sucking potato aphid. These chewing chrysomelids and the aphid readily cohabit the same corn leaf. It was evident that the chewers were much better endowed with the appropriate activities that the aphid. Moreover, the northern corn rootworm with a 330-fold greater susceptibility to aldrin than the western species, had rate-limiting epoxide hydrolase activity, i.e. greater cytochrome P-450 to cis-hydrolase ratio, Table 4. Since the normal course of aldrin detoxification proceeds through the equally or more toxic epoxide, dieldrin (Brooks, 1977), these results suggest the possibility that the northern corn rootworm has insufficient epoxide hydrolase activity to dispose of the excess dieldrin produced by its higher cytochrome P-450 activity. The low susceptibility of the potato aphid to aldrin is not easily explained solely by its low cytochrome P-450 content. Perhaps other toxicodynamic features such as penetration or target action are responsible for this insensitivity. Obviously, more robust comparisons in detoxification-susceptibility profiles would be obtained between species of similar morphologies. Carbofuran toxicity, however, was similar to the rootworms and the aphid by topical application, Table 4.

Levels of general carboxylesterase activity in the aphids, Table 3, were similar to those in other arthropods including both chewing herbivores and carnivores (Mullin, 1985). The use of these enzymes in some aspects of digestion (cf. Mullin and Croft, 1983) may explain why aphids compare favorably with other insects in their ability to hydrolyze potentially toxic carboxylesters.

To further explore host plant - aphid enzyme relationships, we reared potato and green peach aphids on a variety of host plants for at least one generation, and then measured detoxification enzyme activities.

Table 4. Enzyme activities and insecticide susceptibilities of insects feeding on corn leaves.

Species	Enzyme Activity (nmol/min-mg protein)[a]				LD50 (μg/g insect)[b]	
	Epoxidation ($X 10^3$)	Epoxide Hydrolase		Epoxidase/cis-EH		
		trans	cis		Aldrin	Carbofuran
Western corn rootworm	41[d]	24.1[d]	1.46[cd]	28.0	1980[d]	1.16[c]
Northern corn rootworm	156[e]	34.9[d]	2.21[d]	71.0	6.0[c]	1.05[c]
Potato aphid	3.0[c]	0.22[c]	1.03[c]	2.9	>1300[d]	3.1[d]

[a] Assays on S15,000g of gut homogenates for adult rootworms, and whole body homogenates of aphids. Species with same letter are not significantly different at $p < 0.01$.

[b] Rootworms were fed squash and aphids maintained on parafilm sachets containing buffered 35% sucrose (pH 8) during bioassay. Mortality was assessed 24h after topical application.

Significant ($p < 0.05$; Student's two-tailed t test), although moderate, host-related changes of aldrin epoxidation activities were observed; potato aphids reared on potato had two-fold higher activity than when reared on milkweed, _Asclepias syriaca_. Aphicides such as carbofuran and dinoseb were 1.5 to 2.4 times more toxic to milkweed-reared potato aphids than to those reared on potato or corn. This indicates a role for cytochrome P-450 in the detoxification of these aphicides (Mullin and Reeves, unpubl.). Greater host plant effects on the efficacy of aphicides have been observed when chemicals are applied to aphid-infested plant tissue (Richardson and Casanges, 1942; Selander et al., 1972); these results may be due to differences in chemical mobility imposed by the plant tissue and/or the physicochemical properties of the compounds. When aphids were sprayed separate from the plant, more moderate differences were obvious (Juneja and Sharma, 1973).

Celery enhanced levels of esterase in both potato and green peach aphids relative to other host plants, Table 5. A similar esterase responsiveness to umbellifers in the green peach aphid (Wachendorff and Klingauf, 1978) and a chewing-sucking type mite _Tetranychus urticae_ Koch (Acari: Tetranychidae), Table 5, has been found. The potential involvement of coumarins (Berenbaum, 1983), internal cinnamic acid esters common to umbellifers, with these interactions needs to be explored, since substituted coumarins have chemical features that may impart phloem solubility. Both trans- and cis-epoxide hydrolases in potato aphid (seven plant species), and oleander aphid (two plant species) were not significantly modulated by the host plant (Mullin and Reeves, unpubl.). These host-related differences in detoxification enzyme levels are generally much less than those reported in noctuid larvae (Brattsten, 1979a; Yu, 1984) and Coleoptera (Ahmad, 1983). For example, more than a four-fold increase in aldrin epoxidation was found in _Spodoptera eridania_ fed carrot relative to lima bean, while an almost 32-fold increase in glutathione tranferase, another epoxide-metabolizing enzyme, occurred in _S. frugiperda_ fed parsley relative to soybeans, Table 5. Although plant allelochemicals clearly induce detoxification enzymes, they may also impair the overall fitness of the herbivore thereby reducing its ability to cope with additional toxicant exposures (Gordon, 1961; Beck and Reese, 1976; Scriber, Chapter 2 in this text). Even a reduction in lipid reserves may increase the susceptibility of an insect to an insecticide such as DDT due to loss of sequestration sites (cf. Fast, 1964).

5.5. Alternative detoxification strategies

Aphids may resort to strategies other than conventional detoxification enzymes (see above) to deal with dietary toxicants. Indeed, degradation may begin in the host plant via aphid salivary secretions which are known to contain tyrosinase (E.C.1.10.3.1) (Miles, 1972; Sen Gupta and Miles, 1975). Other oxidative enzymes including peroxidases (E.C. 1.11.1.7), laccases (E.C.1.10.3.2), and dioxygenases causing ring-fission (E.C.1.13.11.) may be used by aphids to degrade flavonoids and other phenolics, as is the case in microbes, plants and higher animals (Luckner, 1984; Barz and Koster, 1981; Griffiths, 1982). Similar oxidative enzymes are used to incorporate phenolic compounds into the insect cuticle (Andersen, 1979).

Increased phloem-solubility may be imparted by glycosylation of an allelochemical. Toxicities of flavonoid and phenolic compounds to aphids could be metabolically regulated by interconversion between the usually more toxic aglycone and its glycoside (Juneja et al., 1975). Thus, glycosidases may be important. The ß-glucosidases (E.C.3.2.1.21) are more active in herbivores (Blum, 1981), particularly the piercing-sucking types, than carnivores. We have found low levels of ß-glucosidase and high

Table 5. Umbelliferae-dependent increase in arthropod lipophile-mobilizing enzymes.

Host	% of Activity on nonresponsive diet[a]			
Arthropod Species	Aldrin Epoxidation	General Esterase	Glutathione Transferase	cis-Epoxide Hydrolase
Carrot				
Spodoptera eridania[b]	415*	---	---	---
Spodoptera frugiperda[c]	---	---	164*	---
Tetranychus urticae	---	223*	---	199*
Celery				
Tetranychus urticae[d]	111	236*	---	178*
Macrosiphum euphorbiae[e]	128	166*	---	---
Myzus persicae	100	209*	---	---
Parsley				
Spodoptera eridania[b]	39*	---	---	---
Spodoptera frugiperda[c]	---	---	3186*	---

[a] Relative to artificial or legume-based diets; *Different from non-responsive diet at $p < 0.05$.
[b] Brattsten (1979a).
[c] Yu (1984).
[d] Mullin and Croft (1983).
[e] Mullin, Reeves, and Steighner, unpubl.

levels of α-glucosidase (E.C.3.2.1.20) in potato aphid homogenates by measuring, in vitro, the release of p-nitrophenol from the respective glucosides (Halvorson and Ellias, 1958); these activities have also been detected in the pea aphid, A. pisum (Auclair, 1965; Macleod and Pridham, 1965). Polyphenolic flavonoids, which inhibit α-glycosidases more than β-glycosidases (Iio et al., 1984), may have important implications for co-adaptation of aphids to their host plants. Aphid pigments are phenolic β-glucosides (Cameron and Craik, 1968). Aphid food contains excess sugar relative to nitrogen and they conjugate, presumably, excess sucrose with glucose to form melezitose, a major honeydew component. Therefore, glucose conjugating and releasing enzymes should be more utilized than amino acid and peptide conjugating enzymes such as glutathione transferases in aphids compared to chewing insects. In support of this, we assayed the post-mitochondrial supernatant fraction of potato aphid for glutathione transferase (E.C.2.5.1.18) with the substrate 1,2-dichloro-4-nitrobenzene and found very low activity (<0.9 nmol/min, mg protein).

Biosynthesis of polyketide pigments by aphids and other homopterans are processes more typical of microorganisms and higher plants (Luckner, 1984; Gibson, 1984; Brown, 1975) than animals. Sterol and other lipid biosynthetic processes in aphids have been attributed to prokaryotic symbiotes (Noda and Mittler, 1983; Houk et al., 1976). Perhaps metabolism of flavonoids and related compounds in aphids is mediated by gut flora or intracellular microorganisms within mycetocytes. If aphids and other phloem-sucking insects rely on symbiotes more than chewing herbivores for detoxification, this could have important implications for their selective control.

6. TARGET SITE SELECTIVITIES

The contrast in insecticide susceptibility between sucking and chewing insects may be, in part, due to differing attributes of cholinergic nerve sites. The acetycholinesterase of aphids has a markedly different substrate specificity (Casida, 1955) and sensitivity to sulfhydryl group modifiers (Smissaert, 1976; Brestkin et al., 1985) than that characteristic of chewing herbivores or other insects. Moreover, cholinergic pesticides that are ionizable at physiological pH, and thus can not cross the ion-impermeable membranes ensheathing the insect central nervous system, are normally poor insecticides. Included among these are the cationic organophosphates such as amiton, and the nicotine alkaloids. Schradan, which is converted to polar metabolites that fail to penetrate the nerve sheath, is unexpectedly toxic to some arthropods including mites, aphids and some hemipterans. Attempts to find ion permeability differences in the CNS of an amiton and schradan sensitive (aphid) and insensitive (cockroach) insect failed. The alternative possibility that aphids have peripheral cholinergic sites has been suggested (Toppozada and O'Brien, 1967; O'Brien, 1967). Toxicant sensitivity at target sites may also be partly regulated by detoxification within nerve tissue as is indicated for dieldrin (Brooks, 1977), pyrethroids (Soderlund et al., 1983a) and nicotine (Morris, 1983). Hence, caution is required in assigning target insensitivity based on experiments with crude insect preparations.

7. SUMMARY

It is clear that sucking insects may markedly diverge from other insects in the metabolic strategies used to deal with plant defensive chemistry and chemical control agents. Indeed, phloem-imbibers often become the primary pests in agroecosystems where chemical control of the formerly predominant chewing pests is achieved. Perhaps the contrast can best be seen with aphids. They adapt metabolically to plants in ways that greatly differ from those of other insects, particularly chewing herbivores, as suggested by their unusual chemistry. Aphids synthesize pigmentary phenolic glycosides, trisaccharides such as melezitose, myristic- and sorbic acid-containing triglycerides, and a sesquiterpenoid alarm pheromone that are either unique to this animal group or peculiar to phloem-imbibing insects. Their distinctive chemical ecology is also reflected by the differences in insecticide susceptibility compared to chewing, holometabolous insects. This is perhaps, in part, due to differing attributes of cholinergic nerve sites in aphids relative to other insects.

Plant defensive chemicals are thought to be allocated mostly to specialized organelles or tissues of external structures, and not to vascular tissue. However, phloem loading and translocation of chemicals within plants is poorly understood. Chewing herbivores such as lepidopterous larvae and coleopterans are apparently exposed to higher concentrations of plant toxicants than phloem-sucking counterparts such as aphids. Thus, metabolic adaptations to toxic chemicals should be better developed in chewing relative to sucking herbivores. This is indicated by the generally higher susceptibility of sucking herbivores to conventional insecticides compared to chewing ones. Also, chewers have a much better developed cytochrome P-450 - epoxide hydration metabolic pathway than phloem-sucking insects. This may have important implications for chemical control of herbivorous pests since this detoxification pathway may be blocked be selective inhibition of epoxide hydrolase.

Sucking herbivores, nonetheless, must adapt to plant toxicants. Flavonoids and related phenolics are the major group of allelochemicals that aphids encounter. It is probable that these abundant compounds are predisposed by their weak acidity and polarity to loading into the alkaline phloem. Consumers of phloem should be adaptively distinct from chewing insects which feed on acidic portions of plants. Our studies demonstrate that aphids contain considerable amounts of these compounds, presumably, as a consequence of phloem ingestion. It is likely that aphids resort to alternatives to conventional enzymes for some of their detoxification needs. Possibilities include various polyphenolic oxidases and sequestration. Study in metabolic coadaptations of aphids with their host plants should help identify enzyme systems crucial for aphid herbivory.

8. ACKNOWLEDGEMENT

The technical support of E. M. Reeves, R. J. Steigner and B. J. Morrow and the support of the National Science Foundation (grant BSR-8306008), United States Department of Agriculture (82-CRSR-2-2057) and a Pennsylvania State University Research Initiation Grant are much appreciated. This is paper No. 7323 of the Pennsylvania Agricultural Experiment Station journal series.

9. REFERENCES

Ahmad, S., 1983. Mixed-function oxidase activity in a generalist herbivore in relation to its biology, food plants, and feeding history, Ecology, 64:235-243.

Ajami, A. M., and L. M. Riddiford, 1973. Comparative metabolism of the Cecropia juvenile hormone, J. Insect Physiol., 19:635-645.

Al-Rajhi, D. H., 1982. "Estimating esterase and monoxygenase detoxification by bioassay, electrophoresis and metabolic studies in pea aphid, Acyrthosiphon pisum (Harris) populations (Homoptera: Aphididae)", Ph.D. Thesis, Utah State University, Logan.

Andersen, S. O., 1979. Biochemistry of the insect cuticle, Annu. Rev. Entomol., 24:29-61.

Argandona, V. H., L. J. Corcuera, H. M. Niemeyer, and B. C. Campbell, 1983. Toxicity and feeding deterrency of hydroxamic acids from Graminae in synthetic diets against the greenbug, Schizaphis graminum, Entomol. Exp. Appl., 34:134-138.

Ashton, F. M., and A. S. Crafts, 1981. "Mode of Action of Herbicides", Wiley, New York.

Auclair, J. L., 1963. Aphid feeding and nutrition, Annu. Rev. Entomol., 8:439-490.

Auclair, J. L., 1965. Feeding and nutrition of the pea aphid, Acyrthosiphon pisum (Homoptera: Aphidae), on chemically defined diets of various pH and nutrient levels, Ann. Ent. Soc. Am., 58:855-875.

Ayad, H. M., and T. N. Wheeler, 1984. Synthesis and biological activity of pyrethroids derived from halo-4-alkenoic acids, J. Agric. Food Chem., 32:85-92.

Baker, E. A., M. J. Bukovac, and G. M. Hunt, 1982. Composition of tomato fruit cuticle as related to fruit growth and development, in: "The Plant Cuticle", D. F. Cutler, K. L. Alvin and C. E. Price, eds., pp. 33-44, Academic Press, New York.

Banks, H. J., and D. W. Cameron, 1973. Phenolic glycosides and pterins from the Homoptera, Insect Biochem., 3:139-162.

Barz, W., and J. Koster, 1981. Turnover and degradation of secondary (natural) products, in: "The Biochemistry of Plants", E. E. Conn, ed., Vol. 7, p. 35-84, Academic Press, New York.

Beck, S. D., 1965. Resistance of plants to insects, Annu. Rev. Entomol. 10:207-232.

Beck, S. D., and J. C. Reese. 1976. Insect-plant interactions: nutrition and metabolism, Rec. Adv. Phytochem., 10:41-92.

Bell, W. J., and R. T. Carde, 1984. "Chemical Ecology of Insects", Chapman and Hall, London.

Berenbaum, M., 1983. Coumarins and caterpillars: a case for coevolution, Evolution, 37:163-179.

Bergot, B. J., K. J. Judy, D. A. Schooley, and L. W. Tsai, 1980. Precocene II metabolism: comparatie in vivo studies among several species of insects, and structure elucidation of two major metabolites, Pestic. Biochem. Physiol., 13:95-104.

Bernays, E. A., D. J. Chamberlain, and S. Woodhead, 1983. Phenols as nutrients for a phytophagous insect Anacridium melanorhodon, J. Insect Physiol., 29:535-539.

Bernays, E. A., and S. Woodhead, 1982. Plant phenols utilized as nutrients by a phytophagous insect, Science, 216:201-203.

Blum, M. S., 1981. "Chemical Defenses of Arthropods", Academic Press, New York.

Botha, C. E. J., S. B. Malcolm, and R. F. Evert. 1977. An investigation of preferential feeding habits in four Asclepiadaceae by the aphid, Aphis nerii B. de F., Protoplasma, 92:1-19.

Bowers, W. S., 1982. Endocrine strategies for insect control, Entomol. Exp. Appl., 31:3-14.

Brattsten, L. B., 1979a. Ecological significance of mixed-function oxidations, Drug Metab. Rev., 10:35-58.

Brattsten, L. B., 1979b. Biochemical defense mechanisms in herbivores against plant allelochemicals, in: "Herbivores, Their Interaction with Secondary Plant Metabolites", G. A. Rosenthal and D. H. Janzen, eds., pp. 199-270, Academic Press, New York.

Brattsten, L. B., 1983. Cytochrome P-450 involvement in the interactions between plant terpenes and insect herbivores, in: "Plant Resistance to Insects", P. A. Hedin, ed., pp. 173-195, Symp. Ser. No. 208, Amer. Chem. Soc., Washington.

Brattsten, L. B., and R. L. Metcalf, 1970. The synergistic ratio of carbaryl with piperonyl butoxide as an indicator of the distribution of multifunction oxidases in the Insecta, J. Econ. Entomol., 63:101-104.

Brattsten, L. B., and R. L. Metcalf, 1973. Synergism of carbaryl toxicity in natural insect populations, J. Econ. Entomol., 66:1347-1348.

Brestkin, A. P., E. B. Maizel, S. N. Moralev, K. V. Novozhilov, and I. N. Sazonova, 1985. Cholinesterases of aphids - 1. Isolation, partial purification and some properties of cholinesterases from spring grain aphid Schizaphis graminum (Rond.), Insect Biochem., 15:309-314.

Brooks, G. T., 1977. Epoxide hydratase as a modifier of biotransformation and biological activity, Gen. Pharmac., 8:221-226.

Brooks, G. T., 1979. The metabolism of xenobiotics in insects, in: "Progress in Drug Metabolism", J. W. Bridges and L. F. Chasseaud, eds., Vol. 3, pp. 151-214, Wiley, New York.

Brower, L. P., J. N. Seiber, C. J. Nelson, S. P. Lynch, and M. M. Holland, 1984. Plant-determined variation in the cardenolide content, thin-layer chromatography profiles, and emetic potency of Monarch butterflies, Danaus plexippus L. reared on milkweed plants in California: 2. Asclepias speciosa, J. Chem. Ecol., 10:601-639.

Brown, K. S., 1975. The chemistry of aphids and scale insects, Chem. Soc. Rev., 4:263-288.

Busvine, J. R., 1942. Relative toxicity of insecticides, Nature, 150:208-209.

Cameron, D. W., and J. C. A. Craik, 1968. Colouring matters of the Aphididae. Part 36. The configuration of the glucoside linkage in protoaphins, J. Chem. Soc., (C):3068-3072.

Campbell, B. C., and D. L. Dreyer, 1985. Host-plant resistance of sorghum: Differential hydrolysis of sorghum pectic substances by polysaccharases of greenbug biotypes (Schizaphis graminum, Homoptera: Aphididae), Arch. Insect Biochem. Physiol., 2:203-215.

Campbell, B. C., and W. D. Nes, 1983. A reappraisal of sterol biosynthesis and metabolism in aphids, J. Insect Physiol., 29:149-156.

Campbell, B. C., B. G. Chan, L. L. Creasy, D. L. Dreyer, L. B. Rabin, and A. C. Waiss, Jr., 1984. Bioregulation of host plant resistance to insects, in: "Bioregulators, Chemistry and Uses", R. L. Ory and F. R. Rittig, eds., pp. 193-203, Symp. Ser. No. 257, Amer. Chem. Soc., Washington.

Casida, J. E., 1955. Comparative enzymology of certain insect acetylesterases in relation to poisoning by organophosphorus insecticides, Biochem. J., 60:487-496.

Casida, J. E., 1983. Propesticides: bioactivation in pesticide design and toxicological evaluation, in: "Pesticide Chemistry: Human Welfare and the Environment. Mode of Action, Metabolism and Toxicology", S. Matsunaka, D. H. Hutson, and S. D. Murphy, eds., Vol. 3, pp. 239-246, Pergamon Press, New York.

Casida, J. E., D. W. Gammon, A. H. Glickman, and L. J. Lawrence, 1983. Mechanisms of selective action of pyrethroid insecticides, Annu. Rev. Pharmacol. Toxicol., 23:413-438.

Chamberlain, K., M. M. Burrell, D. N. Butcher, and J. C. White, 1984.

Phloem transport of xenobiotics in Ricinus communis var. Gibsonii., Pestic. Sci., 15:1-8.

Chiang, H.-S., and D. M. Norris, 1984. "Purple Stem", a new indicator of soybean stem resistance to bean flies (Diptera: Agromyzidae), J. Econ. Entomol., 77:121-125.

Classen, D., C. Nozzolillo, and E. Small, 1982. A phenolic-taxometric study of Medicago (Leguminosae), Can. J. Bot., 60:2477-2495.

Cohen, A. J., J. N. Smith, and H. Turbert, 1964. Comparative detoxification 10. The enzymic conjugation of chloro compounds with glutathione in locusts and other insects, Biochem. J., 90:457-464.

Crisp, C. E., and J. E. Larson, 1983. Effect of ring substituents on phloem transport and metabolism of phenoxyacetic acid and six analogues in soybean (Glycine max), in: "Pesticide Chemistry: Human Welfare and the Environment", P. Doyle and T. Fujita, eds. Vol. 1, pp. 213-222, Pergamon Press, New York.

Croft, B. A., and M. E. Whalon, 1982. Selective toxicity of pyrethroid insecticides to arthropod natural enemies and pests of agricultural crops, Entomophaga, 27:3-21.

Cross, A. D., 1960. The chemistry of naturally occurring 1,2-epoxides, Quart. Rev. Chem. Soc., Lond., 14:317-335.

Czeczuga, B., 1976. Investigations on the carotenoids in nineteen species of aphids and their host plants, Zoolog. Polon., 25:27-45.

Dauterman, W. C., 1983. Role of hydrolases and glutathione S-transferases in insecticide resistance, in: "Pest Resistance to Pesticides", G. P. Georghiou and T. Saito, eds., pp. 229-247, Plenum Publ. Corp., New York.

Dean, F. M., 1963. "Naturally Occurring Oxygen Ring Compounds", Butterworths, London.

Devonshire, A. L., and G. D. Moores, 1982. A carboxylesterase with broad substrate specificity causes organophosphorus, carbamate and pyrethroid resistance in peach-potato aphids (Myzus persicae), Pestic. Biochem. Physiol., 18:235-246.

Dixon, A. F. G., 1975. Aphids and translocation, in: "Transport in Plants I. Phloem Transport", M. H Zimmermann and J. A. Milburn, eds., pp. 154-170, Springer Verlag, New York.

Dixon, A. F. G., 1985. "Aphid Ecology", Blackie, London.

Dowd, P. F., C. M. Smith, and T. C Sparks, 1983. Detoxification of plant toxins by insects, Insect Biochem., 13:453-468.

Dreyer, D. L., and K. C. Jones, 1981. Feeding deterrency of flavonoids and related phenolics towards Schizaphis graminum and Myzus persicae: aphid feeding deterrents in wheat, Phytochemistry, 20:2489-2493.

Dreyer, D. L., J. C. Reese, and K. C. Jones, 1981. Aphid feeding deterrents in sorghum. Bioassay, isolation, and characterization, J. Chem. Ecol., 7:273-284.

D'Silva, T. D., J. A. Durden, Jr., A. A. Sousa, and M. H. J. Weiden, 1985.

Novel insecticidal-miticial cyclic dithiacarbamoyloximes, J. Agric. Food Chem., 33:110-115.

Duffey, S. S., 1980. Sequestration of plant natural products by insects, Annu. Rev. Entomol., 25:447-477.

Edwards, J. S., 1965. On the use of gut characters to determine the origin of migrating aphids, Ann. Appl. Biol., 55:485-494.

Elliger, C. A., B. C. Chan, and A. C. Waiss, Jr., 1980. Flavonoids as larval growth inhibitors. Structural features governing toxicity, Naturwiss., 67:358-360.

Fallon, W. E., and Y. Shimizu, 1977. Sorbic acid containing triglycerides in aphids and their fractionation by high pressure liquid chromatography, Lipids, 12:765-768.

Fast, P. G., 1964. Insect lipids: a review, Mem. Entomol. Soc. Can., 37:1-50.

Fast, P. G., 1970. Insect lipids, Prog. Chem. Fats Other Lipids, 11:181-242.

Feeny, P. P., 1976. Plant apparency and chemical defense, Rec. Adv. Phytochem., 10:1-40.

Fife, J. M., and V. L. Frampton, 1936. The pH gradient extending from the phloem into the parenchyma of the sugar beet and its relation to the feeding behavior of Eutettix tenellus, J. Agric. Res., 53:581-593.

Fisk, J., 1980. Effects of HCN, phenolic acids and related compounds in Sorghum bicolor on the feeding behavior of the planthopper, Peregrinus maidis, Entomol. Exp. Appl., 27:211-222.

Forrest, J. M. S., and B. A. Knights, 1972. Presence of phytosterols in the food of the aphid, Myzus persicae, J. Insect Physiol., 18:723-728.

Fukami, J.-I., I. Yamamoto, and J. E. Casida, 1967. Metabolism of rotenone in vitro by tissue homogenates from mammals and insects, Science, 155:713-716.

Geary, R. J., 1953. Development of organic phosphates as systemic insecticides, J. Agric. Food Chem., 1:880-882.

Giaquinta, R. T., 1983. Phloem loading of sucrose, Annu. Rev. Plant Physiol., 34:347-387.

Giaquinta, R. T., 1985. Physiological basis of phloem transport of agrichemicals, in: Bioregulators for Pest Control", P. A. Hedin, ed., Symp. Ser. No. 276, pp. 7-18, Amer. Chem. Soc., Washington.

Gibson, D. T., ed., 1984. "Microbial Degradation of Organic Compounds", Marcel Dekker, New York.

Gibson, R. W., and R. T. Plumb, 1977. Breeding plants for resistance to aphid infestation, in: Aphids as Virus Vectors", K. F. Harris and K. Maramorosch, ed., pp. 473-500, Academic Press, New York.

Goodchild, A. J. P., 1966. Evolution of the alimentary canal in the hemiptera, Biol. Rev., 41:97-140.

Gordon, H. T., 1961. Nutritional factors in insect resistance to chemicals, Annu. Rev. Entomol., 6:27-54.

Gould, F., 1983. Genetics of plant-herbivore systems: interactions between applied and basic study, in: "Variable Plants and Herbivores in Natural and Managed Systems", R. F. Denno and M. S. McClure, eds., pp. 599-653, Academic Press, New York.

Griffiths, L. A., 1982. Mammalian metabolism of flavonoids, in: The Flavonoids: Advances in Research", J. B. Harborne and T. J. Mabry, eds. pp. 681-718, Chapman and Hall, London.

Grundy, D. L., and C. C. Still, 1985. Isolation and identification of the major insecticidal compound of poleo (Lippia stoechadifolia), Pestic. Biochem. Physiol., 23:378-382.

Guthrie, F. E., W. V. Campbell, and R. L. Baron, 1962. Feeding sites of the green peach aphid with respect to its adaptation to tobacco, Ann. Entomol. Soc. Amer., 55:42-46.

Halvorson, H., and L. Ellias, 1958. The purification and properties of an α-glucosidase of Saccharomyces italicus Y 1225, Biochim. Biophys. Acta., 30:28-40.

Hammock, B. D., and G. B. Quistad, 1981. Metabolism and mode of action of juvenile hormone, juvenoids, and other insect growth regulators, Prog. Pestic. Biochem., 1:1-83.

Hammock, B. D., E. Kuwano, A. Ketterman, R. H. Scheffrahn, S. N. Thompson, and D. Sallume, 1978. Acute toxicity and developmental effects of analogues of ethyl α-(4-chlorophenoxy)-α-methylproprionate on two insects, Oncopeltus fasciatus and Tenebrio molitor, J. Agric. Food Chem., 26:166-170.

Harborne, J. B., 1979. Flavonoid pigments, in: "Herbivores, Their Interaction with Secondary Plant Metabolites", G. A. Rosenthal and D. H. Janzen, eds., pp. 619-655, Academic Press, New York.

Harborne, J. B., and T. J. Mabry, eds., 1982. "The Flavonoids: Advances in Research", Chapman and Hall, London.

Harris, M. K., ed., 1979. "Biology and Breeding for Resistance to Arthropods and Pathogens in Agricultural Plants", Texas Agri. Extp. Sta., Misc. Publ. No. 1451.

Hart, S. V., M. Kogan, and J. D. Paxton, 1983. Effect of soybean phytoalexins on the herbivorous insects Mexican bean beetle and soybean looper, J. Chem. Ecol., 9:657-672.

Harwood, J. L., 1980. Plant acyl lipids: Structure, distribution and analysis, in: The Biochemistry of Plants", P. K. Stumpf, ed., Vol. 4, pp. 1-55, Academic Press, New York.

Hedin, P. A., J. N. Jenkins, and F. G. Maxwell, 1977. Behavioral and developmental factors affecting host plant resistance to insects, in: "Host Plant Resistance to Insects", P. A. Hedin, ed., pp. 231-275, Symp. Ser. No. 62, Amer. Chem. Soc., Washington.

Hedin, P. A., J. N. Jenkins, D. H. Collum, W. H. White, W. L. Parrott, and M. W. MacGown, 1983. Cyanidin-3-β-glucoside, a newly recognized basis for resistance in cotton to the tobacco budworm Heliothis virescens (Fab.), Experientia, 39:799-801.

Hemingway, R. W., L. Y. Foo, and L. J. Porter, 1981. Polymeric proanthocyanidins: interflavonoid linkage isomerism in (epicatechin-4)-(epicatechin-4)-catechin procyanidins, J. Chem. Soc. Chem. Commun., 320-322.

Henrick, C. A., 1982. Juvenile hormone analogs: structure-activity relationships, in: Insecticide Mode of Action", J. R. Coats, eds., pp. 315-402, Academic Pres, New York.

Hodgson, E., 1983. The significance of cytochrome P-450 in insects, Insect Biochem.,13:237-246.

Hollingworth, R. M., 1976. The biochemical and physiological basis of selective toxicity, in: "Insecticide Biochemistry and Physiology", C. F. Wilkinson, eds., pp. 431-506, Plenum Publ. Corp., New York.

Holloway, P. J., and A. H. B. Deas, 1973. Epoxyoctadecanoic acids in plant cutins and suberins, Phytochemistry, 12:1721-1735.

Houk, E. J., G. W. Griffiths, and S. D. Beck, 1976. Lipid metabolism in the symbiotes of the pea aphid, Acyrthosiphon pisum, Comp. Biochem. Physiol. 54B:427-431.

Huber, L. L., and G. H. Stringfield, 1940. Strain susceptibility to the European corn borer and the corn leaf aphid in maize, Science, 92:172.

Hussain, A., J. M. S. Forrest, and A. F. G. Dixon, 1974. Sugar, organic acid, phenolic acid and plant growth regulator content of extracts of honeydew of the aphid Myzus persicae and of its host plant, Raphanus sativus, Ann. Appl. Biol., 78:65-73.

Iio, M., A. Yoshioka, Y. Imayoshi, C. Koriyama, and A. Moriyama, 1984. Effects of flavonoids of α-glucosidase and β-fructosidase from yeast, Agr. Biol. Chem., 48:1559-1563.

Isman, M. B., and S. S. Duffey, 1983. Pharmacokinetics of chlorogenic acid and rutin in larvae of Heliothis zea, J. Insect Physiol., 29:295-300.

Jacob, F., and S. Neumann, 1983. Quantitative determination of mobility of xenobiotics and criteria of their phloem and xylem mobility, in: "Pesticide Chemistry: Human Welfare and the Environment", P. Doyle and T. Fujita, eds., Vol. 1, pp. 357-362, Pergamon Press, New York.

Jao, L. T., and J. E. Casida, 1974a. Esterase inhibitors as synergists for (+)-trans-chrysanthemate insecticide chemicals, Pestic. Biochem. Physiol., 4:456-464.

Jao, L. T., and J. E. Casida, 1974b. Insect pyrethroid-hydrolyzing esterases, Pestic. Biochem. Physiol., 4:465-472.

Junde, Q., and K. Lidao, 1984. The influence of secondary plant substances on the growth and development of Myzus persicae of Beijing, Entomol. Exp. Appl., 35:17-20.

Juneja, P. S., S. C. Pearce, R. K. Gholson, R. L. Burton, and K. J. Starks, 1975. Chemical basis for greenbug resistance in small grains. II. Identification of the major neural metabolite of benzyl alcohol in barley, Plant Physiol., 56:385-389.

Juneja, V. K., and J. C. Sharma, 1973. Effect of host plants on the susceptibility of Aphis gossypii Glover to certain insecticides, Indian J. Ent., 35:179-186.

Kain, W. M., and D. R. Biggs, 1980. Effect of pea aphid and bluegreen lucerne aphid (Acyrthosiphon sp.) on coumestrol levels in herbage of lucerne (Medicago sativa), N. Z. J. Agric. Res., 23:563-568.

Kennedy, G. G., 1978. Recent advances in insecticide resistance of vegetable and fruit crops in North America: 1966-1977, Bull. Entomol. Soc. Am., 24:375-384.

Kennedy, J. A., M. H. G. Munro, H. K. J. Powell, L. J. Porter, and L. Y. Foo, 1984. The protonation reactions of catechin, epicatechin and related compounds, Aust. J. Chem., 37:885-892.

Kennedy, J. S., and I. H. M. Fosbrooke, 1972. The plant in the life of an aphid, in: "Insect/Plant Relationships", H. F. van Emden, ed., pp. 129-140, Blackwell Scientific Publ., Oxford.

Klingauf, F., 1971. Die wirkung des glucosids phlorizin auf das wirtswahlverhalten von Rhopalosiphum insertum (Walk.) und Aphis pomi DeGeer (Homoptera: Aphididae), Z. Ang. Ent., 68:41-55.

Klun, J. A., C. L. Tipton, and T. A. Brindley, 1967. 2,4-Dihydroxy-7-methoxy-1,4-benzoxazin-3-one (DIMBOA), an active agent in the resistance of maize to European corn borer, J. Econ. Entomol., 60:1529-1533.

Kogan, M., and J. Paxton, 1983. Natural inducers of plant resistance to insects, in: "Plant Resistance to Insects", P. A. Hedin, ed., pp. 153-171, Symp., Ser. No. 208, Amer. Chem. Soc., Washington.

Kolattukudy, P. E., 1976. Lipid polymers and associated phenols, their chemistry, biosynthesis, and role in pathogenesis, Rec. Adv. Phytochem., 11:185-246.

Kolattukudy, P. E., 1980. Biopolyester membranes of plants: cutin and suberin, Science, 208:990-1000.

Krieger, R. I., P. P. Feeny, and C. F. Wilkinson, 1971. Detoxication enzymes in the guts of caterpillars: an evolutionary answer to plant defenses? Science, 172:579-581.

Kulkarni, A. P., and E. Hodgson, 1984. The metabolism of insecticides: the role of monooxygenase enzymes, Annu. Rev. Pharmacol. Toxicol., 24:19-42.

Kumar, K., and R. Lal, 1966. Comparative toxicity of some recently introduced organic insecticides to some insect pests of crops, Indian J. Entomol., 28:258-264.

Larson, K. C., and R. E. Berry. 1984. Influence of peppermint phenolics and monoterpenes on twospotted spider mite (Acari: Tetranychidae), Environ. Entomol., 13:282-285.

Levin, D. A., 1971. Plant phenolics: an ecological perspective, Amer. Nat., 105:157-181.

Long, B. J., G. M. Dunn, J. S. Bowman, and D. G. Routley, 1977. Relationship of hydroxamic acid content in corn and resistance to the corn leaf aphid, Crop Sci., 17:55-58.

Loper, G. M., 1968. Effect of aphid infestation on the coumestrol content of alfalfa varieties differing in aphid resistance, Crop Sci., 8:104-106.

Lowe, H. J. B., 1967. Interspecific differences in the biology of aphids (Homoptera: Aphididae) on leaves of Vicia faba. I. Feeding behavior, Entomol. Exp. Appl., 10:347-357.

Luck, E., 1972. "Sorbinsaure, Vol. 2, Biochemie-Mikrobiologie", 127 pp., Beh's Verlag, Hamburg.

Luckner, M., 1984. "Secondary Metabolism in Microorganisms, Plants, and Animals", 576 pp., Springer-Verlag, Berlin.

Mabry, T. J., and A. Ulubelen, 1980. Chemistry and utilization of phenylpropanoids including flavonoids, coumarins, and lignans, J. Agric. Food Chem., 28:188-196.

MacGibbon, D. B., and R. M. Allison, 1971. An electrophoretic separation of cabbage aphid and plant glucosinolases, N. Z. J. Sci., 14:134-140.

Macleod, N. J., and J. B. Pridham, 1965. Observations on the translocation of phenolic compounds, Phytochemistry, 5:777-781.

Markham, K. R., 1982. "Techniques of Flavonid Identification", 113 pp., Academic Press, New York.

Marty, M. A., and R. I. Krieger, 1984. Metabolism of uscharidin, a milkweed cardenolide, by tissue homogenates of Monarch butterfly larvae, Danaus plexippus L., J. Chem. Ecol., 10:945-956.

Matile, P., 1984. Das toxische kompartiment der pflanzenzelle, Naturwiss., 71:18-24.

Maxwell, F. G., and P. R. Jennings, 1980. "Breeding Plants Resistant to Insects", 683 pp., Wiley Interscience Publ., New York.

McAllan, J. W., and J. B. Adams, 1961. The significance of pectinase in plant penetration by aphids, Can. J. Zool., 39:305-310.

McClure, J. W., 1975. Physiology and functions of flavonoids, in: "The Flavonoids", J. B. Harborne, T. J. Mabry and H. Mabry, eds., Part 2, pp. 970-1055, Academic Press, New York.

McFarlane, J. E., and M. H. W. Distler, 1982. The effect of rutin on growth, fecundity and food utilization in Acheta domesticus (L.), J. Insect Physiol., 28:85-88.

McKey, D., 1979. The distribution of secondary compounds within plants, in: "Herbivores, Their Interaction with Secondary Plant Metabolites", G. A. Rosenthal and D. H. Janzen, eds., pp. 199-270, Academic Press, New York.

Miles, P. W., 1972. The saliva of Hemiptera, Adv. Insect Physiol., 9, 183-255.

Miles, P. W., 1978. Redox reactions of hemipterous saliva in plant tissues, Entomol. Exp. Appl., 24:534-539.

Montgomery, M. E., and H. Arn, 1974. Feeding response of Aphis pomi, Myzus persicae and Amphorophora agathonica to phlorizin, J. Insect Physiol., 20:413-421.

Morris, C. E., 1983. Uptake and metabolism of nicotine by the CNS of a nicotine-resistant insect, the tobacco hornworm (Manduca sexta), J. Insect Physiol., 29:807-817.

Mullin, C. A., 1985. Detoxification enzyme relationships in arthropods of differing feeding strategies in: "Bioregulators for Pest Control", P. A. Hedin, ed., pp. 267-278, Symp. Ser. No. 276, Amer. Chem. Soc., Washington.

Mullin, C. A., and B. A. Croft, 1983. Host-related alterations of detoxification enzymes in Tetranychus urticae (Acari: Tetranychidae), Environ. Entomol., 12:1278-1282.

Mullin, C. A., and B. A. Croft, 1984. Trans-epoxide hydrolase: a key indicator enzyme for herbivory in arthropods, Experientia, 40:176-178.

Mullin, C. A., B. A. Croft, K. Strickler, F. Matsumura, and J. R. Miller, 1982. Detoxification enzyme differences between a herbivorous and predatory mite, Science, 217:1270-1272.

Mullin, C. A., F. Matsumura, and B. A. Croft, 1984. Epoxide forming and degrading enzymes in the spider mite, Tetranychus urticae, Comp. Biochem. Physiol., 79C:85-92.

Nault, L. R., M. E. Montgomery, and W. S. Bowers, 1976. Ant-aphid association: role of aphid alarm pheromone, Science, 192:1349-1351.

Nielsen, J. K., L. M. Larsen, and H. Sorensen, 1979. Host plant selection of the horseradish flea beetle Phyllotreta armoraciae (Coleoptera: Chrysomelidae): Identification of two flavonol glycosides stimulating feeding in combination with glucosinolates, Entomol. Exp. Appl., 26:40-48.

Noda, H., and T. E. Mittler, 1983. Sterol biosynthesis by symbiotes of aphids and leafhoppers, in: "Metabolic Aspects of Lipid Nutrition in Insects", T. E. Mittler and R. H. Dadd, eds., pp. 41-55, Westview Press, Boulder.

O'Brien, R. D., 1967. "Insecticides: Action and Metabolism", Academic Press, New York.

Osman, D. H., and W. A. Brindley, 1981. Estimating monooxygenase detoxification in field populations: toxicity and distribution of carbaryl in three species of Labops grassbugs, Environ. Entomol., 10:676-680.

Patch, E., 1938. Food plant catalogue of the aphids of the world, including the phylloxeridae, Bull. Maine Agric. Expt. Stn., 393:35-431.

Pathak, M. D., and D. Dale, 1983. The biochemical basis of resistance in host plants to insect pests, in: "Chemistry and World Food Supplies: The New Frontiers", L. W. Shemilt, ed., pp. 129-142, Pergamon Press, New York.

Pratt, G. E., 1983. The mode of action of pro-allatocidins, in: "Natural Products for Innovative Pest Management", D. L. Whitehead and W. S. Bowers, eds., pp. 323-355, Pergamon Press, New York.

Putter, I., J. G. MacConnell, F. A. Preiser, A. A. Haidri, S. S. Ristich, and R. A. Dybas, 1981. Avermectins: novel insecticides, acaricides, and nematocides from a soil microorganism, Experientia, 37:963-964.

Rafaeli-Bernstein, A., and W. Mordue, 1979. The effects of phlorizin, phloretin and ouabain on the reabsorption of glucose by the Malpighian tubules of Locusta migratoria migratorioides, J. Insect Physiol., 25:241-247.

Rappoport, Z., 1967. "Handbook of Tables for Organic Compound Identification", 563 pp., CRC Press, Cleveland.

Raven, J. A., 1983. Phytophages of xylem and phloem: a comparison of animal and plant sap-feeders, Adv. Ecol. Res., 13:135-234.

Richardson, H. H., and A. H. Casanges, 1942. Studies of nicotine as an insect fumigant, J. Econ. Entomol., 35:242-246.

Robinson, D., 1956. The fluorimetric determination of β-glucosidase: its occurrence in the tissues of animals, including insects, Biochem. J., 63:39-44.

Rosenthal, G. A., and D. H. Janzen, eds., 1979. "Herbivores, Their Interaction with Secondary Plant Metabolites", Academic Press, New York.

Rothschild, M., 1972. Secondary plant substances and warning colouration in insects, in: "Insect/Plant Relationships", H. F. van Emden, ed., Royal Entomological Society of London Symposium No. 6, pp. 59-83, Blackwell Scientific Publ., Oxford.

Rothschild, M., J. von Euw, and T. Reichstein, 1970. Cardiac glycosides in the oleander aphid, Aphis nerii, J. Insect Physiol., 16:1141-1145.

Ryan, R. O., M. de Renobales, J. W. Dillwith, C. R. Heisler, and G. J. Blomquist, 1982. Biosynthesis of myristate in an aphid: involvement of a specific acylthioesterase, Arch. Biochem. Biophys., 213:26-36.

Saito, T., 1969. Selective toxicity of systemic insecticides, Residue Revs., 25:175-186.

Sawicki, R. M., A. L. Devonshire, A. D. Rice, G. D. Moores, S. M. Petzing, and A. Cameron, 1978. The detection and distribution of organophosphorous and carbamate insecticide-resistant Myzus persicae (Sulz.) in Britain in 1976, Pestic. Sci., 9:189-201.

Schaller, G., 1968. Biochemische analyse des aphidenspeichels und seine bedeutung fur die gallenbildung, Zool. Jb. Physiol., 74:54-87.

Schmid, P. P. S., and W. Feucht, 1981. Proteins, enzymes and polyphenols in the phloem of cherry shoots in correlation to the seasonal shoot development, Gartenbauwiss., 46:228-233.

Schoental, R., 1976. Carcinogens in plants and microorganisms, in: "Chemical Carcinogens", C. E. Searles, ed., pp. 626-689, Monograph Ser. No. 173, American Chemical Society, Washington.

Schoonhoven, L. M., and I. Derksen-Koppers, 1976. Effects of some allelochemics on food uptake and survival of a polyphagous aphid, Myzus persicae, Entomol. Exp. Appl., 19:52-56.

Schultz, G., 1969. Zurfrage des transports von flavonoiden in oberirdischen organen von Trifolium, Z. Pflanzenphysiol., 61:29-40.

Schultz, J. C., and I. T. Baldwin, 1982. Oak leaf quality declines in response to defoliation by gypsy moth larvae, Science, 217:149-151.

Seaman, F. C., 1982. Sesquiterpene lactones as taxonomic characters in Asteraceae, Bot. Rev., 48:121-592.

Seiber, J. N., S. M. Lee, and J. M. Benson, 1984. Chemical characteristics and ecological significance of cardenolides in Asclepias (milkweed) species, in: "Isopentenoids in Plants: Biochemistry and Function", W. D. Nes, G. Fuller, and L.-S. Tsae, eds., pp. 563-588, Marcel Dekker, New York.

Selander, J., M. Markkula, and K. Tiittanen, 1972. Resistance of the aphids Myzus persicae (Sulz.), Aulacorthum solani (Kalt.) and Aphis gossypii Glov. to insecticides, and the influence of the host plant on this resistance, Ann. Agric. Fenn., 11:141-145.

Self, L. S., F. E. Guthrie, and E. Hodgson, 1964. Metabolism of nicotine by tobacco-feeding insects, Nature, 204:300-301.

Sen Gupta, G. C., and P. W. Miles, 1975. Studies on the susceptibility of varieties of the apple to the feeding of two strains of woolly aphids (Homoptera) in relation to the chemical content of the tissues of the host, Aust. J. Agric. Res., 26:157-168.

Shaver, T. N., and M. J. Lukefahr, 1969. Effect of flavonoid pigments and gossypol on growth and development of the bollworm, tobacco budworm, and pink bollworm, J. Econ. Entomol., 62:643-646.

Slama, K., and M. Romanuk, 1976. Juvenogens, biochemically activated juvenoid complexes, J. Insect Physiol., 6:579-586.

Slama, K., Z. Wimmer, and M. Romanuk, 1978. Juvenile hormone activity of some glycosidic juvenogens, Hoppe-Seyler's Z. Physiol. Chem., 359-1407-1412.

Smissaert, H. R., 1976. Reactivity of a critical sulfhydryl group of the acetylcholinesterase from aphids (Myzus persicae), Pestic. Biochem. Physiol., 6:215-222.

Smith, B. D., 1966. Effect of the plant alkaloid sparteine on the distribution of the aphid, Acyrthosiphon spartii (Koch), Nature, 212:213-214.

Smith, F. A., and J. A. Raven, 1979. Intracellular pH and its regulation, Annu. Rev. Plant Physiol., 30:289-311.

Smith, J. N., 1955. Detoxification mechanisms in insect, Biol. Rev., 30:455-475.

Smith, J. N., 1968. The comparative metabolism of xenobiotics, Adv. Comp. Physiol. Biochem., 3:173-232.

Soderlund, D. M., C. W. Hessney, and D. W. Helmuth, 1983a.

Pharmacokinetics of cis- and trans-substituted pyrethroids in the American cockroach, Pestic. Biochem. Physiol., 30:161-168.

Soderlund, D. M., J. R. Sanborn, and P. W. Lee, 1983b. Metabolism of pyrethrins and pyrethroids in insects, Prog. Pestic. Biochem. Toxicol., 3:401-435.

Sogawa, K., 1976. Studies on the feeding habits of the brown planthopper, Nilaparvata lugens (Stal.) (Hemiptera: Delphacidae). V. Probing stimulatory effect of rice flavonoid, Appl. Ent. Zool., 11:160-164.

Stipanovic, R. D., 1983. Function and chemistry of plant trichomes and glands in insect resistance, in: "Plant Resistance to Insects", P. A. Hedin, ed., pp. 69-100, Symp. Ser. No. 208, Amer. Chem. Soc., Washington.

Sutherland, O. R. W., G. B. Russell, D. R. Biggs, and G. A. Lane, 1980. Insect feeding deterrent activity of phytoalexin isoflavonoids, Biochem. Syst. Ecol., 8:73-75.

Svoboda, J. A., W. R. Lusby, and J. R. Aldrich, 1984. Neutral sterols of representatives of two groups of Hemiptera and their correlation to ecdysteroid content, Arch. Insect. Biochem. Physiol., 1:139-145.

Swain, T., 1979. Phenolics in the environment, Rec. Adv. Phytochem., 12:617-640.

Thompson, S. N., 1973. A review and comparative characterization of the fatty acid compositions of seven insect orders, Comp. Biochem. Physiol., 45B:467-482.

Thurston, R., W. T. Smith, and B. P. Cooper, 1966. Alkaloid secretion by trichomes of Nicotiana species and resistance to aphids, Entomol. Exp. Appl., 9:428-432.

Tingey, W. M., 1984. Glycoalkaloids as pest resistance factors, Amer. Pot. J., 61:157-167.

Todd, G. W., A. Getahum, and D. C. Cress, 1971. Resistance in barley to the greenbug, Schizaphis graminum. I. Toxicity of phenolic and flavonoid compounds and related substances, Ann. Entomol. Soc. Amer., 64:718-722.

Todd, L., 1963. The chemistry of the aphid colouring matters, Pure Appl. Chem., 6:709-717.

Toppozada, A., and R. D. O'Brien, 1967. Permeability of the ganglia of the willow aphid, Tuberolachnus salignus, to organic ions, J. Insect Physiol., 13:941-954.

van Emden, H. F., 1972. Aphids as phytochemists, in: "Phytochemical Ecology", J. B. Harborne, ed., pp. 25-43, Academic Press, New York.

van Emden, H. F., V. F. Eastop, R. D. Hughes, and M. J. Way, 1969. The ecology of Myzus persicae, Annu. Rev. Entomol., 14:197-270.

Wachendorff, U., and F. Klingauf, 1978. Esterasetest zur diagnose von insektizidresistenz bei aphiden, Z. Pflkrankh. Pflschutz., 85:218-227.

Waiss, A. C., B. G. Chan, and C. A. Ellinger, 1977. Host plant resistance to insects, in: "Host Plant Resistance to Pests", P. A. Hedin, ed., p. 115-128, Symp. Ser. No. 62, Amer. Chem. Soc., Washington.

Waller, G. R., and E. K. Nowacki, 1978. "Alkaloid Biology and Metabolism in Plants", Plenum Publ. Corp., New York.

Ware, G. W., 1980. Effects of pesticides on nontarget organisms, Residue Rev., 76:173-201.

Weiden, M. H. J., 1971. Toxicity of carbamates to insects, Bull. Wld. Health Org., 44:203-213.

Wilkinson, C. F., and L. B. Brattsten, 1972. Microsomal drug metabolizing enzymes in insects, Drug. Metab. Rev., 1:153-228.

Williams, W. G., G. G. Kennedy, R. T. Yamamoto, J. D. Thacker, and J. Bordner, 1980. 2-Tridecanone: a naturally occurring insecticide from the wild tomato *Lycopersicon hirsutum* f. *glabratum*, Science, 207:888-889.

Wink, M., T. Hartmann, L. Witte, and J. Rheinheimer, 1982. Interrelationship between quinolizidine alkaloid producing legumes and infesting insects: exploitation of the alkaloid-containing phloem sap of *Cytisus scoparius* by the broom aphid *Aphis cytisorum*, Z. Naturforsch, 37C:1081-1086.

Winteringham, F. P. W., 1969. Mechanisms of selective insecticidal action, Annu. Rev. Entomol., 14:409-442.

Wolfenbarger, D. A., and C. E. Holscher, 1967. Contact and fumigant toxicity of oils, surfactants, and insecticides to two aphid and three beetle species, Fla. Entomol., 50:27-36.

Woodhead, S., D. E. Padgham, and E. A. Bernays, 1980. Insect feeding on different sorghum cultivars in relation to cyanide and phenolic acid content, Ann. Appl. Biol., 95:151-157.

Yu, S. J., 1983. Induction of detoxifying enzymes by allelochemicals and host plants in the fall armyworm, Pestic. Biochem. Physiol., 19:330-336.

Yu, S. J., 1984. Interactions of allelochemicals with detoxification enzymes of insecticide-susceptible and resistant fall armyworms, Pestic. Biochem. Physiol., 22:60-68.

Ziegler, H., 1975. Nature of transported substances, in: "Transport in Plants. I. Phloem transport", M. H. Zimmerman and J. A. Milburn, eds., pp. 59-100, Springer Verlag, New York.

Zucker, W. V., 1982. How aphids choose leaves: the roles of phenolics in host selection by a galling aphid, Ecology, 63:972-981.

FATE OF INGESTED PLANT ALLELOCHEMICALS IN HERBIVOROUS INSECTS

L. B. Brattsten

E. I. DuPont de Nemours & Co., Inc.
Experimental Station, Bldg. 402
Agricultural Products Dept.
Wilmington, DE 19898

1. INTRODUCTION

Insect herbivores and higher plants have coexisted for at least 250 million years and during this time many remarkable mutualisms between them have developed. There are also, of necessity, antagonisms in that herbivorous insects eat plants and plants can kill insects (Courtney, 1981; Raffa, Chapter 9 in this text). Not having available to them motility, tooth, or claw, the plants have developed many morphological and anatomical barriers to insect feeding activities, and also rely prominently on chemical defenses. The insects, in turn, being handicapped by very small sizes which impose limits on their energy available for motility and jaw power, have responded with a diversity of biochemical, physiological, and behavioral adaptations which allow them not only to continue to use the plants as food sources but also to take advantage of the plant defenses for purposes of their own, in many cases. This is emphasized by the many instances in which herbivorous insects use for defense, chemicals that they themselves have biosynthesized from intermediary metabolism precursors by pathways and enzyme assemblages similar to those found in plants (Duffey, 1977; Rotschild and Marsh, 1978; Eisner et al., 1986).

Ingested non-nutrients that are not utilized may be either of no consequence, toxic with acutely lethal effects, or, more commonly, acting subacutely by decreasing growth, feeding, or reproductive rates. Insects may escape injury from ingested toxicants by several mechanisms including extra rapid excretion, metabolism to an excretable product (Ahmad et al., Chapter 3 in this text), extensive metabolism so that, in the fashion of bacteria and plants, elements can be used, sequestration and storage sometimes in special body compartments, or by a protected or modified target with which the toxicant can not interfere (Berenbaum, Chapter 7 in this text).

I will in the following briefly review some examples of these adaptive mechanisms with examples representing several of the major classes of allelochemicals. Excellent recent reviews provide detailed analyses of cases where the fate of a compound or class of compounds has been studied in insects. In chapter three of this text, Ahmad et al. describe the enzymes involved in the metabolism of plant allelochemicals and in several cases illustrate the enzyme action with some of the few well documented cases available on the metabolic fate of such compounds in insects. Therefore I shall use other examples as far as possible and make frequent references to chapter three.

2. TERPENES

The terpenes constitute a biosynthetically related group of compounds with extreme molecular diversity and a fairly matching diversity of effects on insect herbivores (Brattsten, 1983).

2.1. Monoterpenes

2.1.1. Bark beetle pheromones. The most spectacular case of monoterpene metabolism in insects is their use as pheromone precursors in bark beetles (Coleoptera: Scolytidae). An early recognition of the importance of host plant components in bark beetle aggregation was that it seemed necessary, in some cases, for the beetles to feed before they could attract conspecifics. Since the beetles use volatile monoterpenes that could be taken in by inhalation also, some species may only need to bore into healthy tree tissue. It is also possible that monoterpenes that larvae and pupae are exposed to, being highly lipophilic, may be sequestered in the pupal tissues and later released. Hughes (1975) showed that exposing pupae to α-pinene vapors resulted in a significantly higher pheromone concentration in the emerging adults than in those emerging from untreated pupae.

The major pheromones isolated from beetle hindguts and frass are cis- and trans-verbenol, verbenone, ipsdienol, and ipsenol. Adult beetles can produce these compounds from host plant monoterpenes such as α-pinene, β-pinene, and myrcene (Renwick et al., 1973, 1976a; Hughes, 1974, 1975). Frontalin, brevicomin, and other pheromones of restricted occurrence may also be derived from host terpenes. In experiments with optically purified α-pinene, Renwick et al., 1976b) showed that adults of *Ips paraconfusus* (Lanier) metabolized (+)-α-pinene to trans-verbenol and (+)-myrtenol and (−)-α-pinene to the geometrical isomer cis-verbenol and (+)-myrtenol:

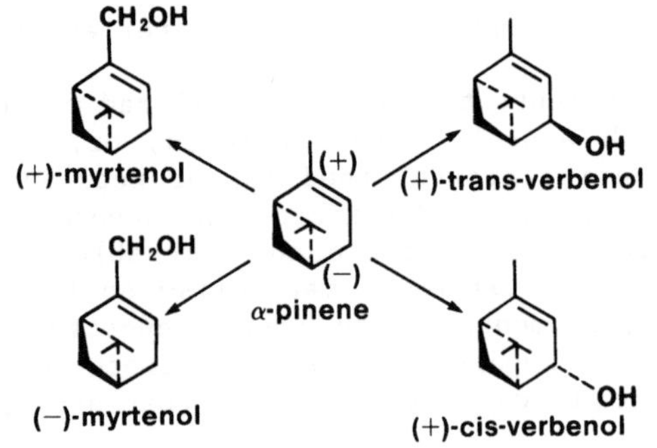

Only cis-verbenol is part of the pheromone blend in this species; as the authors point out, the relative content of not just a single monoterpene, but the enantiomer of one, can thus strongly influence the ability of the beetles to colonize trees. It may take two or even four oxidase isoenzymes to produce the verbenols and myrtenols from the two α-pinene isomers.

Myrcene is used by male *Ips* and *Dendroctonus* (Coleoptera: Scolytidae) beetles to make ipsdienol, a common pheromone component in both species; some species of *Ips* convert the ipsdienol further to ipsenol, a component in their pheromone blend (Hughes, 1974):

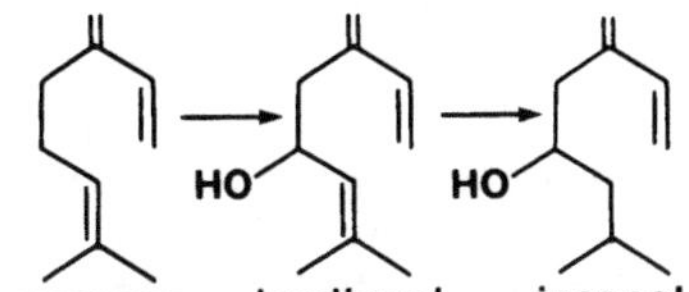

myrcene ipsdienol ipsenol

Both male and female *D. brevicomis* LeConte use myrcene to make myrcenol, but only males produce ipsdienol (Byers, 1982). In *D. frontalis* Zimmerman, α-pinene is converted to trans-verbenol in both males and females but only males convert the trans- verbenol further to verbenone (Renwick et al., 1973). In this species, males produce more kinds of metabolites from α-pinene; females appear restricted to making alcohols whereas males can make further oxidation products.

Microorganisms in the guts of bark beetles convert α-pinene to cis- and trans-verbenol and further to verbenone (Brand et al., 1975, 1976). *I. paraconfusus* males fed the antibiotic streptomycin could convert α-pinene to cis-verbenol but could not convert myrcene into ipsdienol (Byers and Wood, 1981). It was thus unclear whether the beetles themselves or their gut microbes produced the beetle aggregation pheromones. If the microbes were tree pathogens they would have a "vested interest" in promoting beetle aggregation and may be a somewhat likely source of the pheromones. However, axenically reared *D. ponderosae* Hopkins and *I. paraconfusus* were able to convert host log vapors to pheromone components (Conn et al., 1984). Without exposure to α-pinene the axenic beetles produced about half the amounts of the wild ones, but after α-pinene-exposure the axenic beetles produced up to 20 times more trans-verbenol than the wild ones. Conn et al. (1984) suggest that gut microbes may competitively consume some of the log monoterpenes and/or the beetle-produced pheromone intermediate. This idea is supported by work on bacterial cytochrome P-450, particularly in *Pseudomonas putida* which has a cytochrome P-450 that oxidizes camphor for use as a nutrient and can be induced to very high activity by camphor (Katagiri et al., 1968). Certain *Pseudomonas* strains use α-pinene as their sole carbon source (Gibbon and Pirt, 1971; Gibbon et al., 1972). This is also consistent with the finding of Brandt et al. (1975) that bacteria isolated from bark beetle guts oxidize α-pinene. In analogy, excessive bacterial contamination in laboratory cultures of the boll weevil, *Anthonomus grandis* Boheman (Coleoptera: Curculionidae) inhibits its pheromone production (Gueldner et al., 1977).

Juvenile hormone III stimulated trans-verbenol production almost 700-fold in wild *D. ponderosae* but only about eight-fold in axenic beetles (Conn et al. (1984). This difference was attributed to an age difference in the wild and axenic beetles; this study also revealed a 23-fold increase in trans-verbenol production between the fifth and the 35th day after eclosion in axenic female *D. ponderosae*. Borden et al. (1969), Hughes and Renwick (1977a, and 1977b), and Bridges (1982) also found a stimulation of bark beetle pheromone production by exposure to juvenile hormone. This effect combined with other experiments (Hughes and Renwick 1977a, 1977b) strongly indicates that pheromone biosynthesis is regulated by juvenile hormone; this seems plausible, in particular, considering the otherwise multiple effects of juvenile hormone.

The oxidations of host monoterpenes to pheromones are catalyzed by cytochrome P-450 (Ahmad et al., Chapter 3 in this text). Several host tree monoterpenes are toxic to invading bark beetles (Raffa, Chapter 9 in this text). Monoterpenes are detoxified primarily by cytochrome P-450-catalyzed oxidations. A well documented case is limonene metabolism in mammals (Igimi et al., 1974; Ariyoshi et al., 1975) because of the therapeutic use of limonene for dissolving gall bladder stones. It is, thus, likely that ancestral detoxification products that happened to be volatile came to serve as behavioral signals. This idea is supported by the existence of bark beetles that do not fabricate aggregation pheromones but instead are attracted directly to host resin vapors (Kangas et al., 1965). Some bark beetles that do produce oxidation products from host monoterpenes are also primarily attracted by resin vapors. Hughes (1973) suggested that this may represent a primitive stage in the evolution of bark beetle communication.

The increase in pheromone production after eclosion in the bark beetles could result from an increase in cytochrome P-450 content with adult age such as observed in the house fly, Musca domestica L. (Diptera: Muscidae), the house cricket, Acheta domesticus (L.) (Orthoptera: Gryllidae), and the Madagascar cockroach, Gromphadorhina portentosa Schaum (Orthoptera: Blattidae) (Wilkinson and Brattsten, 1972; Ahmad et al., Chapter 3 in this text). Binding of α-pinene to cytochrome P-450 from midguts of the southern armyworm, Spodoptera eridania (Cramer) (Lepidoptera: Noctuidae) has been documented as well as its oxidation by cytochrome P-450 measured by substrate-generated NADPH oxidation (Brattsten et al., 1984). It is also interesting to consider that both α-pinene (Brattsten et al., 1977; Yu, Chapter 4 in this text) and juvenile hormone (Terriere and Yu, 1973) are inducers of cytochrome P-450 in insects. White et al. (1979, 1980) showed that α-pinene induces cytochrome P-450 and its own oxidation in rat liver; moreover, α-pinene was oxidized by a microsomal fraction prepared from D. terebrans (Olivier), indicating that cytochrome P-450 catalyzed the reaction. It may, thus, be possible that the host trees contribute, by cytocrhome P-450 induction, to aggregation pheromone production in the beetles; this would seem self-defeating, but may have implications for forest self-rejuvenation (Raffa and Berryman, 1986).

2.1.2. Pyrethrins. The pyrethrins are botanical insecticides extracted from the mature flowers of Chrysanthemum cinerariaefolium (Compositae). Having been of considerable commercial value in household pest control before they were superceeded by the synthetic pyrethroids, the pyrethrins were studied intensively to elucidate their mode of action and metabolism. The acute toxicity of pyrethroids by interference with the nerve membrane sodium channel is discussed by Berenbaum (Chapter 7 in this text).

The pyrethrin molecule consists of an acid and an alcohol moiety held together by an ester bond. The acid part of the molecule is considered to arise by an unusual variant of the isoprenoid biosynthetic pathway resulting in a rearranged monoterpene derivative. The pathway leading to the alcohol moiety is incompletely known (Casida, 1973; Mabry and Gill, 1979).

The pyrethrin molecule is converted both by cytochrome P-450 and carboxylesterases to nontoxic products. One of the oxidation products is shown in Chapter 3; other points where the pyrethrins can be oxidized both by insects (Yamamoto et al., 1969) and mammals (Elliott et al., 1972) are indicated by the arrows:

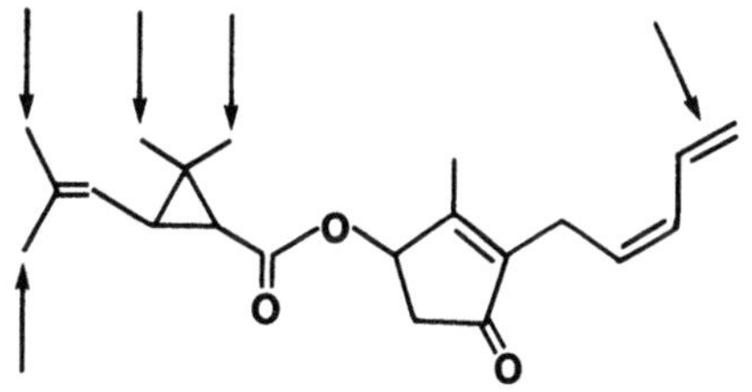

The major reason for the relative safety of the pyrethrins to mammals is that the rates of oxidation and, particularly, hydrolysis of the ester bond are much (probably about 10-fold) higher in mammals than in insects (Hollingworth, 1976). The major oxidation products are conjugated in the house fly to glucose (Yamamoto et al., 1969) and, apparently, to glucuronic acid in the rat (Elliott et al., 1972).

2.1.3. Iridoid glycosides. The iridoid glycosides are glucosylated monoterpene derivatives. The aglycone iridoidal was first found in Iridomyrmex ants (Cavill et al., 1956). The iridoids also serve as precursors for several kinds of alkaloids, e. g. the indole alkaloids found in Catharanthus roseus (Apocynaceae) (Vickery and Vickery, 1981). They are extremely bitter compounds and thus good candidates for antifeedants. Iridoid glycosides in the nectar of catalpa (Catalpa speciosa, Bignoniaceae) are distasteful and toxic to would-be nectar thieves including several species of ants and the common skipper, Poanes hobomok Harris (Lepidoptera: Hesperiidae) but not to species adapted to use catalpa nectar such as several species of bumble bees, carpenter bees, and the catalpa sphinx, Ceratomia catalpae (Boisduval) (Lepidoptera: Sphingidae) (Stephenson, 1982).

The catalpa sphinx larvae and buckeye butterfly, Junonia coenia Hubner (Lepidoptera: Nymphalidae), larvae have a feeding preference for iridoid glycoside-containing plants and sequester the compounds from their food plants (Bowers, 1984; Bowers and Puttick, 1986). They may use the compounds, which have detrimental effects also on vertebrates, for defense against bird predators; this is particularly effective if the insects have a conspicuous pattern of warning colors and live gregariously.

The entire life history of the baltimore checkerspot butterfly, Euphydryas phaeton (Drury) (Lepidoptera: Nymphalidae), and probably all other North American checkerspots is closely tied to the iridoid glycosides (Bowers, 1980); the females oviposit on plants containing the compounds, notably those belonging to Scrophulariaceae, the larvae use them as feeding cues and sequester them. Unlike the catalpa sphinx and buckeye, the checkerspots retain the compounds through the metamorphosis. They are then redistributed in all tissues of the adults where they probably serve as defense against predators (Bowers, 1980, 1981).

Bowers and Puttick (1986) found that the meconium of the buckeye and the catalpa sphinx contain iridoid glycosides that the larvae had ingested. They propose that retaining the compounds through to the newly emerged adult when the meconium and its content is discarded, may be a predator protection device for the very sensitive period during which the exoskeleton and wings harden; the meconium content can be forcefully ejected at an attacker.

The iridoid glycoside composition in insects sequestering the compounds and that in the food plant sometimes differ, indicating selective excretion and/or biotransformation. Bowers and Puttick (1986) found that baltimore adults from larvae fed Chelone glabra (Scrophulariaceae) with more catalpol than aucubin, true to the plant, contained more catalpol than aucubin. However, adults from larvae fed Plantago lanceolata (Plantaginaceae) with more aucubin than catalpol, nevertheless, contained more catalpol than aucubin. Since they could not detect either iridoid glycoside in the frass from larvae fed either plant, it is possible that the aucubin was converted to catalpol by the insect:

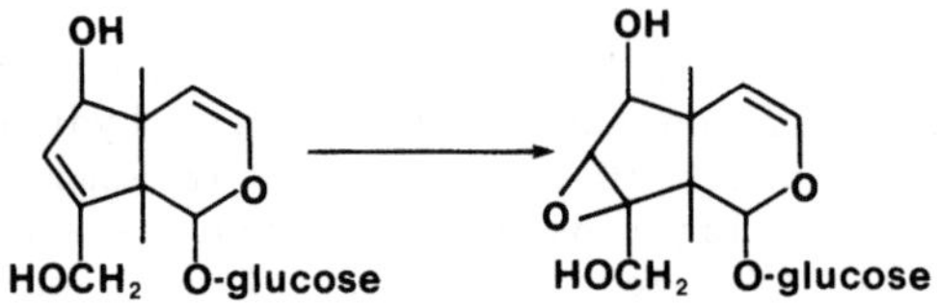

This would require an epoxidation, a reaction typically catalyzed by cytochrome P-450. Stermitz et al. (1986) showed that E. anicia (Doubleday & Hewitson) (Lepidoptera: Nymphalidae) adults contain iridoid glycosides presumably ingested by the caterpillars feeding on Castilleja and Besseya (Scrophulariaceae) plants. In some cases the adults contained an opposite ratio of the two major compounds, catalpol and aucubin, than the plants; this was particularly evident in male butterflies. In some cases the plants contained esterified iridoid glycosides that can be converted to catalpol by ester hydrolysis; the butterflies did not contain any of these esterified compounds but may have hydrolyzed them to catalpol.

Apart from being bitter and generally toxic, there is apparently nothing known about the molecular effects of these compounds on any organism or how they may be metabolized. They could perhaps also protect caterpillars from parasitoids.

2.2. Sesquiterpenes

2.2.1. Juvenile hormones. The possibly most important sesquiterpene derivatives for insects are the juvenile hormones (JH) of which there are at least four different variants:

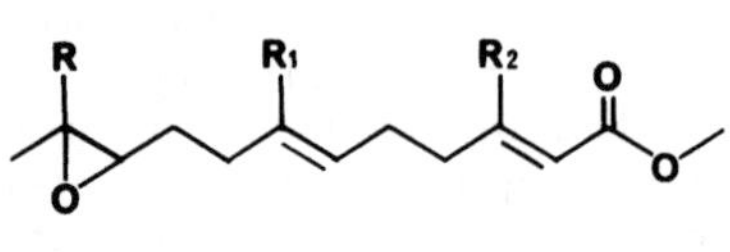

JH III in which R = R1 = R2 = methyl is the most widely occurring among insect orders; in addition JH 0 (R = R1 = R2 = ethyl), JH I (R = R1 = ethyl, R2 = methyl), and JH II (R = ethyl, R1 = R2 = methyl) have been found in Lepidoptera (Schooley and Baker, 1985). The juvenile hormones are biosynthesized in the corpora allata of insects from the universal sesquiterpene precursor, farnesyl pyrophosphate. Several insect enzymes participate in the conversion of autogenous farnesyl pyrophosphate to juvenile hormones, among them microsomal cytochrome P-450 (see Ahmad et al., Chapter 3 in this text). Even though many plants contain farnesyl pyrophosphate so that the compound likely is ingested by insects, this is not known to upset their hormone balance. The compound, being an intermediate in plant secondary metabolism, may not accumulate to interferingly high concentrations or gut enzymes may metabolize it so that it can no longer serve as a juvenile hormone precursor.

Redundant juvenile hormone is, like all other hormones and pheromones, recognized as a foreign compound and metabolically inactivated. A hemolymph esterase that is highly specific for juvenile hormone is most important in its degradation (Abdel-Aal and Hammock, 1986). It is also inactivated by the action of an extra-allatal epoxide hydrolase of unknown specificity, and a cytochrome P-450 also of unknown specificity is active in a minor degradation pathway:

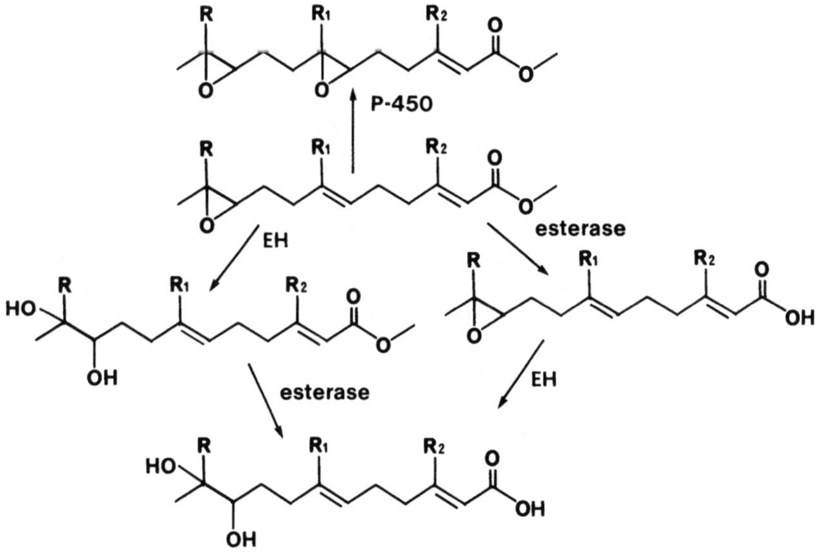

In the southern armyworm, the sulfate conjugate of the JH diol acid is the major breakdown product (Slade and Wilkinson, 1974). The elimination of injected juvenile hormone is quite rapid in some insects; the vagrant grasshopper, Schistocerca vaga (Scudder) (Orthoptera: Acrididae), converts 45 percent of injected JH I to sulfate and possibly glucose conjugates in two hours (Slade and Zibitt, 1972). For excellent recent reviews on juvenile hormone biosynthesis and degradation, see Schooley and Baker (1985), Feyereisen (1985), and Hammock (1985).

2.2.2. Sesquiterpene antifeedants. Sesquiterpenes e. g. warburganal (Frazier, Chapter 1 in this text) and their derivatives such as the sesquiterpene lactones are often powerful feeding inhibitors for herbivorous insects and mammals; they are some of the most bitter compounds in nature. Glaucolide A, a sesquiterpene lactone in Vernonia spp. (Compositae):

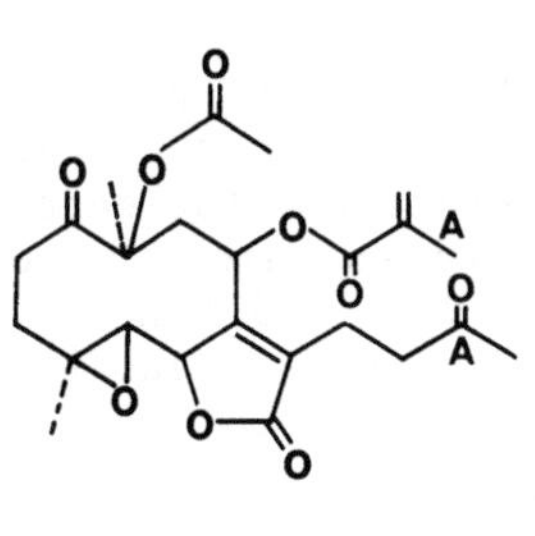

inhibited feeding of the polyphagous larvae of the southern and fall armyworm, Spodoptera frugiperda (J. E. Smith) (Lepidoptera: Noctuidae) and the saddleback caterpillar, Sibine stimulea (Clemens) (Lepidoptera: Limacodidae) in laboratory feeding tests (Burnett et al., 1978). These species were not seen feeding on Vernonia species with high glaucolide A concentration in nature. It also inhibited feeding in the yellow-striped armyworm, Spodoptera ornithogalli Guenee (Lepidoptera: Noctuidae) even though this species feeds on high glaucolide A plants in nature. It did, however, not inhibit feeding by cabbage looper, Trichoplusia ni Hubner (Ibid.), larvae, or yellow woollybear, Spilosoma virginica F. (Lepidoptera: Arctiidae), larvae both of which also use Vernonias with high levels of the compound (Mabry and Gill, 1979).

The mode of feeding inhibiting action of this and similar compounds is not known (see Frazier, Chapter 1 in this text). It is not known if they have post-ingestive toxic effects and, if so, how those effects are circumvented in adapted species.

Other sesquiterpene lactones including conchosin B, parthenium, tetraneurin A, and coronopilin were toxic to males of the migratory grasshopper, Melanoplus sanguinipes F. (Orthoptera: Acrididae) when injected into the hemocoel but not when ingested or contacted topically (Isman, 1985). Both gut and integument tissues in insects contain detoxifying enzymes, particularly carboxylesterases which may make the lipophilic sesquiterpene lactones too polar to penetrate the integument or gut cell membranes.

2.3. Diterpenes

Gossypol and the related heliocides characteristically occur in the glands of cotton, Gossypium (Malvaceae), plants and defend the plants from herbivory by the tobacco budworm, Heliothis virescens F. (Lepidoptera: Noctuidae), other Heliothis species and the pink bollworm, Pectinophora gossypiella (Saunders) (Lepidoptera: Gelechiidae). Interestingly, these compounds are not toxic to the boll weevil which is specialized to cotton and a few related plants (Stipanovic et al., 1977). Both gossypol and the heliocides are rather highly hydroxylated and may undergo conjugation and subsequent excretion. Not much is known about the behavior of conjugating enzymes in the presence of a specific selection pressure. However, the cotton budworm is a severe pest of cotton and easily develops resistance to synthetic insecticides (Brattsten, 1986a); this resistance may simultaneously decrease its sensitivity to the cotton diterpenes. Nothing is known about the metabolism or disposition of these compounds in insects.

2.4. Triterpenes

The cucurbitacins are extremely bitter, oxygenated tetracyclic triterpenes and characteristic of cucumbers, squashes, and other plants in the Cucurbitaceae family. They are highly toxic to many mammals and deter feeding in many nonadapted insect herbivores. The squash beetle, Epilachna borealis (F.) (Coleoptera: Coccinellidae), may be an arbiter elegantarium among the insects; in its apparent dilemma of depending on a highly toxic compound as a feeding cue, it has developed a unique and highly sophisticated behavioral defense against the cucurbitacins (Tallamy, Chapter 8 in this text). In contrast, diabroticite (Coleoptera: Chrysomelidae) beetles to which the cucurbitacins are extremely strong feeding stimulants and arrestants (Metcalf et al., 1980, 1982) have solved the same problem by biochemical and physiological means. Metcalf et al. (1980) estimated a median lethal dose much in excess of 2000 mg per kg of cucurbitacin B to adults of the southern, Diabrotica undecimpunctata howardi Barber, and western corn rootworm, D. virgifera LeConte, whereas the median lethal dose of the same compound is about one mg per kg, intraperitoneally, to the mouse (David and Vallance, 1955).

Ferguson et al. (1985) studied the disposition of cucurbitacin B (cuc B) in adults of five species of diabroticite beetles. They found that most of the ingested dose was excreted and the beetles sequestered up to three percent in the form of a conjugate. The western corn rootworm, a grass specialist, excreted most of the dose unchanged whereas the three generalist feeders, the southern corn rootworm, the banded cucumber beetle, *Diabrotica balteata* LeConte, and *D. cristata* Harris metabolized the compound more extensively; up to 90 percent of the excreted material from the banded cucumber beetle was polar, probably conjugated metabolites. The fifth species, the striped cucumber beetle, *Acalymma vittatum* F., a cucurbit specialist, like the grass specialist tended to excrete more of the dose unchanged than did the generalists. This supports the idea that generalist feeders may need to have more active and versatile metabolic defenses than specialist ones (Gordon, 1961). The excreted polar and/or conjugated products were not identified, but the portion of the ingested dose sequestered into the hemolymph as a conjugate had been deacetylated by the beetles before conjugation:

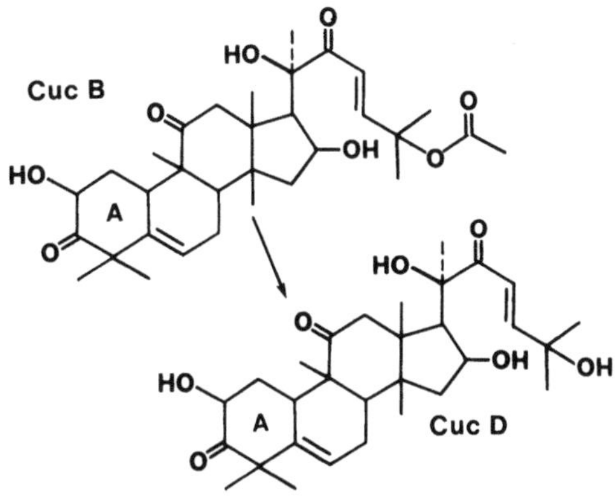

When the conjugate was hydrolyzed to restore the free metabolite, it was still a feeding stimulant for the beetles having merely been converted into another cucurbitacin (cuc D); the steric configuration of the oxygens at the A ring which match the beetles' receptors (Metcalf et al., 1980) had been preserved intact. Presumably, they were also still feeding deterrents for nonadapted herbivores. Remarkably, none of the excreted metabolites stimulated feeding. This implies that the beetles "used" different metabolic routes for storing versus eliminating the ingested cucurbitacin. Polar and/or conjugated cucurbitacin products were not translocated to the hemolymph whereas some portion of those without an acetoxy group were, and perhaps subsequently conjugated by fat body enzymes. The stored material can easily be reactivated by a β-glycosidase.

This difference in metabolic routes provides the beetles with a genuine opportunity to use these plant-derived compounds for their own defense and that of their offspring because of the deterrent effect of the intact cucurbitacin. A conjugate identical to that stored was also found in eggs (Ferguson et al., 1985). Although there were no acutely toxic consequences of the beetles' eating huge amounts of cucurbitacins, they usually suffered a somewhat shortened life span, particularly the males (Ferguson et al., 1985). A chinese mantid, Tenodera aridifolia sinensis Saussure (Orthoptera: Mantidae), rejected diabroticite beetles that had been fed cucurbitacin-containing squash fruits but not those that had been fed a cucurbitacin-free diet (Ferguson and Metcalf, 1985).

3. STEROIDS

Throughout the animal kingdom, steroids are used as membrane building blocks and hormones. Cholesterol is a very common membrane component and is required in substantial concentrations; many insects out of those studied have sterols other than cholesterol in their membranes. The steroid hormones, on the other hand, occur in minute concentrations subject to precision timing and titer regulating mechanisms and exist in a considerable array of molecular variants often with specific effects. Insects and some other invertebrates, lacking a crucial enzyme in the steroid biosynthetic pathway perhaps as a consequence of originating from carnivorous ancestor(s), rely on their food source for a supply of the basic steroid nucleus. Many herbivorous insects, thus, need to bulk convert phytosteroids to membrane-compatible sterols, particularly, cholesterol and then selectively convert cholesterol to 20-hydroxyecdysone and ecdysone, the molting hormones for many insects. They also need mechanisms for removing the active hormones from their systems when not needed.

3.1. Phytosteroids

Sitosterol, campesterol, and stigmasterol are the three major steroids in higher plants and mostly utilized by herbivorous insects for cholesterol synthesis (Svoboda and Thompson, 1985). All three are characterized by an alkyl substituent on C24 that needs to be removed:

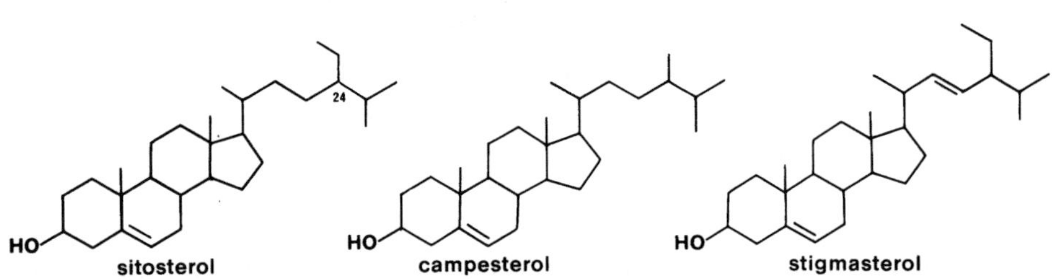

Instead of squalene oxidocyclase and the enzymes involved in the conversion of lanosterol to cholesterol, herbivorous insects had to aquire the enzymes to accomplish the dealkylation of the phytosterol side chain. These enzymes should, thus, be unique to insect herbivores and offer a target site for selective insecticides that would spare natural enemies, since carnivorous insects do not have the ability to dealkylate the side chain. Attempts have been made to develop such an insecticide but so far without success; most of the enzymes involved have not been identified. The evolutionary implications of phytosterol utilization instead of cholesterol de novo biosynthesis are mysterious and tantalizing.

In the tobacco hornworm, Manduca sexta (L.) (Lepidoptera: Sphingidae), sitosterol is converted to fucosterol, stigmasterol to 5,22,24-cholestatrienol, and campesterol to 24-methylenecholesterol:

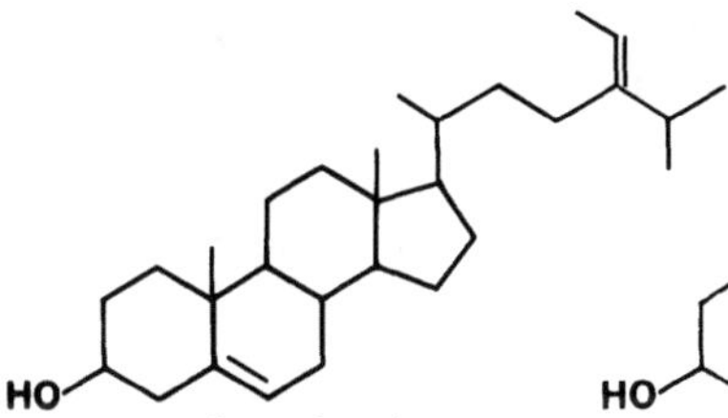

fucosterol

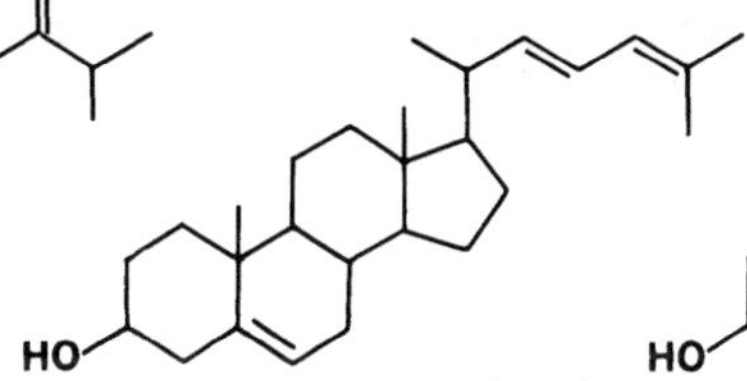

5,22,24-cholestatrienol

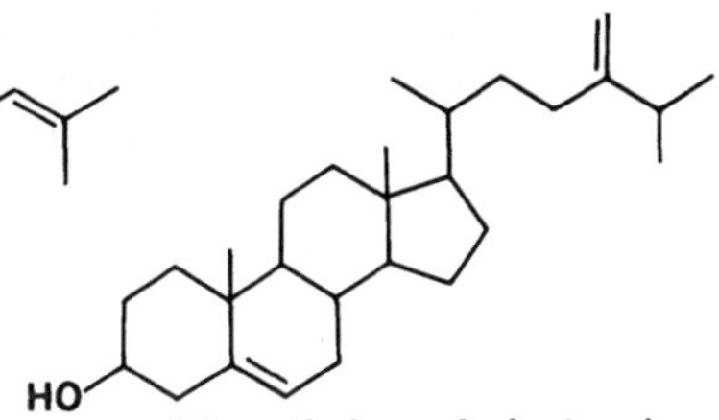

24-methylenecholesterol

All of these are converted into desmosterol, the immediate precursor of cholesterol:

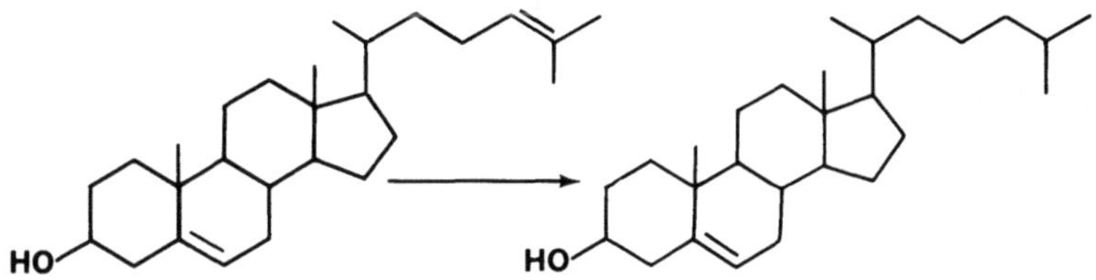

desmosterol cholesterol

The above biotransformations were clarified and reviewed by Svoboda and Thompson (1983) and Svoboda et al. (1978). There is considerable information about this pathway and the variants of it that have been found in other insect species. In the confused flour beetle, *Tribolium confusum* J. du Val (Coleoptera: Tenebrionidae), half of the membrane sterol content consists of 7-dehydrocholesterol:

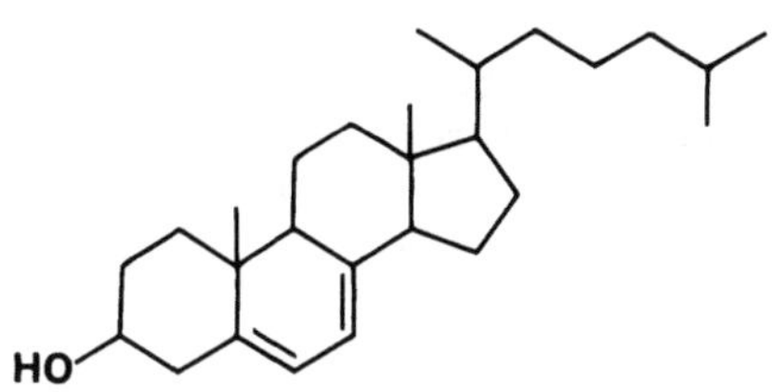

7-dehydrocholesterol

a likely intermediate between desmosterol and cholesterol but of unknown significance (Svoboda et al., 1972). Whereas the predatory ladybeetle, *Coccinella septempunctata* L. (Coleoptera: Coccinellidae), directly utilizes the cholesterol of its prey, the Mexican bean beetle, *Epilachna varivestis* Mulsant (Ibid.) saturates stigmasterol and other plant sterols and, apparently, uses the saturated analogs as membrane components (Svoboda et al., 1974). Svoboda and Thompson (1983) and Svoboda et al. (1978) review several other equally intriguing deviations from the tobacco hornworm pattern. The implications of variations in membrane sterol composition for membrane-bound detoxifying enzymes such as cytochrome P-450 is worth investigating.

Although there is considerable information about the pathways and the intermediates, very little is known about the enzymes involved in the utilization of phytosterols. The dealkylation of sitosterol occurs in several steps the first of which is a desaturation to fucosterol (Rees, 1985; Svoboda et al., 1971). Nothing is known about the properties of this 24,28-desaturase. The fucosterol is then epoxidized across the newly formed double bond. The enzyme involved in this step is a cytochrome P-450 of unknown subcellular location and specificity; the reaction requires molecular oxygen and reduced nicotinamide adenine dinucleotide phosphate

(NADPH) (Fujimoto et al., 1980). The next step is a complex opening of the 24,28-epoxide during which a hydrogen from the neighboring carbon, C25, migrates to C24, in analogy with the "NIH-shift" observed during epoxide hydration of polycyclic aromatic hydrocarbons (Daly et al., 1969). The extra hydrogen on C24 destabilizes the C24,C28 bond and the result is dealkylation to desmosterol (Rees, 1985):

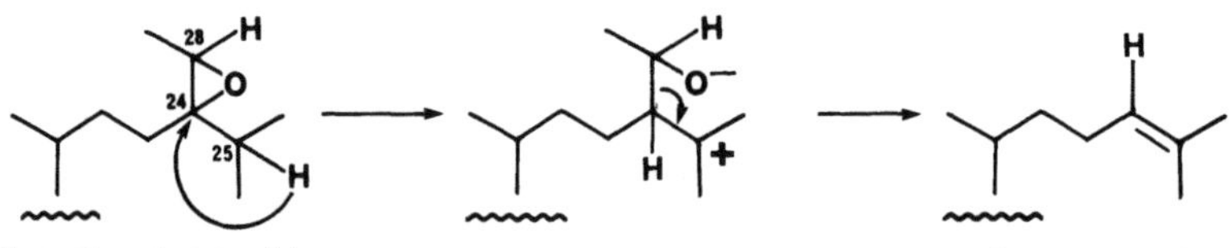

Prestwich et al. (1985) demonstrated conversion of one of the possible stereoisomer, 24R,28R-fucosterol epoxide, by an enzyme, fucosterol epoxide lyase, in the post-mitochondrial supernatant of tobacco hornworm gut cells; the activity was not affected by sulfhydryl reagents or nucleotide cofactors but remains to be further characterized. The saturation of the C24-C25 double bond by an unidentified enzyme then results in cholesterol.

3.2. Ecdysones

During another multiple step process, cholesterol is converted to ecdysone. None of the enzymes involved in this pathway seem to have been characterized. Ecdysone is the active molting hormone in some arthropods, but 20-hydroxyecdysone is the more common active hormone in insects. Ecdysone is converted to 20-hydroxyecdysone by a cytochrome P-450 (see Ahmad et al., Chapter 3 in this text):

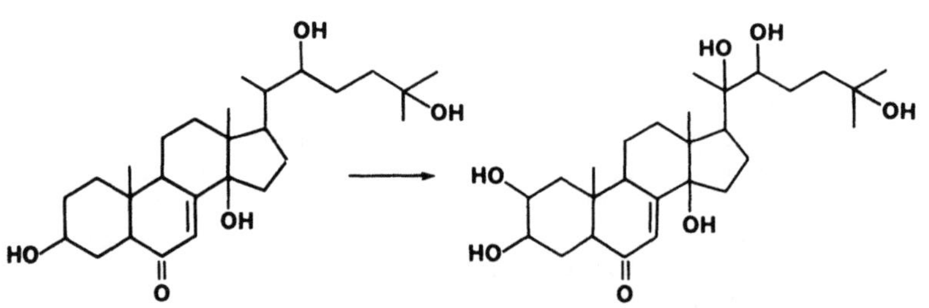

The titer of molting hormone varies characteristically throughout the development of insects. It peaks immediately preceding a molt, and in some insects there is a second smaller peak midway through an immature stage for unknown reasons (Smith, 1985). This variability in endogenous ecdysteroid titer is attributed to biosynthesis regulated by prothoracotropic hormones, and breakdown. The 20-hydroxyecdysone molecule is susceptible to oxidations and conjugations in many places and all changes of the molecule remove the hormonal activity:

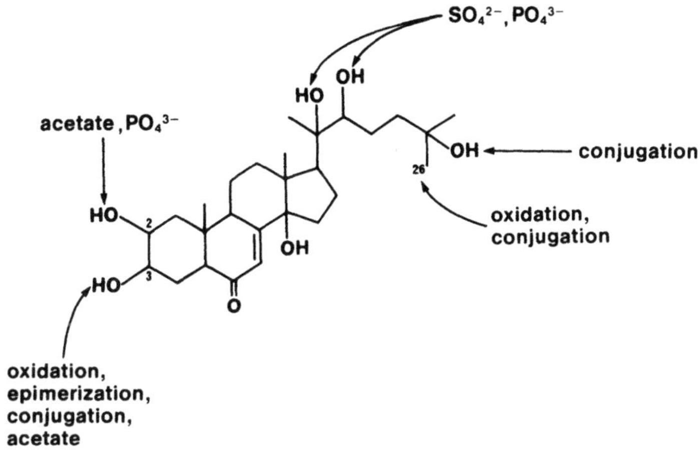

The major inactivation mechanism probably varies with species, perhaps even with life stage. Hydroxylation occurs primarily at C26 and C20 (Lafont et al., 1983; Thompson et al., 1985). The hydroxylated C26 can be further oxidized to the carboxyl acid. Both the hydroxylated metabolites and the acidic product, ecdysonoic acid, can then undergo conjugation. These products may be for disposal since their reversal to the active hormone would not easily occur. Other excretion products may be formed by oxidation followed by epimerization at C3 (Karlson and Koolman, 1975; Mayer et al., 1979).

Conjugation can also take place directly via the hydroxyl groups at C3, C4, C22, C23 and C25. These conjugates which can relatively easily be restored to the active aglycone (22-hydroxyecdysone) may be storage forms. The major conjugates are sulfates and phosphates (Koolman and Karlson, 1985). Acetates (Isaac et al., 1984) and nucleotide containing metabolites (Tsoupras et al., 1983) have also been reported. No ecdysteroid glycosides have been found, and insects are not known to utilize glucuronidation as a conjugating mechanism (Ahmad et al., Chapter 3 in this text). Interestingly, Yang and Wilkinson (1972) and Yang et al. (1973) found that sulfotransferase activity in southern armyworm larvae was high between molts and declined prior to molts at the same time as arylsulfatase activity increased (Wilkinson, 1986). This supports the hypothesis that a sulfate conjugate may be a hormone storage form. The arylsulfatase would restore hormonal activity to an inactive sulfate storage form produced earlier in the instar by the sulfotransferase. Insects may also store conjugates of phytosterols for use after release of the aglycone at the appropriate time.

A variety of higher plants contain phytoecdysteroids, phytosterols with molting hormone activity, notably ecdysone and 20-hydroxyecdysone. Some of the phytoecdysteroids, e.g., ponasterone (Kubo and Klocke, 1986) are toxic at relatively high concentration to lepidopterous larvae that are generalist feeders; most phytoecdysteroids are harmless to herbivorous insects at the concentrations at which they occur in plants (Jones and Firn, 1978). This is not surprising considering the many mechanisms insects have for inactivating endogenous ecdysteroids that, presumably, would also work on ingested ones.

3.3. Cardenolides

The cardenolides are C3 glycosylated C23 steroid derivatives, often with unusual hexoses attached. They occur in several plant families, notably, Apocynaceae and Asclepiadaceae. They are notoriously very toxic to mammals. By inhibiting Na,K-dependent ATPases, they disrupt the ratio of sodium to potassium to which the heart muscle is particularly sensitive; the effect is a decreased rate of heart beat with increased intensity and is usually lethal except when the compound is used in minute, therapeutic doses. The inhibitory effect appears to be related to the steric relationship between the unsaturated lactone substituent and the C14 hydroxyl group.

The large milkweed bug, Oncopeltus fasciatus Dallas (Hemiptera: Lygaeidae), has ATPases with reduced sensitivity to cardenolide inhibition (Moore and Scudder 1986). The gut ATPases of monarch, Danaus plexippus L. (Lepidoptera: Danaidae), larvae also have diminished sensitivity to ouabain, the most commonly therapeutically used cardenolide (Jungreis and Vaughan, 1977). This target site insensitivity is likely the major factor which enables the large milkweed bug and the monarch to store the compounds (Berenbaum, Chapter 7 in this text) presumably for their own protection. Other insects including certain grasshoppers and arctiids are also known to store cardenolides obtained from their host plants, but the sensitivity of their ATPases to cardenolide inhibition has not been investigated.

It is well established that insects store somewhat different cardenolides from those found in their host plants in many cases. This has been shown with the large milkweed bug (Moore and Scudder, 1985), the monarch (Roeske et al., 1976), and the grasshopper Poikilocerus bufonius (von Euw et al., 1967). The selective storage could be due to selective uptake, metabolism, or both. Scudder and Meredith (1982) showed that the relatively polar non-milkweed cardenolide ouabain was very slowly sequestered through the gut; three percent of the dose was translocated to the hemolymph in 30 minutes. In contrast, a relatively non-polar non-milkweed cardenolide, digitoxin, was taken up rapidly; 77 percent of the dose was recovered in the hemolymph after 30 minutes. As with synthetic insecticides which like the cardenolides penetrate membranes by passive diffusion along a concentration gradient, it is likely that the cardenolides should have some optimum polarity to be efficiently sequestered (Brooks, 1976; Shah and Guthrie, 1985). It is also likely that a selective sequestration process is obscured by subsequent metabolism.

In the large milkweed bug, a second sequestration process occurs to accumulate the cardenolides in the modified, highly vacuolized inner epidermal layer of their cuticle, the "dorsolateral spaces" (Scudder et al., 1986). The uptake into the dorsolateral spaces differs from that through the gut; the cardenolides are accumulated against a concentration gradient by an unknown mechanism. Moore and Scudder (1985) found predominantly relatively nonpolar cardenolides in the dorsolateral spaces. A possibility would be a metabolic mechanism in the epidermal cells whereby, like in the gut cells, cardenolides of optimum polarity in the hemolymph are allowed in and, inside, converted to less polar metabolites that would, by lipophilic interaction, accumulate in the vacuole membrane and from there spill over into the vacuole. This possibility could be approached experimentally.

Unlike the large milkweed bug and P. bufonius which store cardenolides in special morphological structures, epidermal vacuoles and poison "glands", the monarch has no special storage space. In adult monarchs, the cardenolides are accumulated in the wings, abdomen, and thorax (Brower and Glazier, 1975). The monarch, thus, does not have the problem to take up cardenolides through a second membrane barrier. It does, however, store different cardenolides in different body parts; those stored in the abdomen, although present in lower concentration, are more emetic when tested with blue jays than those in the wings (Roeske et al., 1976). This could be related to the relative lipid content in the body parts that cardenolides with different polarities would have affinity to.

The stereoisomers calactin and calotropin have been found in P. bufonius (von Euw et al., 1967) and monarchs (Reichstein et al., 1967). Seiber et al. (1980) found calactin and calotropin in monarch larvae fed uscharidin and no other cardenolide. They, thus, demonstrated in vivo metabolism of uscharidin to calactin and calotropin. In agreement with this, Marty and Krieger (1984) showed that soluble carbonyl reductases (Ahmad et al., Chapter 3 in this text) prepared from midgut and fatbody tissues of monarch larvae convert uscharidin to calactin and calotropin in vitro. The fatbody was less active than the gut and produced only calotropin, whereas both stereoisomers were produced by the midgut enzymes.

4. PHENYLPROPANOIDS

Phenylpropanes are considered derivatives of a six-carbon aromatic unit and a three-carbon unit. Most of these, including many plant phenols, coumarins, and lignans, arise biosynthetically via the shikimic acid pathway (Vickery and Vickery, 1981). The chromenes bear structural similarity to the coumarins, but unlike the latter, their biosynthetic origin seems to be undocumented.

4.1. Precocenes

The two chromenes, precocene I (7-methoxy-2,2-dimethylchromene) and precocene II (6,7-dimethoxy-2,2-dimethylchromene) isolated from Ageratum houstonianum disrupt the immature molts of certain insects (Bowers 1976). In susceptible insects such as the large milkweed bug, this effect derives from specific and complete destruction of the corpora allata (Nair et al., 1981), the site of juvenile hormone synthesis. Most of the work done to elucidate the mode of action of the precocenes has employed precocene II, the less volatile and more toxic of the two; it provides a first- rate example of the great importance of the concerted action of several different xenobiotic-metabolizing enzymes to offset the inherent hazards in cytochrome P-450-catalyzed reactions that produce polar and therefore in some cases more reactive metabolites.

Juvenile hormones are made by epoxidation of methyl farnesoate (Hammock and Mumby, 1978) in the corpora allata, although they can also arise from methylation of 10,11-epoxyfarnesoic acid (Schooley and Baker, 1985). To produce enough juvenile hormone it is necessary to either have a large excess of the lipophilic precursor, an unusually active epoxidase, or a high concentration of epoxidase enzyme. Corpora allata cells are characterized by extensive endoplasmic reticulum membranes (Cassier, 1979; Feyereisen et al., 1981), and so appear to have taken the route of having very high concentrations of cytochrome P-450 to catalyze the necessary epoxidation of methyl farnesoate. The corpus allatum is "an endocrine gland with highly dedicated biochemistry" (Ellis-Pratt, 1983). There is no evidence of any cytochrome P-450 reaction other than epoxidation going on in it, and no evidence of epoxide hydrolase or group transferase activities (see Ahmad et al., Chapter 3 in this text). Worse, there seems to be a shortage or perhaps even an absence of water, dispensible nucleophilic groups such as glutathione, or storage and transport proteins in the corpora allata.

A lipophilic precocene will readily enter this environment and be accepted as a substrate by cytochrome P-450; it is converted to the 3,4-epoxide which is extremely reactive and readily forms adducts with nearby macromolecules:

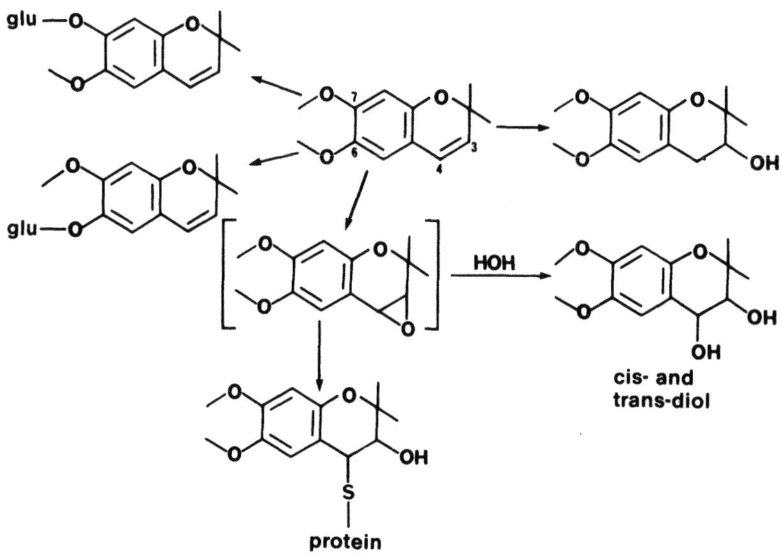

The only metabolites found in the corpora allata are the trans- and cis-3,4-diols that likely formed by spontaneous reaction with water (perhaps from the buffer) of the cytochrome P-450-produced 3,4-epoxide (Ellis-Pratt, 1983). Epoxide hydrolase inhibitors do not prevent the diol formation (Burt et al., 1978). The diols are not formed and there is no necrosis of corpora allata in the presence of cytochrome P-450 inhibitors (Brooks et al., 1979a, 1979b). Aizawa et al. (1985) showed that the 3,4-epoxide rapidly forms adducts with thiols such as thiophenol and cysteine and also with some amino-compounds e. g., morpholine.

Precocene II is more extensively metabolized in other insect tissues. In the cabbage looper and the European corn borer, Ostrinia nubilalis Hubner (Ibid.), the fat body was most active in metabolizing precocene II followed by the gut; enzymes in Malpighian tubules silk gland, head, and cuticle also metabolized the compound (Burt et al., 1978). The 3,4-diols were the major metabolites and hydroxylation at C3 was also substantial in vitro, as well as demethylation at C6 and C7 (Soderlund et al., 1980). Bergot et al. (1980) showed that glucose conjugates are formed in vivo in several insect species from 6- and 7-monodemethylated precocene II. There may be substantial differences in the rate of overall metabolism between sensitive and insensitive species. The large milkweed bug (sensitive) had metabolized 10 percent of a 50 ng dose in 20 hours, whereas the insensitive corn earworm, Heliothis zea Boddie (Lepidoptera: Noctuidae) had in the same time metabolized 90 percent of the dose (Haunerland and Bowers, 1985).

In addition to extensive metabolism in extra-allatal tissues in insensitive species, notably lepidopterans, the uptake of precocene may differ in sensitive and insensitive species. After topical application, more than 80 percent of the dose was found in the gut of the corn earworm, whereas more than 80 percent of the dose accumulated rapidly in the fat-body of the large milkweed bug from where it was slowly released (Haunerland and Bowers, 1985). Insects also have special hemolymph proteins that bind xenobiotics either for storage or for transport. Precocene II binds to lipophorin and arylphorin in corn earworm larval hemolymph (Haunerland and Bowers, 1986). Lipophorin is a highly lipophilic protein with a high percent of lipids that nonspecifically binds xenobiotics of intermediate polarity; arylphorin is a large glycoprotein that may be important in carrying sequestered xenobiotics over from the larval to the adult stage; it appears in late instar larvae and occurs at high concentrations in pupae but is absent from adult hemolymph (Haunerland and Bowers, 1986).

4.2. Furanocoumarins

Berenbaum (1981) has shown that the phototoxic furanocoumarins are lethal to generalist insect herbivores such as the southern armyworm, although adapted specialists such as black swallowtail, Papilio polyxenes F. (Lepidoptera: Papilionidae), larvae tolerate them and use them as feeding cues. This seems to be a case of metabolic resistance enhanced by rapid excretion. In black swallowtail larvae, the linear furanocoumarin xanthotoxin is rapidly metabolized to two nontoxic products by enzymes in their gut tissues (Ivie et al., 1983; Bull et al., 1984):

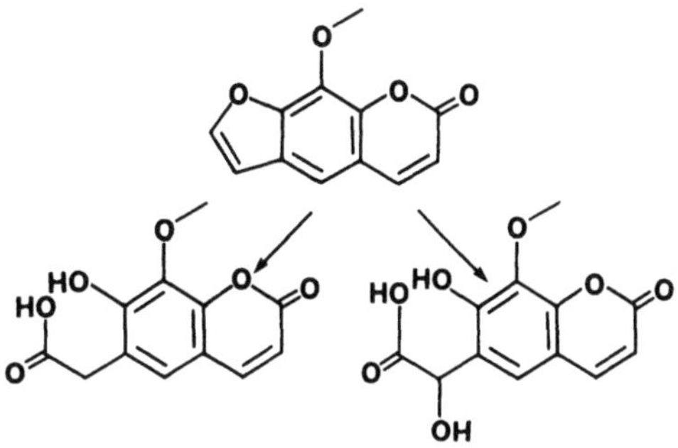

Fall armyworm larvae are generalist feeders and susceptible to the toxicity of xanthotoxin. Nevertheless, fall armyworm larvae produce the identical two metabolites but at a 33-fold slower rate. As a consequence, the toxic parent compound accumulates in their body tissues at a 60-fold higher concentration than in black swallowtail larvae. The latter had also excreted 50 percent of the dose in 90 minutes, whereas the fall armyworm larvae had only excreted one percent of the dose in the same time (Ivie et al., 1983; Bull et al., 1984).

The xanthotoxin metabolites are produced by gut cytochrome P-450; they are not formed by gut homogenates or fractions in vitro in the presence of piperonyl butoxide (Bull et al., 1986). This is in agreement with the increased toxicity in vivo of xanthotoxin to corn earworm larvae when fed to them together with myristicin (Berenbaum and Neal, 1985). Myristicin is a lignan with the methylenedioxyphenyl group characteristic of cytochrome P-450 inhibitors; it occurs along with xanthotoxin in many umbellifers. Bull et al. (1986) also showed that last instar black swallowtail larvae that had been feeding on parsley since hatching, epoxidize aldrin extremely rapidly. This is in agreement with the unusually high titer of cytochrome P-450 and rapid N-demethylation of p-chloro N-methylaniline (Brattsten, 1979) in last instars fed carrot foliage since hatching. The rate of microsomal aldrin epoxidation in the swallowtail larvae was 2.6-fold higher than in parsley-fed fall armyworm larvae, a sufficient molecular difference to result in a large difference in susceptibility to a toxicant; this is known to be the case also for synthetic insecticides (Brattsten, 1986b).

Angular furanocoumarins are considered less phototoxic than the linear ones such as xanthotoxin owing to their ability to form DNA-adducts with one end of the molecule but not cross links. Both linear and angular furanocoumarins may have light-independent toxic effects as well. Berenbaum (Chapter 7 in this text) pointed out that it would be difficult or impossible to rely on target site insensitivity, even though that has clear advantages over reliance on detoxification, for feeders of plants with an array of allelochemicals with different toxic modes of action. It may, likewise, be difficult or impossible to develop target site insensitivity to an allelochemical with multiple toxic effects. In such such situations, efficient metabolic degradation may be the most "economic" defense. Even though black swallowtail larvae easily contend with linear furanocoumarins, they suffer reduced growth and fecundity from ingesting angular furanocoumarins (Berenbaum and Feeny, 1981). This toxicity may have its explanation in their slower rate of metabolic degradation. Ivie et al. (1986) showed that 90 minutes after a dose of either psoralen (linear) or isopsoralen (angular), black swallowtail larval body tissues contained twice as much unmetabolized isopsoralen as psoralen. They also showed that isopsoralen was metabolized two to three times more slowly than the linear analog.

4.3. Phenolics

Phenolics are extremely abundant plant allelochemicals often associated with feeding deterrency and/or growth inhibiting effects (Feeny, 1970 and 1976; Steinly and Berenbaum, 1985). A case of their nutritional utilization by insects is the incorporation of gallic acid and other phenolics into the sclerotized exoskeleton of a tree locust, *Anacridium melanorhodon* (Walker) (Orhtoptera: Acrididae) and the silkworm *Bombyx mori* (L.) (Lepidoptera: Bombycidae) (Bernays and Woodhead, 1982a, 1982b; Bernays et al., 1983). In this case the ingested plant phenols may spare the insect phenylalanine which would otherwise be used for sclerotization and which may, as well as other amino acid and protein nutrients, be in short supply for tree-feeding insects (Bernays and Woodhead, 1984).

Ingested phenolics may also be used as precursors for insect defensive chemicals as exemplified by the transformation of salicin to salicylaldehyde, a strong repellent to ants and other insect predators, in chrysomeline larvae (Pasteels et al., 1983 and 1984). In some cases the salicin obtained by the adult from their willow (Salicaceae) food plants are transmitted to the eggs; the neonate larvae convert it to salicylaldehyde before they have fed (Pasteels et al., 1986). Salicylaldehyde was not found in the eggs, implying that the full complement of enzymes necessary for the conversion develops very rapidly in newly hatched larvae.

5. FLAVONOIDS

The 2000 or so identified flavonoids are classified into 12 groups depending on the oxidation level of their central ring (Harborne, 1979). Just one of the groups, the anthocyanins, may be responsible for their perhaps most spectacular role in flowering plants, the colors of flowers. Flavonoids occur in the form of glycosides in plants; there are often several different glycosides for each aglycone either by the attachment of different sugars or by their attachment in different positions. The flavonoid aglycones are polyhydroxylated molecules and sometimes referred to as polyphenolics. With the exception of the isoflavonoids, the flavonoids are not known as highly toxic compounds. This could be a function of efficient metabolism in animals ingesting them. The commonly occurring flavonol aglycone quercetin interferes with trans-membrane ion transport by inhibiting a calcium-dependent ATPase pump (Racker et al.,

1980) and other flavonols are also cytotoxic. As with other classes of allelochemicals with a potential for feeding deterrency and growth inhibition, herbivorous insects have in many cases adapted to them or utilize them. Quercetin and several of its glycosides deterred feeding by tobacco budworm and pink bollworm larvae and were also toxic at higher concentrations (Shaver and Lukefahr, 1969). Rutin, another of the quercetin glycosides, stimulated growth and fecundity in the house cricket (McFarlane and Distler, 1982). Rutin did not affect the food consumption rate in the crickets but improved the digestion of the ingested food. The reasons for this effect are unknown.

It has long been known that flavonoids are stored by certain lepidopterans; Thomson (1926), Ford (1941, 1944), and Morris and Thomson (1964) identified anthocyanins in the wings of butterflies. The marbled white butterfly, Melanargia galathea L. (Lepidoptera: Satyridae), use flavonoids in their larval food plants for wing pigmentation (Wilson, 1985a). As with insects storing cardenolides, the Melanargia larvae selectively sequester and metabolize the host plant flavonoids so that the wings consistently contain a pattern of flavonoids that is only in part identical to that in the grass ingested by the larvae. The larvae seemed to selectively sequester tricin and also convert it to the 4'-glycoside which was absent in the larval food plants. By comparing flavonoid wing patterns in wild collected butterflies and in butterflies from larvae reared on 25 species of grasses, Wilson (1985a) could also pinpoint the host range of the species; instead of marbled whites being generalist grass feeders as assumed, they seem to be specialized to Festuca rubra and a few related grasses; the flavonoid pattern in the wild butterflies consistently corresponded most closely to that in butterflies from larvae reared on F. rubra. The larvae also grew and developed best on F. rubra. Moreover, analysis of several other Melanargia species upheld the constancy and specificity of the wing flavonoid pattern and its relation to the larval food plants (Wilson, 1985b).

The isoflavone rotenone is probably the only flavonoid whose primary metabolism has been studied in detail in insects; the obvious reason for this attention its the insecticidal properties and use of rotenone which is oxidized

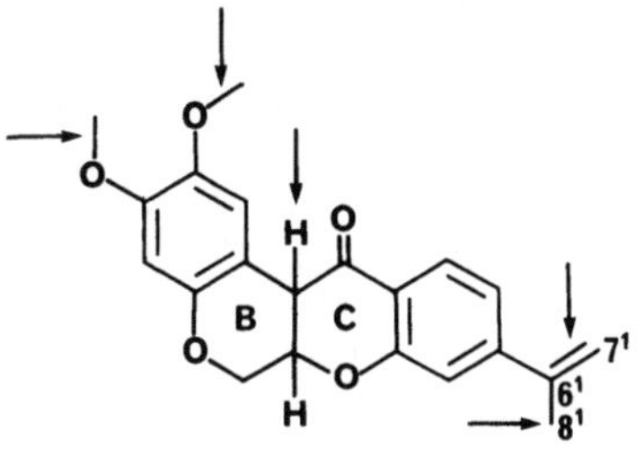

in several different positions in house flies, German, Blattella germanica (L.) (Orthoptera: Blattidae), American, Periplaneta americana (L.) (Ibid.), cockroaches, and mice (Fukami et al., 1967 and 1969). The 6',7'-dihydro-6',7'-dihydroxy rotenone was the major metabolite in all species and formed in quantity only in the presence of NADPH. This metabolite is produced by a cytochrome P-450-catalyzed epoxidation followed by hydration of the epoxide (see Ahmad et al., Chapter 3 in this text). The metabolite production was inhibited by several typical cytochrome P-450 inhibitors such as piperonyl butoxide. This metabolite was three to four times less toxic than rotenone to male mice. Hydroxylation between the A and B rings most effectively eliminated the ability of rotenone to inhibit mitochondrial respiration and thereby its toxicity. It seems likely that the oxidation products of rotenone undergo conjugation and subsequent excretion.

There is no information on the secondary metabolism of rotenone; this is the case most often with work undertaken to elucidate the detoxification of insecticides.

Several isoflavones are estrogen agonists and can thereby interfere with reproduction in birds and mammals (see Harborne, 1979). Insects are not known to be affected in any such way. Insects, however, have estrogens and androgens of unknown function (Mechoulam et al., 1984).

6. AMINO ACIDS

More than 200 nonprotein amino acids are known from higher plants and several of these are known for their toxicity or lethality to herbivorous insects. Only the interactions of L-canavanine with insect biochemistry have been investigated, however. Canavanine is the major nitrogen storage form of seeds of certain neotropical legumes (Rosenthal and Bell, 1979) which are eaten by extremely few insects, the exceptions being some beetle larvae. Larvae of *Caryedes brasiliensis* (Coleoptera: Bruchidae) develop in the seeds of *Diochlea megacarpa* and *Sternechus tuberculatus* (Coleoptera: Curculionidae) larvae may be specialized to develop in seeds of *Canavalia brasiliensis* (Rosenthal, 1986); both plants have very high concentrations of canavanine in their seeds.

Canavanine is toxic to insects by virtue of being "mistakenly" incorporated into proteins instead of arginine; this results in aberrant and malfunctioning proteins (Rosenthal and Dahlman, 1986; Rosenthal, 1986). Insects would be especially sensitive to canavanine at times when critical proteins are synthesized, for example during molts and egg production. Canavanine causes lethal morphological malformations in tobacco hornworm pupae, by this mechanism (Rosenthal and Dahlman, 1986; Rosenthal, 1986). It also causes sterility in adult migratory locusts, *Locusta migratoria migratorioides* (Orthoptera: Acrididae) (Pines et al., 1981) and in adult *Dysdercus koenigii* (Hemiptera: Pyrrhocoridae) (Koul, 1983).

Caryedes brasiliensis grubs are resistant to the toxic effects of canavanine by a target site insensitivity mechanism (Rosenthal, 1986; see Berenbaum, Chapter 7 in this text) which allows them to utilize this toxic amino acid to satisfy their own nitrogen requirements. They do this by metabolic mechanisms that also effectively detoxify the compound. It is not possible to do justice here to all the exquisite and well documented mechanisms whereby *C. brasiliensis* accomplishes this. The success of the beetle is related to a simple property of canavanine: it is sufficiently dissimilar to arginine to be, at best, a very marginal substrate for the arginyl tRNA-synthetase of the larvae, but sufficiently similar to be accepted as a substrate by their arginase.

Arginase converts canavanine to canaline and urea. Canaline is a neurotoxic amino acid which the grubs must also have a defense against; it is toxic to tobacco hornworm moths (Rosenthal and Dahlman, 1975). In *C. brasiliensis*, canaline is detoxified by conversion to ammonia and homoserine, the latter being cycled into the amino acid pool and the ammonia being utilized for nitrogen (Rosenthal et al., 1978). The conversion of canaline to homoserine and ammonia is catalyzed by a dehydrogenase and requires energy in the form of reduced nicotinamide adenine dinucleotide (NADH). This energy is, however, regenerated in subsequent reactions of homoserine.

The urea is converted to ammonia and carbon dioxide by urease which is highly active in *C. brasiliensis* (Rosenthal et al., 1977) but presumed to be generally rare in insects (Dowd et al., 1983). Baldwin (1964), however, states that "urease [is] commonly present among invertebrates". Uricolytic enzymes have undergone "negative specialization" in the animal kingdom and disappear "up the evolutionary scale" as it progresses from aquatic to terrestrial life styles. If not the urease protein itself, the genetic competence for it may still exist at least sporadically in terrestrial insects.

The relatively large amount of ammonia thus generated also poses a threat of toxicity; ammonia kills rabbits at very low concentrations (Baldwin, 1964) and southern armyworm moths (Brattsten, unpubl. observation). The beetle larvae actually excrete very little ammonia and instead cycle most of it into the amino acid metabolic network. They may use glutamic acid dehydrogenase and glutamine synthetase (Rosenthal and Janzen, 1985). The former enzyme uses ammonia and 2-oxoglutaric acid to form glutamic acid, and glutamine synthetase uses ammonia and glutamic acid to form glutamine. Both enzymes are particularly active in C. brasiliensis larvae, compared to other insects and even adult beetles (Rosenthal and Janzen, 1985), and their activities could fully account for ammonia detoxification in the grubs.

C. brasiliensis is an unusually interesting and well investigated case of canavanine resistance in a highly specialized insect; the biochemical pathways used by the beetle larvae may be identical to those used by their host plants to protect them from canavanine autointoxication (Rosenthal et al., 1978). Such resistance may also occur in other insect herbivores since all the special adaptations are based on enzymes, or possibly at least genes for enzymes, already endowed in the intermediary metabolism of most organisms. It may then be most likely in generalist plant-feeders. The tobacco budworm, the corn earworm, and the fall armyworm, all generalists, are also relatively insensitive to the toxic effects of canavanine (Berge et al., 1986).

A canavanine dose of five grams per kilogram was taken up but cleared from the hemolymph of tobacco budworm larvae in 12 hours; the hemolymph half-life was a short 135 minutes. Nevertheless, less than one percent of a dose ingested over three days was excreted unmetabolized (Berge et al., 1986). This insect, unlike the tobacco hornworm which is an alkaloid specialist, but like C. brasiliensis, incorporates very little canavanine into its proteins. The ratio of canavanine to arginine incorporation into proteins was one in 365 for C. brasiliensis, one in 100 for the tobacco budworm, but one in 5.6 for the tobacco hornworm (Rosenthal, 1986). It thus appears that the tobacco budworm must also have the enzymes to effectively metabolize canavanine or, at least the genetic capacity to respond to dietary presence of canavanine. Comparative enzyme studies, however, did not reveal extraordinary arginase utilization of canavanine in the tobacco budworm as in C. brasiliensis (Rosenthal and Janzen, 1983), or comparably high activities of either glutamic acid dehydrogenase or glutamine synthetase (Rosenthal and Janzen, 1985). It appears that the tobacco budworms used for these studies had not been exposed to dietary canavanine prior to the enzyme assays to allow possible induction of the activities to take place. Other possibilities for detoxifying canavanine may be glycoside or dipeptide (amino acid conjugation, see Ahmad et al., Chapter 3 in this text) formation (Chen, 1985). An amino acid would be a highly unusual substrate for cytochrome P-450.

7. CYANOGENIC GLYCOSIDES

Cyanide was once the dominating component of the earth's atmosphere and is, perhaps as a consequence, still essential in trace quantities in many biological and biochemical systems. It is, however, highly toxic to most extant organisms at very low concentrations. It is generated, mostly in the form of glycosides, by some bacteria, many plants, and by a few arthropods including polydesmid millipedes, some chrysomelid beetles and a few lepidopterans, in some cases in enough quantity to kill small mammalian predators. Plants and arthropods produce cyanogenic substances, presumably for defense. Larvae of Zygaena trifolii Esper (Lepidoptera: Zygaenidae) make linamarin and lotaustralin from valine and isoleucine (Wray et al., 1983). The cyanogenic glycosides may be biosynthesized in the fat body; they occur in the hemolymph from where they are translocated and stored in epidermal cavities in the dorsal and lateral parts of the cuticle (Franzl et al., 1986). The glycosides are emitted through ruptures in thin cuticle sections (Povolny and Weyda, 1981; Franzl and Naumann, 1985) when the larvae are disturbed. This arrangement bears remarkable resemblance to the cardenolide-storing dorso-lateral spaces of the large milkweed bug (see above). The insects use identical biosynthetic pathways to those of plants to produce their cyanogens (Duffey, 1981; Davis and Nahrstedt, 1985). Many insects are insensitive to cyanide and may employ defense mechanisms also identical to those of plants.

Cyanogenic plants have their cyanogenic glycosides compartmentalized to separate them from the enzymes that liberate the cyanide. Dhurrin, a phenylalanine-derived cyanogenic glycoside in sorghum, occurs only in the epidermal plant tissues, whereas the two enzymes necessary for its degradation reside in the mesophyll (Conn, 1979 and references therein). When an insect crushes the leaf during feeding the tissues are mixed and cyanide is liberated. The insect, thus, ingests cyanide or hydrocyanic acid.

The cyanide ion is a highly reactive nucleophile and can inhibit some 90 enzymes; the irreversible inhibition of cytochrome oxidase, the terminal oxidase in the mitochondrial respiratory pathway constitutes the toxic lesion by the overall inhibition of energy (ATP) production. Some plants, e. g. the skunk cabbage, have an alternative enzyme, an autooxidizable flavoprotein that does not bind cyanide, to perform the terminal oxidation (Henry and Nyns, 1975). This target site insensitivity may occur also in some insects (see Berenbaum, Chapter 7 in this text).

Cyanide can be eliminated in two major ways: by being incorporated into the sulfane sulfur pool, or by incorporation into the amino acid pool.

Thiocyanate is formed by the combination of cyanide and a sulfur atom attached to another sulfur as in disulfides, persulfates, polysulfates, protein persulfides, thiosulfates, and others (Westley, 1981). Thiocyanate is formed in vertebrates, invertebrates, plants and fungi, and many kinds of microorganisms. At least two enzymes can catalyze the formation of thiocyanate. The thiosulfate: cyanide sulfurtransferase also called rhodanese is the most familiar one (see Ahmad et al., Chapter 3 in this text). Rhodanese is a constitutive, mitochondrial enzyme that may be important in the turnover of mitochondrial non-heme iron respiratory proteins and is probably ubiquitous in organisms. Its importance in cyanide detoxification is probably limited by its location exclusively in the mitochondrial matrix whence the thiosulfate ion can penetrate only slowly, whereas cyanide is an extremely fast acting poison (unusual for a non-nerve poison). Mercaptopyruvate sulfurtransferase (E.C. 2.8.1.2) is another enzyme that can catalyze the formation of thiocyanate (Westley, 1981) by the concomittant conversion of mercaptopyruvate to pyruvate. This enzyme has been found in animals, fungi and heterotrophic bacteria; it has not been studied in insects and perhaps not in plants. It appears to be a cytosolic enzyme which would help in a cyanide emergency if sufficient mercaptopyruvate were available. It also seems that there is appreciable spontaneous reaction to form thiocyanate. The enzyme blanks in the rhodanese assays of Long and Brattsten (1982) were never completely colorless; the activities were based on the difference in absorbance between the incubations and the appropriate blanks. Therefore, a possible defense would be for organisms exposed to cyanide to store a sufficiently high concentration of sulfane sulfur compounds in their body fluids for the cyanide to be bound by. The extreme reactivity of the cyanide ion may make such a mechanism feasible. It would account for the thiocyanate formation observed in vivo and would be reflected in the activities of enzymes which contribute to the maintenance of the sulfane sulfur pool. There are more reasons than cyanide toxicity for studying sulfur metabolism in insects, a virtually unknown area.

Incorporation of cyanide into the amino acid pool seems to be the route plants have taken for eliminating excess cyanide. The first step is the combination of cyanide with cysteine to form β-cyanoalanine. This reaction was first discovered if not explicitly described in an insect. Bond (1961) found the major portion of [^{14}C] cyanide inhaled by the granary weevil, Sitophilus granarius (L.) (Coleoptera: Curculionidae), excreted as labeled amino acids and incorporated in the aspartyl residue of a polypeptide. The reaction and pathway were described from plants (Blumenthal-Goldschmidt et al., 1963) and have also been found in bacteria (Castric, 1981), millipedes (Conn, 1979; Duffey, 1981) and in several more insects (Duffey, 1981; Davis and Nahrstedt, 1985). The reaction is catalyzed by β-cyanoalanine synthetase (E.C. 4.4.1.9). This enzyme shows some of the adaptations useful for detoxification. The activity is higher in cyanogenic plants than in acyanogenic ones (Miller and Conn, 1980). The activity is present in feeding larvae of Heliconius melpomene (Lepidoptera: Nymphalidae) but not in pupae (Davis and Nahrstedt, 1985). The product, β-cyanoalanine, is a neurotoxic amino acid that can kill vertebrates (Rosenthal, 1979) and insects ; it caused lethal water loss in the migratory locust (Schlesinger, 1976) and killed larvae of Callosobruchus maculatus (Coleoptera: Bruchidae) (Janzen et al., 1977). The β-cyanoalanine synthesis will work as a detoxification mechanism only if the insect also has an active β-cyanoalanine hydrase for converting the β-cyanoalanine to asparagine which is subsequently deaminated to aspartic acid. Aspartic acid can then be converted to other amino acids. The granary weevil has both enzymes since labeled carbon from inhaled cyanide turned up in aspartic as well as other amino acids.

8. GLUCOSINOLATES

Cruciferous plants are characterized by an unusual type of glycosides, the glucosinolates or mustard oil glycosides. These compounds have glucose attached by a β-bond via a sulfur atom. The plants also contain a thioglycosidase (thioglucoside glucohydrolase E.C. 3.2.3.1) commonly called myrosinase (see Ahmad et al., Chapter 3 in this text). The enzyme is physically separated from the glucosinolates in the plant tissues (van Etten and Tookey, 1979 and references therein) as in the case of the cyanogenic glycosides and the cyanide-liberating β-glycosidases and nitrile lyases in cyanogenic plants. When the tissues are crushed, myrosinase hydrolyzes the β-thioglucosidic bond and the mustard oils are released:

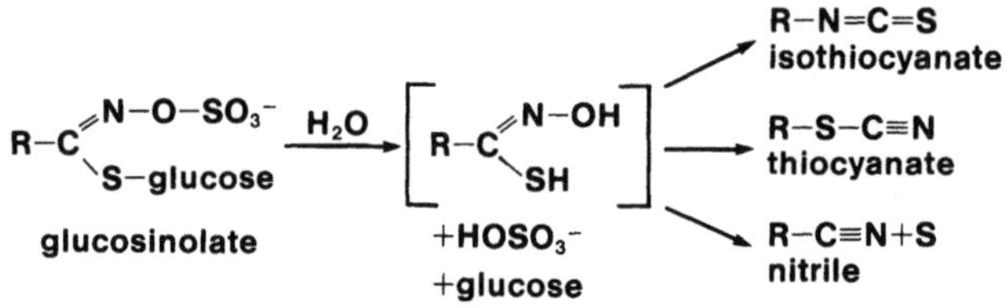

These are lipophilic straight or branched aliphatic or hydroxylated carbon chains or aromatic or heterocyclic moieties. The sulfate group is also disconnected during the thioglucosidase reaction. The organic products are unstable and rearrange to isothiocyanates, nitriles (organic cyanide), or thiocyanates which are ingested. The most common end products are isothiocyanates which make radishes, watercress and other crucifers taste hot.

Myrosinase occurs in many bacteria including intestinal ones (Oginsky et al., 1965) but has been reported from only one insect, the cabbage aphid, Brevicoryne brassicae (L.) (Homoptera: Aphidae) (MacGibbon and Allison, 1971). This may be a thioglucosidase that serves some special function in the biochemistry of aphids (see Mullin, Chapter 5 in this text) and also hydrolyses glucosinolates. This aphid specializes to feed on glucosinolate-containing plants and does not crush the plant tissues during feeding, an advantage considering the toxic effects of the aglucone products. Unless the cabbage aphid has some special use for isothiocyanates, thiocyanates, or nitriles, it would be better off without this enzyme.

The isothiocyanates are toxic to nonadapted insect herbivores. Blau et al. (1978) showed that black swallowtail larvae suffered heavy mortality from concentrations of allyl glucosinolate, possibly converted to allyl isothiocyanate, that did not at all affect larvae of the imported cabbageworm, Pieris rapae (L.) (Lepidoptera: Pieridae) and inhibited growth in southern armyworm larvae only mildly. The imported cabbageworm is a specialist on glucosinolate-containing plants; its detoxifying enzymes have not been studied and the mechanism by which it is insensitive is unknown. The southern armyworm is a generalist feeder with relatively high activities of the enzymes involved in foreign compound metabolism. The swallowtail is known to have unusually high cytochrome P-450 levels but its other detoxifying enzyme capacities are unknown.

Thiocyanates interfere with iodine uptake by the thyroid and other glands and isothiocyanates have a similar effect owing to their in vivo conversion to thiocyanates (van Etten and Tookey, 1979 and references therein). Ingestion of glucosinolate-containing plants can thus cause disease in vertebrate herbivores and people. Allyl and benzyl isothiocyanate were conjugated to glutathione in rat liver and kidney in vitro, and the corresponding mercapturic acid was excreted in vivo by rats (Brusewitz et al., 1977; Scheline, 1978). The glutathione conjugate reacted with a further reduced glutathione and oxidized diglutathione, hydrogen cyanide and the thiol derivative of the thiocyanate were formed:

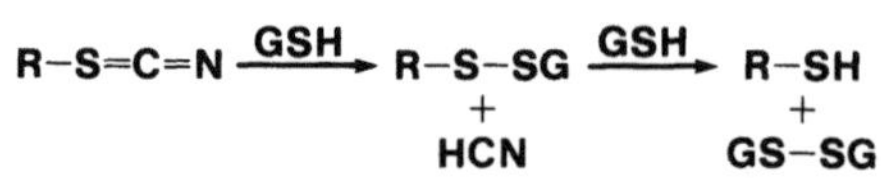

A resistance mechanism based on glutathione transferases may be operating in the southern armyworm larvae; this remains to be investigated.

Cabbage white butterflies, Pieris napi macdunnoughii Remington (Lepidoptera: Pieridae) deposit their eggs on several different species of crucifers (Chew, 1975, 1977). The larvae are very small and unable to move any distance further than a few centimeters. Of the plants the females oviposited on, the larvae developed best on Descurainia richardsonii and were killed by Thlaspi arvense. A subsequent analysis (Rodman and Chew, 1980) showed marked qualitative differences in glucosinolate content between the plants that could influence their suitability as larval food plants; this could also be influenced by quantitative differences between species or individuals of plants and/or by synergistic interactions between the glucosinolates. Differences in non-glucosinolate allelochemicals could also be important but Usher and Feeny (1983), studying the effects of isoquinoline alkaloids, cucurbitacins, and cardenolides occurring in crucifers on larvae of the imported cabbageworm, found no evidence for this.

The ovipositing females apparently detect glucosinolates but do not discriminate between them. This seems to be the case also with females of Anthocharis cardamines ovipositing on Alliaria petiolata on which the larvae survive well and Hesperis matronalis on which they they suffer heavy mortality (Courtney, 1981). Louda and Rodman (1983) found that adapted crucifer-feeders among chrysomelid beetles and psyllids preferred plants with a lower glucosinolate content to those with higher concentration.

The lack of suitability of some of the plants for the larvae that the females select for oviposition sites could derive from a combination of toxic allelochemicals, nutritional, and environmental factors. The considerable quantitative and qualitative differences among the plants in terms of glucosinolate content may counteract the development of adaptations in the insects that feed on them. This situation also indicates the complexity of the adaptation process when both adult behavior and larval xenobiotic and/or nutritional metabolism must coevolve.

9. ALKALOIDS

The true alkaloids are nitrogen-heterocyclic compounds biosynthetically derived from amino acids. Pseudoalkaloids including the insecticidal caffeine are made with purine bases as starting materials (Robinson, 1979; Vickery and Vickery, 1981). Alkaloids occur in most or perhaps all plant families with the exception of the ferns the evolutionary origin of which is earlier than that of all other plants. Not all plants with the biosynthetic capacity for alkaloid production store the compounds and so are not known as "alkaloid plants". Most alkaloids have biological activity. Considering how many alkaloids have been and still are used as drugs, there is surprisingly little detailed information about their metabolism and disposition. The exceptions may be the tobacco alkaloids and the pyrrolizidine alkaloids. Also, considering the common and worldwide occurrence of alkaloids, the information on their effects on and interactions with insects is not overwhelming.

9.1. Tobacco alkaloids

Nicotine is the most toxic of the tobacco alkaloids. It kills nonadapted insects and vertebrates at very low doses by being an acetylcholine mimic (see Berenbaum, Chapter 7 in this text). Insects adapted to feed on tobacco plants some of which contain very high concentrations of nicotine have several effective defenses ranging from metabolic to behavioral. The green peach aphid, Myzus persicae Sulzer (Homoptera: Aphididae), and other aphids feed in the phloem of plants where nicotine may be absent or present in minute concentrations (see Mullin, Chapter 5 in this text).

Tobacco hornworm larvae are protected by several mechanisms, the major one being extra rapid excretion. Very little, if any metabolism occurs; instead, 93 percent of an ingested 0.5 mg dose was excreted unmetabolized in two hours (Self et al., 1964a). This is very fast compared to adult house flies which excrete 10 percent of a dose in 18 hours (Self et al., 1964b). In the hornworm, a very small fraction of the ingested nicotine accumulates (briefly) in the hemolymph from where it is absorbed by an alkaloid-specific transport mechanism located in the Malpighian tubules (Maddrell and Gardiner, 1976). This transport system is saturable and independent of ions; it concentrated nicotine in a bathing fluid to a more than four times higher concentration in the tubule output. Its specificity for alkaloids was demonstrated by competitive inhibition of nicotine transport by other alkaloids such as atropine and morphine. It likely corresponds to a renal transport mechanism for organic bases occurring in vertebrates. The alkaloid-specific transport system was found in several insects but not in all. It was found in tobacco hornworm and Pieris brassicae L. (Lepidoptera: Pieridae) caterpillars but not in the adults of these species. It was also found in the blood-sucking adults of Rhodnius prolixus Stal (Heteroptera: Reduviidae) and in the detritivorous adults of Calliphora erythrocephala (Meigen) (Diptera: Calliphoridae) and the house fly (Maddrell and Gardiner, 1976).

Only about two percent of nicotine exists as a free base at the pH, 6.6, of the hornworm hemolymph (Self et al., 1964a). In the event that some of this nicotine should penetrate the ion-impermeable sheath surrounding the insect nervous system, the nerve cells have cytochrome P-450 which readily detoxifies nicotine (Morris, 1983) in the free base form. Finally, as the sixth line of defense, the nerves of the tobacco hornworm are relatively insensitive to nicotine compared to those of the American cockroach (Morris, 1984; see Berenbaum, Chapter 7 in this text). Thus, even in the almost complete absence of gut metabolism, the tobacco hornworm has no problem eating nicotine-laden tobacco plants.

Self et al. (1964b) found that other lepidopterous larvae such as the tobacco budworm and the cabbage looper, which also invade tobacco plantations, also escape nicotine poisoning by very fast excretion of the unmetabolized compound. Except for the ion impermeable nerve sheath which all herbivorous insects are likely to have (Chapman, 1982), it is not known if they also have all the other adaptations found in the tobacco hornworm for eliminating the toxic effects of nicotine. It is known that these three lepidopterous larvae have a moderately to highly active cytochrome P-450 system in their gut tissues (Gould and Hodgson, 1978; Kuhr, 1970; Tate et al., 1980). It is debatable why nicotine escapes metabolism. One possibility would be that there is less overall metabolic expense in shoving the ingested food very quickly through the gut than in using the NADPH-requiring detoxification process at the rate needed. Another possibility would be that the "right" form of cytochrome P-450 is missing from lepidopterans.

Other tobacco-feeding insects including the tobacco wireworm, Conoderus vespertinus (F.) (Coloptera: Elateridae), the cigarette beetle, Lasioderma serricorne (F.) (Coleoptera: Anobiidae) and the differential grasshopper, Melanoplus differentialis (Thomas) (Orthoptera: Acrididae) as well as adult house flies metabolize nicotine (Self et al., 1964b). The major metabolite in all was cotinine (see Ahmad et al., Chapter 3 in this text). Cotinine is not toxic and is formed by cytochrome P-450-catalyzed hydroxylation of the pyridine carbon adjacent to the nitrogen followed by further oxidation catalyzed by an aldehyde dehydrogenase (Hucker et al., 1960). Cytochrome P-450 inhibitors including the drug extender SKF 525-A inhibited cotinine formation by liver slices, and cyanide, a known inhibitor of aldehyde dehydrogenase activity, also eliminated cotinine formation. Cotinine is the major nicotine metabolite in vertebrates. It is also the major metabolite in German and American cockroaches (Guthrie et al., 1957); few other metabolites were found.

In the southern armyworm which freely feeds on tobacco plants without ill effects, Guthrie et al., (1957) found nine metabolites none of which was positively identified; the metabolites were detected in the frass within five hours after ingestion of the nicotine. The metabolites were probably at least in part produced by cytochrome P-450 which occurs in southern armyworm larvae (Krieger and Wilkinson, 1969); the toxicity of nicotine to southern armyworm larvae was increased by coapplication with the cytochrome P-450 inhibitor piperonyl butoxide and decreased by pretreatment with α-pinene (Brattsten et al., 1977), an inducer of cytochrome P-450 activity (see Yu, Chapter 4 in this text).

9.2. Pyrrolizidine alkaloids

The pyrrolizidine alkaloids are composed of a nitrogen-containing "necine" base esterified to one or two straight or branched aliphatic acids that may be hydroxylated. They occur conspicuously in plants belonging to the genera Senecio (Compositae) and Crotalaria (Leguminosae) and in many other plant families, as well. Pyrrolizidine alkaloids can cause cumulative liver damage and death in people and livestock and have therefore received considerable attention. A pyrrolizidine alkaloid such as monocrotaline is detoxified in the vertebrate liver by esterases and oxidases, probably other than cytochrome P-450 (see Ahmad et al., Chapter 3 in this text):

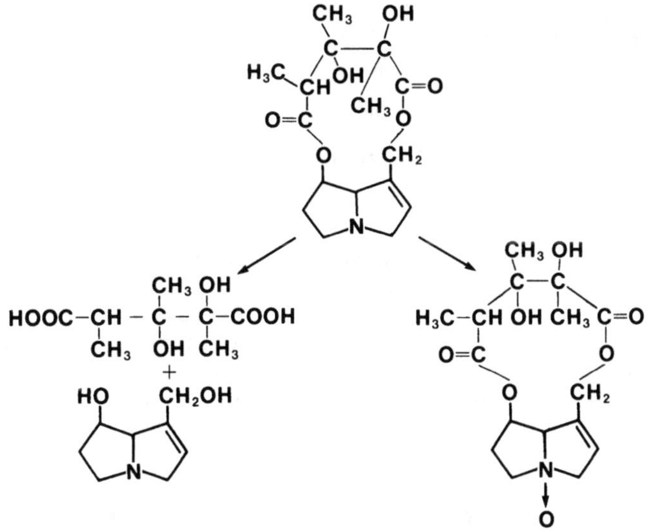

The hydrolysis rate is low (Bull et al., 1968) and the products are readily excreted, as is the N-oxide (Mattocks, 1968). The toxicity of the pyrrolizidine alkaloids is associated with the esterified C1,C2 double bond and depends on metabolic activation.

The liver damage results from alkylation of macromolecules by a reactive intermediate. Mattocks and White (1971) demonstrated the formation in vitro by rat liver microsomes of a highly reactive pyrrole derivative. After phenobarbital pretreatment which induces cytochrome P-450, the rate of the pyrrole formation increased and it was inhibited by the synergist SKF-525A. This corresponded to observed toxicity in vivo of the alkaloids. After incubation with liver slices, the slices turned from colorless to reddish when exposed to Ehrlich's reagent which forms a colored complex with pyrroles; the color could not be washed out completely implying tight binding to the tissue (Mattocks, 1968). The pyrrole is formed by a cytochrome P-450-catalyzed hydroxylation of the carbon adjacent to the nitrogen followed by rearrangement for which the C1,C2 double bond of the parent compound is necessary.

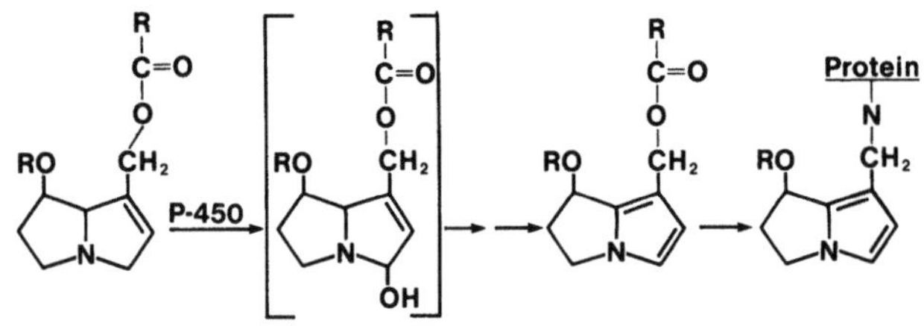

A second, unconfirmed, reactive metabolite may be the trans-4-hydroxy-2-hexenal reported to be formed in incubations with liver microsomes (Segall et al., 1985). This reactive metabolite may result from a highly complex reductive fission of the unesterified pyrrole. Bacteria in the rumen of sheep can accomplish this (Lanigan, 1970; Scheline, 1978). It is unclear how an in vitro liver microsomal system can accomplish the fission. If the pyrrole escaped from the liver, intestinal microbes could probably convert it to the hydroxylated hexenal which would then be expected to cause local damage:

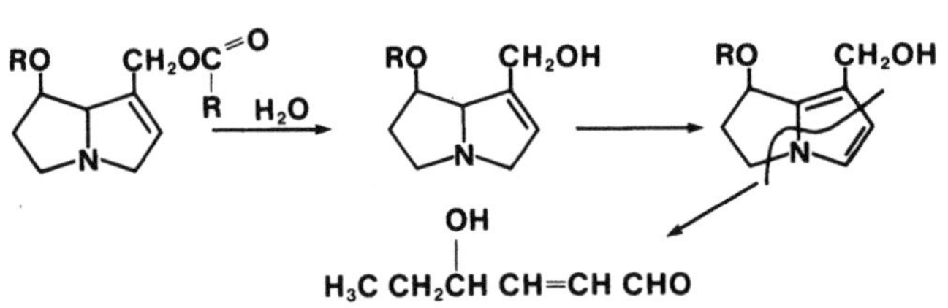

Nothing is known about the general metabolism of pyrrolizidine alkaloids in insects. They are, however, essential for the reproduction of certain danaids and arctiids. Males have special organs, hair pencils or coremata, containing either danaidone or hydroxydanaidal, aphrodisiacal pheromones biosynthesized from ingested pyrrolizidine alkaloids as precursors (Meinwald and Meinwald, 1966; Meinwald et al., 1969; Pliske and Eisner, 1969; Schneider et al., 1975; Conner et al., 1981; Bell et al., 1984).

hydroxydanaidal   danaidone

Apparently, the same metabolic reactions as in the vertebrate liver take place in the insect. If hydrolysis occurs first, a cytochrome P-450, perhaps located in the pheromone gland, could convert the necine base of the alkaloid to a pyrrole which could then be further oxidized to the pheromone. It is not known if insects can make pheromones from pyrrolizidine alkaloids lacking the 1,2 double bond.

Monocrotaline and usaramine in Crotalaria (Leguminosae) plants are feeding cues for larvae of Utetheisa ornatrix (L.) (Lepidoptera: Arctiidae) (Conner et al., 1981). The alkaloids are sequestered by the larvae and carried over to the adults which store them in all tissues; they are also found in the eggs and serve as protection against predators for all life stages. They are probably also oviposition cues.

This is remarkable considering the seemingly unavoidable and cumulative cytotoxicity of the alkaloids following their metabolic activation by cytochrome P-450. An explanation may be that Utetheisa larvae have no cytochrome P-450 form capable of catalyzing the α-hydroxylation of the alkaloid base; they have, however, a measurable level in their gut tissues of a cytochrome P-450 capable of hydroxylating the carbon of p-chloro N-methylaniline adjacent to the amine nitrogen (Brattsten, 1979). It is possible that some small portion of the ingested alkaloid is continually hydroxylated; if the reactive product thus produced alkylates the nearest protein molecule which would be the cytochrome P-450 molecule that produced it, further hydroxylation of incoming alkaloids would not occur due to the irreversible inhibition of cytochrome P-450. As long as the larvae are not exposed to any other xenobiotics needing to be detoxified by cytochrome P-450, its inhibition would be inconsequential, and leave the insects with the opportunity to store the intact alkaloid and to use a fraction of it that may have undergone hydrolysis for pheromone production. There is no experimental evidence for these potential interactions. Cytochrome P-450 in sixth instar southern armyworm larvae that had fed on seeds of Crotalaria spectabilis for 24 hours had a significantly reduced ability to hydroxylate p-chloro N-methylaniline (Brattsten, 1979; see Ahmad et al., Chapter 3 in this text); this provides some circumstantial support for the idea.

Males of Creatonotos gangis and C. transiens (Lepidoptera: Arctiidae) have coremata of different morphological complexity depending on which plant their polyphagous larvae had fed on (Schneider et al., 1982; Boppre and Schneider, 1985). The coremata contained the pheromone hydroxydanaidal in direct proportion to the total amount of pyrrolizidine alkaloid ingested by the larvae. This is not surprising. However, the morphogenic effects of a plant allelochemical on the pheromone gland but on no other aspect of the larval or adult morphology is unique and implies some specific interaction between the genes in that organ and the pheromone, the pyrrolizidine alkaloid, or some intermediate metabolite. Certain intact pyrrolizidine alkaloids, those with an esterified 1,2 double bond are strong mutagens in Drosophila, a grasshopper, marsupials, humans, and a plant (Clark, 1959; Alderson and Clark, 1966; Bull et al., 1968). Low concentrations of these alkaloids cause antimitosis with the consequence that cell division but not cell growth is inhibited (Bull et al., 1968). Some such effect may lead to coremata consisting of giant cells in the Creatonotos moths. The question then becomes why this effect is limited to the coremata of the Creatonotos moths.

One may conclude that, apparently, the behavioral advantages (Eisner, 1980; Conner et al., 1981) gained from the ingestion of pyrrolizidine alkaloids are gained at considerable molecular risks. This may apply to other insect-plant interactions involving toxic allelochemicals, as well.

10. SUMMARY

Herbivorous insects must always avoid the toxic effects of plant allelochemicals. In many cases they use these aquired chemicals for the protection of their own species. They also use plant allelochemicals as feeding cues, oviposition signals, and as pheromone precursors. Many more plant allelochemicals than have been discussed in this paper are toxic to herbivorous insects and/or used by them. In avoiding acute poisoning the insects rely on metabolic transformations of the allelochemicals either to a product stage that allows their elimination or storage, or to the extent of reusing the components of the allelochemical to satisfy their own nutritional requirements. Insects which store quantities of toxic allelochemicals probably always also have an insensitive target site, but reliance on an insensitive target site would not necessarily imply storage, just allow it. Extra rapid excretion of metabolized or unmetabolized toxicants is a known "physiological" defense the biochemical mechanism(s) of which has not been clarified.

It is debatable which one, if any one, of the biochemical and physiological defense mechanisms is the most widespread or effective or important or the earliest one to appear. In natural interactions involving toxic plant allelochemicals where the insect relies on biochemical or physiological defenses, it seems that more than one defense is the rule rather than the exception. A good example is the multiple defenses of the tobacco hornworm against nicotine. In most such cases metabolism appears to be an important factor even when the insect has an insensitive target site. In cases where the insect relies on a behavioral adaptaion for avoiding poisoning, for example the trenching behavior of the squash beetle, it is uncertain that any biochemical or physiological defenses are also involved.

It is also debatable whether a behavioral adaptation allows subsequent development of biochemical defensive specializations to evolve but seems more likely that a biochemical adaptation would allow the insect subsequent behavioral advantages. It would depend on the degree to which learning is involved in insect behavior compared to how much of it is genetically programmed.

The present state of knowledge of the molecular aspects of insect-plant interactions raises many questions and invites speculations. For these not to be completely idle, the most important business at hand is, clearly, continued experimental work.

11. ACKNOWLEDGEMENT

I thank many of my friends and colleagues for stimulating and enlightening discussions of topics included in this review and for help with finding the appropriate references.

12. REFERENCES

Abdel-Aal, Y. A. I. and B. D. Hammock, 1986. Transition state analogs as ligands for affinity purification of juvenile hormone esterase, Science, 233:1073-1076.

Aizawa, H., W. S. Bowers and D. M. Soderlund, 1985. Reactions of precocene II epoxide with model nucleophiles, J. Agric. Food Chem., 33:406-411.

Alderson, T. and A. M. Clark, 1966. Interlocus specificity for chemical mutagens in Aspergillus nidulans, Nature, 210:593-595.

Ariyoshi, T., M. Arakaki, K., Ideguchi, Y. Ishizuka, K. Noda and H. Ide, 1975. Studies on the metabolism of d-limonene (p-mentha-1,8-diene). III. Effects of d-limonene on the lipids and drug-metabolizing enzymes in rat livers, Xenobiotica, 5:33-38.

Baldwin, E., 1964. "An Introduction to Comparative Biochemistry", 4th edition, Cambridge Univ. Press, Cambridge.

Bell, T. W., M. Boppre, D. Schneider and J. Meinwald, 1984. Stereochemical course of pheromone biosynthesis in the arctiid moth, Creatonotos transiens, Experientia, 40:713-714.

Berenbaum, M. R., 1978. Toxicity of a furanocoumarin to armyworms: a case of biosynthetic escape from insect herbivores, Science, 201:532-534.

Berenbaum, M. R., 1981, Effects of linear furanocoumarins on an adapted specialist insect (Papilio polyxenes), Ecol. Entomol., 6:345-351.

Berenbaum, M. R. and P. P. Feeny, 1981. Toxicity of angular furanocoumarins to swallowtail butterflies: escalation in a coevolutionary arms race? Science, 212:927-929.

Berenbaum, M. R. and J. J. Neal, 1985. Synergism between myristicin and xanthotoxin, a naturally coocurring plant toxicant, J. Chem. Ecol., 11:1349-1358.

Berge, M. A., G. A. Rosenthal and D. L. Dahlman, 1986. Tobacco budworm, Heliothis virescens [Noctuidae] resistance to L-canavanine, a protective allelochemical, Pestic. Biochem. Physiol., 25:319-326.

Bergot, B. J., K. J. Judy, D. A. Schooley and L. W. Tsai, 1980. Precocene II metabolism: comparative in vivo studies among several species of insects, and structure elucidation of two major metabolites, Pestic. Biochem. Physiol., 13:95-104.

Bernays, E. A. and S. Woodhead, 1982a. Plant phenols utilized as nutrients by a phytophagous insect, Science, 216:201-203.

Bernays, E. A. and S. Woodhead, 1982b. Incorporation of dietary phenols into the cuticle in the tree locust Anacridium melanorhodon, J. Insect Physiol., 28:601-606.

Bernays, E. A. and S. Woodhead, 1984. The need for high levels of phenylalanine in the diet of Schistocerca gregaria nymphs, J. Insect Physiol., 30:489-493.

Bernays, E. A., D. J. Chamberlain and S. Woodhead, 1983. Phenols as nutrients for a phytophagous insect Anacridium melanorhodon, J. Insect Physiol., 29:535-539.

Blau, P. A., P. Feeny, L. Contardo and D. S. Robson, 1978. Allylglucosinolate and herbivorous caterpillars: a contrast in toxicity and tolerance, Science, 200:1296-1298.

Blumenthal-Goldschmidt, S., G. W. Butler and E. E. Conn, 1963. Incorporation of hydrocyanic acid labelled with carbon-14 into asparagine in seedlings, Nature, 197:718-719.

Bond, E. J., 1961. The action of fumigants on insects. III. The fate of hydrogen cyanide in Sitophilus granarius (L.), Can. J. Biochem. Physiol., 39:1793-1802.

Boppre, M. and D. Schneider, 1985. Pyrrolizidine alkaloids quantitatively regulate both scent organ morphogenesis and pheromone biosynthesis in male Creatonotos moths (Lepidoptera: Arctiidae), J. Comp. Physiol., A157:569-577.

Borden, J. H., K. K. Nair and C. E. Slater, 1969. Synthetic juvenile hormone: induction of sex pheromone production in Ips confusus, Science, 166:1626-1627.

Bowers, M. D., 1980. Unpalatability as a defense strategy of Euphydryas phaeton (Lepidoptera: Nymphalidae), Evolution, 34:586-600.

Bowers, M. D., 1981. Unpalatability as a defense strategy of western checkerspot butterflies (Euphydryas, Nymphalidae), Evolution, 35:367-375.

Bowers, M. D., 1984. Iridoid glycosides and host-plant specificity in larvae of the buckeye butterfly, Junonia coenia (Nymphalidae), J. Chem. Ecol., 10:1567-1577.

Bowers, M. D. and G. M. Puttick, 1986. Fate of ingested iridoid glycosides in lepidopteran herbivores, J. Chem. Ecol., 12:169-178.

Bowers, W. S., 1976. Discovery of insect antiallatotropins, in: "The Juvenile Hormones", L. I. Gilbert, ed., pp. 394-408, Plenum Publ. Corp., New York.

Brand, J. M., J. W. Bracke, A. J. Markovetz, D. L. Wood and L. E. Browne, 1975. Production of verbenol pheromone by a bacterium isolated from bark beetles, Nature, 254:136-137.

Brand, J. M., J. W. Bracke, L. N. Britton, A. J. Markovetz and S. J. Barras, 1976. Bark beetle pheromones: production of verbenone by a mycangial fungus of Dendroctonus frontalis, J. Chem. Ecol., 2:195-199.

Brattsten, L. B., 1979. Ecological significance of mixed-function oxidations, Drug Metab. Rev., 10:35-58.

Brattsten, L. B., 1983. Cytochrome P-450 involvement in the interactions between plant terpenes and insect herbivores, in: "Plant resistance to insects", P. A. Hedin, ed., pp. 173-195, Symp. Ser. No. 208. Amer. Chem. Soc., Washington.

Brattsten, L. B., 1986a. Metabolic insecticide defenses in the boll weevil compared to those in a resistance-prone species, Pestic. Biochem. Physiol., in press.

Brattsten L. B., 1986b. The role of plant allelochemicals in the development of insecticide resistance, in: "Indirect effects of plant allelochemicals on insect herbivores", P. Barbosa, ed., Wiley and Sons, New York, in press.

Brattsten, L. B., C. F. Wilkinson and T. Eisner, 1977. Herbivore- plant interactions: mixed-function oxidases and secondary plant substances, Science, 196:1349-1352.

Brattsten, L. B., C. K. Evans, S. Bonetti and L. H. Zalkov, 1984. Induction by carrot allelochemicals of insecticide-metabolising enzymes in the southern armyworm (Spodoptera eridania), Comp. Biochem. Physiol., 77C:29-37.

Bridges, J. R., 1982. Effects of juvenile hormone on pheromone synthesis in Dendroctonus frontalis, Environ. Entomol., 11:417-420.

Brooks, G. T., 1976. Penetration and distribution of insecticides, in: "Insecticide Biochemistry and Physiology", C. F. Wilkinson, ed., pp. 3-58, Plenum Publ. Corp., New York.

Brooks, G. T., A. F. Hamnett, R. C. Jennings, A. P. Ottridge and G. E. Pratt, 1979a. Aspects of the mode of action of precocenes on milkweed bugs (Oncopeltus fasciatus) and locusts (Locusta migratoria), in: Proc. Brit. Crop Prot. Conf.: Pests and Diseases, pp. 273-279.

Brooks, G. T., G. E. Pratt and R. C. Jennings, 1979b. The actions of precocenes in milkweed bugs (Oncopeltus fasciatus) and locusts (Locusta migratoria), Nature, 281:570-572.

Brower, L. P. and S. C. Glazier, 1975. Localization of heart poisons in the monarch butterfly, Science, 188:19-25.

Brusewitz, G., B. D. Cameron, L. F. Chasseaud, K. Gorler, D. R. Hawkins, H. Koch and W. H. Mennicke, 1977. The metabolism of benzyl isothiocyanate and its cysteine conjugate, Biochem. J., 162:99-107.

Bull, D. L., G. W. Ivie, R. C. Beier, N. W. Pryor and E. H. Oertli, 1984. Fate of photosensitizing furanocoumarins in tolerant and sensitive insects, J. Chem. Ecol., 10:893-911.

Bull, D. L., G. W. Ivie, R. C. Beier and N. W. Pryor, 1986. In vitro metabolism of a linear furanocoumarin (8-methoxypsoralen, xanthotoxin) by mixed-function oxidases of larvae of black swallowtail butterfly and fall armyworm, J. Chem. Ecol., 12:885-892.

Bull, L. B., C. C. J. Culvenor and A. T. Dick, 1968. "The Pyrrolizidine Alkaloids, their Chemistry, Pathogenicity and other Biological Properties", John Wiley & Sons, New York.

Burnett, W. C., Jr., S. B. Jones, Jr. and T. J. Mabry, 1978. The role of sesquiterpene lactones in plant-animal coevolution, in: "Biochemical Aspects of Plant and Animal Coevolution", J. B. Harborne, ed., pp. 233-257, Academic Press, New York.

Burt, M. E., R. J. Kuhr and W. S. Bowers, 1978. Metabolism of precocene II in the cabbage looper and European corn borer, Pestic. Biochem. Physiol., 9:300-303.

Byers, J. A., 1982. Male-specific conversion of the host plant compound, myrcene, to the pheromone, (+)-ipsdienol, in the bark beetle, Dendroctonus brevicomis, J. Chem. Ecol., 8:363-371.

Byers, J.A. and D. L. Wood, 1981. Antibiotic-induced inhibition of pheromone synthesis in a bark beetle, Science, 213:763-764.

Casida, J. E., 1973. Biochemistry of the pyrethrins, in: "Pyrethrum, the Natural Insecticide", J. E. Casida ed., pp. 101-120, Academic Press, New York.

Cassier, P., 1979. The corpora allata of insects, Int. Rev. Cytol., 57:1-73.

Castric, P. A., 1981. The metabolism of hydrogen cyanide by bacteria, in: "Cyanide in Biology", B. Vennesland, E. E. Conn, C. J. Knowles, J. Westley and F. Wissing, eds, pp. 233-261, Academic Press, New York.

Cavill, G. W. K., D. L. Ford and H. D. Locksley, 1956. The chemistry of ants. I. Terpenoid constituents of some Australian Iridomyrmex species, Austr. J. Chem., 9:288-293.

Chapman, R. F., 1982. "The Insects, Structure and Function", Harvard University Press, Cambridge.

Chen, P. S., 1985. Amino acid and protein metabolism, in: "Comprehensive Insect Physiology, Biochemistry, and Pharmacology", G. A. Kerkut and L. I. Gilbert, eds, Vol. 10, pp. 177-217, Pergamon Press, New York.

Chew, F. S., 1975. Coevolution of pierid butterflies and their cruciferous foodplants. I. Relative quality of available resources, Oecologia (Berl.), 20:117-127.

Chew, F. S., 1977. Coevolution of pierid butterflies and their cruciferous foodplants. II. The distribution of eggs on potential foodplants, Evolution, 31:568-579.

Clark, A. M., 1959. Mutagenic activity of the alkaloid heliotrine in Drosophila, Nature, 163:731-732.

Conn, E. E., 1979. Cyanide and cyanogenic glycosides, in: "Herbivores, their Interaction with Secondary Plant Metabolites", G. A. Rosenthal and D. H. Janzen, eds, pp. 387-412, Academic Press, New York.

Conn, E. E., J. H. Borden, D. W. A. Hunt, J. Holman, H. S. Whitney, O. J. Spanier, H. D. Pierce, Jr. and A. C. Oehlschlager, 1984. Pheromone production by axenically reared Dendroctonus ponderosae and Ips paraconfusus (Coleoptera: Scolytidae), J. Chem. Ecol., 10:281-290.

Conner, W. E., T. Eisner, R. K. Vander Meer, A. Guerrero and J. Meinwald, 1981. Precopulatory sexual interaction in an arctiid moth (Utetheisa ornatrix): role of a pheromone derived from dietary alkaloids, Behav. Ecol. Sociobiol., 9:227-235.

Courtney, S. P., 1981. Coevolution of pierid butterflies and their cruciferous foodplants. III. Anthocharis cardamines (L.) survival, development and oviposition on different hostplants, Oecologia (Berl.), 51:91-96.

Daly, J., D. Jerina, J. Farnsworth and G. Guroff, 1969. The migration of deuterium during aryl hydroxylation. II. Effect of induction of microsomal hydroxylases with phenobarbital or polycyclic aromatic hydrocarbons, Arch. Biochem. Biophys., 131:238-244.

David, A. and D. K. Vallance, 1955. Bitter principles of Cucurbitaceae, J. Pharm. Pharmacol., 7:295-296.

Davis, R. H. and A. Nahrstedt, 1985. Cyanogenesis in insects, in: "Comprehensive Insect Physiology, Biochemistry, and Pharmacology", G. A. Kerkut and L. I. Gilbert, eds, Vol. 11, pp. 635-654, Pergamon Press, New York.

Dowd, P. F., C. M. Smith and T. C. Sparks, 1983. Detoxification of plant toxins by insects, Insect Biochem., 13:453-468.

Duffey, S. S., 1977. Arthropod allomones: chemical effronteries and antagonists, Proc. XV Int. Congr. Entomol., Washington, D. C., Aug. 1976, 15:323-394.

Duffey, S. S., 1981. Cyanide and arthropods, in: "Cyanide in Biology", B. Vennesland, E. E. Conn, C. J. Knowles, J. Westley and F. Wissing, eds, pp. 385-414, Academic Press, New York.

Eisner, T., 1980. Chemistry, defense, and survival: case studies and selected topics, in: "Insect Biology in the Future", M. Locke and D. S. Smith, eds, pp. 847-878, Academic Press, New York.

Eisner, T., M. Goetz, D. Aneshansley, G. Ferstandig-Arnold and J. Meinwald, 1986. Defensive alkaloid in blood of Mexican bean beetle (Epilachna varivestis), Experientia, 42:204-207.

Elliott, M., N. F. Janes, E. C. Kimmel and J. E. Casida, 1972. Metabolic fate of pyrethrin I, pyrethrin II, and allethrin administered orally to rats, J. Agric. Food Chem., 20:300-312.

Ellis-Pratt, G., 1983. The mode of action of pro-allatocidins, in: "Natural Products for Innovative Pest Management", D. L. Whitehead and W. S. Bowers, eds, pp. 323-355, Pergamon Press, New York.

Feeny, P. P., 1970. Seasonal changes in oak leaf tannins and nutrients as a case of spring feeding by winter moth caterpillars, Ecology, 51:565-581.

Feeny, P. P., 1976. Plant apparency and chemical defense, Rec. Adv. Phytochem., 10:1-40.

Ferguson, J. E. and R. L. Metcalf, 1985. Cucurbitacins, plant-derived defense compounds for diabroticites (Coleoptera: Chrysomelidae), J. Chem. Ecol., 11:311-318.

Ferguson, J. E., R. L. Metcalf and D. C. Fischer, 1985. Disposition and fate of cucurbitacin B in five species of diabroticites, J. Chem. Ecol., 11:1307-1321.

Feyereisen, R., 1985. Regulation of juvenile hormone titer: synthesis, in: "Comprehensive Insect Physiology, Biochemistry, and Pharmacology, G. A. Kerkut and L. I. Gilbert eds, Vol 7, pp. 391-429, Pergamon Press, New York.

Feyereisen, R., G. Johnson, J. Koener, B. Stay and S. S. Tobe, 1981. Precocenes as pro-allatocidins in adult female Diploptera punctata: a functional and ultrastructural study, J. Insect Physiol., 27:855-868.

Ford, E. B., 1941. Studies on the chemistry of pigments in the lepidoptera with reference to their bearing on systematics. 1. The anthoxanthins, Proc. R. Entomol. Soc., London,(A) 16:65-90.

Ford, E. B., 1944. Studies on the chemistry of pigments in the Lepidoptera, with reference to their bearing on systematics. 4. The calssification of Papilionidae, Trans. R. Entomol. Soc., London, (A) 19:92-106.

Franzl, S. and C. M. Naumann, 1985. Cyanoglucoside storing cavities in the cuticle of Zygaena larvae (Insecta: Lepidoptera): morphology and ultrastructure of the integument, Tiss. Cell, 17:267-278.

Franzl, S., A. Nahrstedt and C. M. Naumann, 1986. Evidence for the site of biosynthesis and transport of the cyanoglucosides linamarin and lotaustralin in larvae of Zygaena trifolii (Insecta: Lepidoptera), J. Insect Physiol., 32:705-709.

Fujimoto, Y., M. Kimura, A. Takasu, F. A. M. Khalifa, M. Morisaki and N. Ikekawa, 1984. Mechanism of stigmasterol dealkylation in insect, Tetrahedron Lett., 25:1501-1504.

Fukami, J.-I., I. Yamamoto and J. E. Casida, 1967. Metabolism of rotenone in vitro by tissue homogenates from mammals and insects, Science, 155:713-716.

Fukami, J.-I., T. Shishido, K. Fukunaga and J. E. Casida, 1969. Oxidative metabolism of rotenone in mammals, fish, and insects and its relation to selective toxicity, J. Agric. Food Chem., 17:1217-1226.

Gibbon, G. H. and S. J. Pirt, 1971. The degradation of α-pinene by Pseudomonas PX1, FEBS Letters, 18:103-105.

Gibbon, G. H., N. F. Millis and S. J. Pirt, 1972. Degradation of α-pinene by bacteria, in: "Fermentation Technology Today", Proc. IV Int. Ferment. Symp., G. Terui ed., pp. 609-612, Japan.

Gordon, H. T., 1961. Nutritional factors in insect resistance to chemicals, Annu. Rev. Entomol., 6:27-54.

Gould F. and E. Hodgson, 1978. Mixed-function oxidase and glutathione transferase activity in last instar Heliothis virescens larvae, Pestic. Biochem. Physiol., 13:34-40.

Gueldner, R. C., P. P. Sikorowski and J. M. Wyatt, 1977. Bacterial load and pheromone production in the boll weevil, Anthonomus grandis, J. Invert. Pathol., 29:397-398.

Guthrie, F. E., R. L. Ringler and T. G. Bowery, 1957. Chromatographic separation and identification of some alkaloid metabolites of nicotine in certain insects, J. Econ. Entomol., 50:821-825.

Hammock, B. D., 1985. Regulation of juvenile hormone titer: degradation, in: "Comprehensive Insect Physiology, Biochemistry, and Pharmacology", G. A. Kerkut and L. I. Gilbert, eds, Vol. 7, pp. 431-472, Pergamon Press, New York.

Hammock, B. D. and S. M. Mumby, 1978. Inhibition of epoxidation of methyl farnesoate to juvenile hormone III by cockroach corpus allatum homogenates, Pestic. Biochem. Physiol., 9:39-47.

Harborne, J. B., 1979. Flavonoid pigments, in: "Herbivores, their Interaction with Secondary Plant Metabolites", G. A. Rosenthal and D. H. Janzen, eds, pp. 619-655, Plenum Publ. Corp., New York.

Haunerland, N. H. and W. S. Bowers, 1985. Comparative studies on pharmacokinetics and metabolism of the anti-juvenile hormone precocene II, Arch. Insect Biochem. Physiol., 2:55-63.

Haunerland, N. H. and W. S. Bowers, 1986. Binding of insecticides to lipophorin and arylphorin, two hemolymph proteins of Heliothis zea, Arch. Insect Biochem. Physiol., 3:87-96.

Henry, M. F. and E. J. Nyns, 1975. Cyanide-insensitive respiration. An alternative mitochondrial pathway, Subcellular Biochem., 4:1-66.

Hollingworth, R. M., 1976. The biochemical and physiological basis of selective toxicity, in: "Insecticide Biochemistry and Physiology", C. F. Wilkinson, ed., pp. 431-506. Plenum Publ. Corp., New York.

Hucker, H. B., J. R. Gillette and B. B. Brodie, 1960. Enzymatic pathway for the formation of cotinine, a major metabolite of nicotine in rabbit liver, J. Pharmacol. Exp. Ther., 129:94-100.

Hughes, P. R., 1973. Dendroctonus: production of pheromones and related compounds in response to host monoterpenes, Z. Angew. Entomol., 73:294-312.

Hughes, P. R., 1974. Myrcene: a precursor of pheromones in Ips beetles, J. insect Physiol., 20:1271-1275.

Hughes, P. R., 1975. Pheromones of Dendroctonus: origin of alpha-pinene oxidation products present in emergent adults, J. Insect Physiol., 21:687-691.

Hughes, P. R. and J. A. A. Renwick, 1977a. Neural and hormonal control of pheromone biosynthesis in the bark beetle, Ips paraconfusus, Physiol. Entomol., 2:117-123.

Hughes, P. R. and J. A. A. Renwick, 1977b. Hormonal and host factors stimulating pheromone synthesis in female western pine beetles, Dendroctonus brevicomis, Physiol. Entomol., 2:289-292.

Igimi, H., M. Nishimura, R. Kodama and H. Ide, 1974. Studies on the metabolism of d-limonene (p-mentha-1,8-diene). I. The absorption, distribution and excretion of d-limonene in rats, Xenobiotica, 4:77-84.

Isaac, R. E., H. P. Desmond and H. H. Rees, 1984. Isolation and identification of 3-acetylecdysone 2-phosphate, a metabolite of ecdysone, from developing eggs of Schistocerca gregaria, Biochem. J., 217:239-243.

Isman, M. B., 1985. Toxicity and tolerance of sesquiterpene lactones in the migratory grasshopper, Melanoplus sanguinipes, Pestic. Biochem. Physiol., 24:348-354.

Ivie, G. W., D. L. Bull, R. C. Beier, N. W. Pryor and E. H. Oertli, 1983. Metabolic detoxification: mechanism of insect resistance to plant psoralens, Science, 221:374-376.

Ivie, G. W., D. L. Bull, R. C. Beier and N. W. Pryor, 1986. Comparative metabolism of [^{3}H]psoralen and [^{3}H]isopsoralen by black swallowtail (Papilio polyxenes Fabr.) caterpillars, J. Chem. Ecol., 12:871-884.

Janzen, D. H., H. B. Juster and E. A. Bell, 1977. Toxicity of secondary compounds to the seed-eating larvae of the bruchid beetle Callosobruchus maculatus, Phytochem., 16;223-227.

Jones, C. G. and R. D. Firn, 1978. The role of phytoecdysteroids in bracken fern, Pteridium aquilinum (L.) Kuhn as a defense against phytophagous insect attack, J. Chem. Ecol., 4:117-138.

Jungreis, A. M. and G. L. Vaughan, 1977. Insensitivity of lepidopteran tissues to ouabain: absence of ouabain binding and Na-K ATPases in larval and adult midgut, J. Insect Physiol., 23:503-509.

Kangas, E., V. Perttunen, H. Oksanen and M. Rinne, 1965. Orientation of Blastophagus piniperda L. (Col., Scolytidae) to its breeding material. Attractant effect of -terpineol isolated from pine rind, Ann. Entomol. Fenn., 31:61-73.

Katagiri, M., B. N. Ganguli and I. C. Gunsalus, 1968. A soluble cytochrome P-450 functional in methylene hydroxylation, J. Biol. Chem., 243:3543-3546.

Koolman, J. and P. Karlson, 1975. Ecdysone oxidase, an enzyme from the blowfly Calliphora erythrocephala (Meigen), Hoppe-Seyler's Z. Physiol. Chem., 356:1131-1138.

Koolman, J. and P. Karlson, 1985. Regulation of ecdysteroid titer: degradation, in: Comprehensive Insect Physiology, Biochemistry, and Pharmacology, G. A. Kerkut and L. I. Gilbert, eds, Vol. 7, pp. 343-361, Pergamon Press, New York.

Koul, O., 1983. L-canavanine, an antigonadal substance for Dysdercus koenigii, Entomol. Exp. Appl., 34:297-300.

Krieger, R. I. and C. F. Wilkinson, 1969. Microsomal mixed-function oxidases in insects: localization and properties of an enzyme system effecting aldrin epoxidation in larvae of the southern armyworm (Prodenia eridania), Biochem. Pharmacol., 18:1403-1415.

Kubo, I. and J. Klocke, 1986. Insect ecdysis inhibitors, in: "Natural Resistance of Plants to Pests, Roles of Allelochemicals", M. B. Green and P. A. Hedin, eds, pp. 206-219, Symp. Ser. No. 296, Amer. Chem. Soc., Washington.

Kuhr, R. J., 1970. Metabolism of carbamate insecticide chemicals in plants and insects, J. Agric. Food Chem., 18:1023-1030.

Lafont, R., C. Blais, P. Beydon, J. F. Modde, U. Enderle and J. Koolman, 1983. Conversion of ecdysone and 20-hydroxyecdysone into 26-oic derivatives is a major pathway in larvae and pupae of species from three insect orders, Arch. Insect Biochem. Physiol., 1:41-58.

Lanigan, G. W., 1970. Metabolism of pyrrolizidine alkaloids in the ovine rumen. II. Factors affecting the rate of alkaloid breakdown by rumen fluid in vitro, Aust. J. Agric. Res., 21:633-639.

Long, K. Y. and L. B. Brattsten, 1982. Is rhodanese important in the detoxification of dietary cyanide in southern armyworm (Spodoptera eridania Cramer) larvae? Insect Biochem., 12:367-375.

Louda, S. M. and J. E. Rodman, 1983. Ecological patterns in the glucosinolate content of a native mustard, Cardamine cordifolia, in the Rocky Mountains, J. Chem. Ecol., 9:397-422.

Mabry, T. J. and J. E. Gill, 1979. Sesquiterpene lactones and other terpenoids, in: "Herbivores, their Interaction with Secondary Plant Metabolites", G. A. Rosenthal and D. H. Janzen, eds, pp. 501-537, Academic Press, New York.

MacGibbon, D. B. and R. M. Allison, 1971. An electrophoretic separation of cabbage aphid and plant glucosinolases, N. Z. J. Sci., 14:134-143.

Maddrell, S. H. P. and B. O. C. Gardiner, 1976.Excretion of alkaloids by Malpighian tubules of insects, J. Exp. Biol., 64:267-281.

Marty, M. A. and R. I. Krieger, 1984. Metabolism of uscharidin, a milkweed cardenolide, by tissue homogenates of monarch butterfly larvae, Danaus plexippus L., J. Chem. Ecol., 10:945-956.

Mattocks, A. R., 1968. Toxicity of pyrrolizidine alkaloids, Nature, 217:723-728.

Mattocks, A. R. and I. N. White, 1971. The conversion of pyrrolizidine alkaloids to N-oxides and to dihydropyrrolizine derivatives by rat liver microsomes in vitro, Chem. Biol. Interactions, 3:383-396.

Mayer, R. T., J. L. Durrant, G. M. Holman, G. F. Weirich and J. A. Svoboda, 1979. Ecdysone 3-epimerase from the midgut of Manduca sexta (L.), Steroids, 34:555-562.

McFarlane, J. E. and M. H. W. Distler, 1982. The effect of rutin on growth, fecundity and food utilization in Acheta domesticus (L.), J. Insect Physiol., 28:85-88.

Mechoulam, R., R. W. Brueggemeier and D. L. Denlinger, 1984. Estrogens in insects, Experientia, 40:942-944.

Meinwald, J. and Y. C. Meinwald, 1966. Structure and synthesis of the major components in the hairpencil secretion of a male butterfly, Lycorea ceres ceres (Cramer), J. Am. Chem. Soc., 88:1305-1310.

Meinwald, J., Y. C. Meinwald and P. H. Mazzocchi, 1969. Sex pheromone of the queen butterfly: chemistry, Science, 164:1174-1175.

Metcalf, R. L., R. A. Metcalf and A. M. Rhoades, 1980. Cucurbitacins as kairomones for diabroticite beetles, Proc. Natl. Acad. Sci., USA, 77:3769-3772.

Metcalf, R. L., A. M. Rhoades, R. A. Metcalf, J. E. Ferguson, E. R. Metcalf and P.-Y. Lu, 1982. Cucurbitacin content and diabroticite (Coleoptera: Chrysomelidae) feeding upon Cucurbita spp., Environ. Entomol., 11:931-937.

Miller, J. M. and E. E. Conn, 1980. Metabolism of cyanide by higher plants, Plant Physiol., 65:1199-1202.

Moore, L. V. and G. G. E. Scudder, 1985. Selective sequestration of milkweed (Asclepias sp.) cardenolides in Oncopeltus fasciatus (Dallas) (Hemiptera: Lygaeidae), J. Chem. Ecol., 11:667-687.

Moore, L. V. and G. G. E. Scudder, 1986. Ouabain-resistant Na,K-ATPases and cardenolide tolerance in the large milkweed bug, Oncopeltus fasciatus, J. Insect Physiol., 32:27-33.

Morris, C. E., 1983. Uptake and metabolism of nicotine by the CNS of a nicotine-resistant insect, the tobacco hornworm (Manduca sexta), J. Insect Physiol., 29:807-817.

Morris, C. E., 1984. Electrophysiological effects of cholinergic agents on the CNS of a nicotine-resistant insect, the tobacco hornworm (Manduca sexta), J. Exp. Zool., 229:361-374.

Morris, S. J. and R. H. Thomson, 1963. The flavonoid pigments of the marbled white butterfly (Melanargia galathea Seltz.), J. Insect Physiol., 9:391-399.

Nair, K. K., G. C. Unnithan, D. R. Wilson and C. J. Kooman, 1981. Cytochemistry of corpora allata, fatbody and follicle cells of precocene-treated Schistocerca gregaria, Sci. Pprs Inst. Org. Phys. Chem. Wroclaw Tech. Univ., 22:389-404.

Pasteels, J. M., M. Rowell-Rahier, J. C. Braekman and A. DuPont, 1983. Salicin from host plant as precursor of salicylaldehyde in defensive secretion of chrysomeline larvae, Physiol. Entomol., 8:307-314.

Pasteels, J. M., M. Rowell-Rahier, J. C. Braekman and D. Daloze, 1984. Chemical defense in leaf beetles and their larvae: the ecological, evolutionary and taxonomic significance, Biochem. Syst. Ecol., 12:395-406.

Pasteels, J. M., D. Daloze and M. Rowell-Rahier, 1986. Chemical defence in chrysomelid eggs and neonate larvae, Physiol. Entomol., 11:29-37.

Pines, M., G. A. Rosenthal and S. W. Applebaum, 1981. In vitro incorporation of L-canavanine into vitellogenin of the fat body of the migratory locust, Locusta migratoria migratorioides, Proc. Natl. Acad. Sci., USA, 78:5480-5483.

Pliske, T. E. and T. Eisner, 1969. Sex pheromone of the queen butterfly: biology, Science, 164:1170-1172.

Povolny, D. and F. Weyda, 1981. On the glandular character of larval integument in the genus Zygaena (Lepidoptera: Zygaenidae), Acta Entomol. Bohemoslav., 73:273-279.

Prestwich, G. D., M. Angelastro, A. De Palma and M. A. Perino, 1985. Fucosterol epoxide lyase of insects: synthesis of labeled substrates and development of a partition assay, Anal. Biochem., 151:315-326.

Racker, E., J. A. Belt, W. W. Carey and J. H. Johnson, 1980. Studies on anion transporters, Ann. N. Y. Acad. Sci., 341:27-36.

Raffa, K. F. and A. A. Berryman, 1986. Interacting selective pressures in conifer - bark beetle systems: a basis for reciprocal adaptations? Amer. Nat., in press.

Rees, H. H., 1985. Biosynthesis of ecdysone, in: "Comprehensive Insect Physiology, Biochemistry, and Pharmacology", G. A. Kerkut and L. I. Gilbert, eds, Vol 7, pp. 249-293, Pergamon Press, New York.

Reichstein, T., J. von Euw, J. A. Parsons and M. Rothschild, 1968. Heart poisons in the monarch butterfly, Science, 161:861-866.

Renwick, J. A. A., P. R. Hughes and T. D. Ty, 1973. Oxidation products of pinene in the bark beetle, Dendroctonus frontalis, J. Insect Physiol., 19:1753-1740.

Renwick, J. A. A., P. R. Hughes, G. B. Pitman and J. P. Vite, 1976a. Oxidation products of terpenes identified from Dendroctonus and Ips bark beetles, J. Insect Physiol., 22:725-727.

Renwick, J. A. A., P. R. Hughes and I. S. Krull, 1976b. Selective production of cis- and trans-verbenol from (-)- and (+)-α-pinene by a bark beetle, Science, 191:199-201.

Robinson, T., 1979. The evolutionary ecology of alkaloids, in: "Herbivores, their Interaction with Secondary Plant Metabolites", G. A. Rosenthal and D. H. Janzen, eds, pp. 413-448, Academic Press, New York.

Rodman, J. E. and F. S. Chew, 1980. Phytochemical correlates of herbivory in a community of native and naturalized Cruciferae, Biochem. Syst. Ecol., 8:43-50.

Roeske, C. N., J. N. Seiber, L. P. Brower and C. M. Moffitt, 1976. Milkweed cardenolides and their comparative processing by monarch butterflies (Danaus plexippus L.), Rec. Adv. Phytochem., 10:93-167.

Rosenthal, G. A., 1986. Biochemical insight into insecticidal properties of L-canavanine, a higher plant protective allelochemical, J. Chem. Ecol., 12:1145-1156.

Rosenthal, G. A. and E. A. Bell, 1979. Naturally occurring, toxic nonprotein amino acids, in: "Herbivores, their Interaction with Secondary Plant Metabolites", G. A. Rosenthal and D. H. Janzen, eds, pp. 353-385, Plenum Publ. Corp., New York.

Rosenthal, G. A. and D. L. Dahlman, 1975. Non-protein amino acid-insect interactions. II. Effect of canaline-urea cycle amino acids on growth and development of the tobacco hornworm, Manduca sexta (L.) [Sphingidae], Comp. Biochem. Physiol., 52A:105-108.

Rosenthal, G. A. and D. L. Dahlman, 1986. L-Canavanine and protein synthesis in the tobacco hornworm Manduca sexta, Proc. Natl. Acad. Sci., USA, 83:14-18.

Rosenthal, G. A. and D. H. Janzen, 1983. Arginase and L-canavanine metabolism by the bruchid beetle, Caryedes brasiliensis, Entomol. Exp. Appl., 34:336-338.

Rosenthal, G. A. and D. H. Janzen, 1985. Ammonia utilization by the bruchid beetle, Caryedes brasiliensis [Bruchidae], J. Chem. Ecol., 11:539-544.

Rosenthal, G. A., D. L. Dahlman and D. H. Janzen, 1978. L-Canaline detoxification: a seed predator's biochemical mechanism, Science, 202:528-529.

Rosenthal, G. A., D. H. Janzen and D. L. Dahlman, 1977. Degradation and detoxification of canavanine by a specialized seed predator, Science, 196:658-660.

Rotschild, M. and N. Marsh, 1978. Some peculiar aspects of danaid/plant relationships, Entomol. Exp. Appl., 24:437-450.

Scheline, R. R., 1978. "Mammalian Metabolism of Plant Xenobiotics", Academic Press, New York.

Schlesinger, H. M., S. W. Applebaum and Y. Birk, 1976. Effect of β-cyano-L-alanine on the water balance of Locusta migratoria, J. Insect Physiol., 22:1421-1425.

Schneider, D., M. Boppre, H. Schneider, W. R. Thompson, C. J. Boriak, R. L. Petty and J. Meinwald, 1975. A pheromone precursor and its uptake in male Danaus butterflies, J. Comp. Physiol., 97:245-256.

Schneider, D., M. Boppre, J. Zweig, S. B. Horsley, T. W. Bell, J. Meinwald, K. Hansen and E. W. Diehl, 1982. Scent organ development in Creatonotos moths: regulation by pyrrolizidine alkaloids, Science, 215:1264-1265.

Schooley, D. A. and F. C. Baker, 1985. Juvenile hormone biosynthesis, in: "Comprehensive Insect Physiology, Biochemistry, and Pharmacology", G. A. Kerkut and L. I. Gilbert, eds, Vol. 7, pp. 363-389, Pergamon Press, New York.

Scudder, G. G. E. and J. Meredith, 1982. The permeability of the midgut of three insects to cardiac glycosides, J. Insect Physiol., 28:689-694.

Scudder, G. G. E., L. V. Moore and M. B. Isman, 1986. Sequestration of cardenolides in Oncopeltus fasciatus: morphological and physiological adaptations, J. Chem. Ecol., 12:1171-1187.

Segall, H. J., D. W. Wilson, J. L. Dallas and W. F. Haddon, 1985. Trans-4-hydroxy-2-hexenal: a reactive metabolite from the macrocyclic pyrrolizidine alkaloid senecionine, Science, 229:472-475.

Seiber, J. N., P. M. Tuskes, L. P. Brower and C. J. Nelson, 1980. Pharmacodynamics of some individual milkweed cardenolides fed to larvae of the monarch butterfly, J. Chem. Ecol., 6:321-339.

Self, L. S., F. E. Guthrie and E. Hodgson, 1964a. Adaptations of tobacco hornworms to the ingestion of nicotine, J. Insect Physiol., 10:907-914.

Self, L. S., F. E. Guthrie and E. Hodgson, 1964b. Metabolism of nicotine by tobacco-feeding insects, Nature, 204:300-301.

Shah, A. H. and F. E. Guthrie, 1985. Role of polarity of insecticides in penetration through the isolated midgut of insects and mammals, Ind. J. Agric. Chem., 18:153-156.

Shaver, T. N. and M. J. Lukefahr, 1969. Effect of flavonoid pigments on growth and development of the bollworm, tobacco budworm, and pink bollworm, J. Econ. Entomol., 62:643-646.

Slade, M. and C. F. Wilkinson, 1974. Degradation and conjugation of cecropia juvenile hormone by the southern armyworm (Prodenia eridania), Comp. Biochem. Physiol., 49B:99-103.

Slade, M. and C. H. Zibitt, 1972. Metabolism of cecropia juvenile hormone in insects and mammals, in: "Insect Juvenile Hormones: Chemistry and Action", J. J. Menn and M. Beroza, eds, pp. 155-176, Academic Press, New York.

Smith, S. L., 1985. Regulation of ecdysteroid titer: synthesis, in: Comprehensive Insect Physiology, Biochemistry, and Pharmacology", G. A. Kerkut and L. I. Gilbert, eds, Vol 7, pp. 295-341, Pergamon Press, New York.

Soderlund, D. M., A. Messeguer and W. S. Bowers, 1980. Precocene II metabolism in insects: synthesis of potential metabolites and identification of initial in vitro biotransformation products, J. Agric. Food Chem., 28:724-731.

Steinly, B. A. and M. R. Berenbaum, 1985. Histopathological effects of tannins on the midgut epithelium of Papilio polyxenes and Papilio glaucus, Entomol. Exp. Appl., 39:3-9.

Stephenson, A. G., 1982. Iridoid glycosides in the nectar of Catalpa speciosa are unpalatable to nectar thieves, J. Chem. Ecol., 8:1025-1034.

Stermitz, F. R., D. R. Gardner, F. J. Odendaal and P. R. Ehrlich, 1986. Euphydryas anicia (Lepidoptera: Nymphalidae) utilization of iridoid glycosides from Castilleja and Besseya (Scrophulariaceae) host plants, J. Chem. Ecol., 12:1459-1468.

Stipanovic, R. D., A. A. Bell and M. J. Lukefahr, 1977. Natural insecticides from cotton (Gossypium), in: "Host Plant Resistance to Pests", P. A. Hedin, ed., pp. 197-214, Symp. Ser. No. 62, Amer. Cem. Soc., Washington.

Svoboda, J. A. and M. J. Thompson, 1983. Comparative sterol metabolism in insects, in: "Metabolic aspects of lipid nutrition in insects", T. E. Mittler and R. H. Dadd, eds, pp. 1-16, Westview Press, Boulder.

Svoboda, J. A. and M. J. Thompson, 1985. Steroids, in: "Comprehensive Insect Physiology, Biochemistry, and Pharmacology", G. A. Kerkut and L. I. Gilbert, eds, Vol. 10, pp.137-175, Pergamon Press, New York.

Svoboda, J. A. M. J. Thompson and W. E. Robbins, 1971. Identification of fucosterol as a metabolite and probable intermediate in conversion of sitosterol to cholesterol in the tobacco hornworm, Nature New Biol., 230:57-58.

Svoboda, J. A., W. E. Robbins, C. F. Cohen and T. J. Shortino, 1972. Phytosterol utilization and metabolism in insects: recent studies with Tribolium confusum, in: "Insect and Mite Nutrition", J. G. Rodriguez, ed., pp. 505-516, North Holland, Amsterdam.

Svoboda, J. A., M. J. Thompson, T. C. Elden and W. E. Robbins, 1974. Unusual composition of sterols in a phytophagous insect, Mexican bean beetle reared on soybean plants, Lipids, 9:752-755.

Svoboda, J. A., M. J. Thompson, W. E. Robbins and J. N. Kaplanis, 1978. Insect steroid metabolism, Lipids, 13:742-753.

Tate, L. G., S. S. Nakat and E. Hodgson, 1982. Comparison of detoxication activity in midgut and fat body during fifth instar development of the tobacco hornworm, Manduca sexta, Comp. Biochem. Physiol., 72C:75-81.

Terriere, L. C. and S. J. Yu, 1973. Insect juvenile hormone: induction of detoxifying enzymes in the house fly and detoxication by house fly enzymes, Pestic. Biochem. Physiol., 3:96-107.

Thompson, M. J., G. F. Weirich, H. H. Rees, J. A. Svoboda, M. F. Feldlaufer and K. R. Wilzer, 1985. New ecdysteroid conjugate: isolation and identification of 26-hydroxyecdysone and 26-phosphate from eggs of the tobacco hornworm, Manduca sexta (L.), Arch. Insect Biochem. Physiol., 2:227-236.

Thomson, D. L., 1926. The pigments of butterflies' wings. 1. Melanargia galathea, Biochem. J., 20:73-75.

Tsoupras, G., C. Hetru, B. Luu, E. Constantin, M. Lagueux and J. Hoffman, 1983. Identification and metabolic fate of ovarian 22-adenosine monophosphoric ester of 2-deoxyecdysone in ovaries and eggs of an insect, Locusta migratoria, Tetrahedron, 39:1789-1796.

Usher, B. F. and P. Feeny, 1983. Atypical secondary compounds in the family Cruciferae: tests for toxicity to Pieris rapae, an adapted crucifer-feeding insect, Entomol. Exp. Appl., 34:257-262.

van Etten, C. H. and H. L. Tookey, 1979. Chemistry and biological effects of glucosinolates, in: "Herbivores, their Interaction with Secondary Plant Metabolites", G. A. Rosenthal and D. H. Janzen, eds, pp. 471-500, Academic Press, New York.

Vickery, M. L. and B. Vickery, 1981. "Secondary Plant Metabolism", University Park Press, Baltimore.

von Euw, J., L. Fishelson, J. A. Parsons, T. Reichstein and M. Rothschild, 1967. Cardenolides (heart poisons) in a grasshopper feeding on milkweeds, Nature, 214:35-39.

Westley, J., 1981. Cyanide and sulfane sulfur, in: "Cyanide in Biology", B. Vennesland, E. E. Conn, C. J. Knowles, J. Westley and F. Wissing, eds, pp. 61-76, Academic Press, New York.

White, R. A., R. T. Franklin and M. Agosin, 1979. Conversion of α-pinene to α-pinene oxide by rat liver and the bark beetle Dendroctonus terebrans microsomal fractions, Pestic. Biochem. Physiol., 10:233-242.

White, R. A., Jr., M. Agosin, R. T. Franklin and J. W. Webb, 1980. Bark beetle pheromones: evidence for physiological synthesis mechanisms and their ecological implications, Z. Angew. Entomol., 90:255-274.

Wilson, A., 1985a. Flavonoid pigments in marbled white butterfly (Melanargia galathea) are dependent on flavonoid content of larval diet, J. Chem. Ecol., 11:1161-1179.

Wilson, A., 1985b. Flavonoid pigments of butterflies in the genus Melanargia, Phytochem., 24:1685-1691.

Wilkinson, C. F., 1986. Xenobiotic conjugation in insects, in: "Xenobiotic conjugation chemistry", G. D. Paulson, J. Caldwell, D. H. Hutson and J. J. Menn, eds, pp. 48-61, Symp. Ser. No. 299, Amer. Chem. Soc., Washington.

Wilkinson, C. F. and L. B. Brattsten, 1972. Microsomal drug-metabolizing enzymes in insects, Drug Metab. Rev., 1:153-228.

Wray, V., R. H. Davis and A. Nahrstedt, 1983. Biosynthesis of cyanoglucosides in butterflies and moths: incorporation of valine and isoleucine into linamarin and lotaustralin by Zygaena and Heliconius species (Lepidoptera), Z. Naturf., 38C:583-588.

Yamamoto, I., E. C. Kimmel and J. E. Casida, 1969. Oxidative metabolism of pyrethroids in house flies, J. Agric. Food Chem., 17: 1227-1236.

Yang, R. S. H. and C. F. Wilkinson, 1972. Sulfotransferases and phosphotransferases in insects, Comp. Biochem. Physiol., 46B:717-726.

Yang, R. S. H. , J. G. Pellicia and C. F. Wilkinson, 1973. Age-dependent arylsulfatase and sulfotransferase activities in the southern armyworm: a possible endocrine regulatory mechanism? Biochem. J., 136:817-820.

TARGET SITE INSENSITIVITY IN INSECT-PLANT INTERACTIONS

May R. Berenbaum

Department of Entomology
320 Morrill Hall
University of Illinois Urbana-Champaign
505 S. Goodwin Avenue
Urbana, IL 61801

1. INTRODUCTION

By acting as a selective source of mortality in insect populations, insecticides, both natural and man-made, have wrought extraordinary changes in the genetic composition and physiology of insects. Among these changes are several fundamentally different resistance mechanisms. One of the less well understood forms of resistance is target site insensitivity (TSI), defined as the failure of a toxicant to bind to the target due to alteration in the structure or accessibility of that target site (Brooks 1976). Studies of TSI have been severely hampered by the fact that, in order to understand TSI as a resistance mechanism, it is necessary first to know what the target site and mode of action are. This is decidedly not the case for the majority of plant allelochemicals; it is, however, true for a few synthetic organic insecticides, and the phenomenon of target site insensitivity was first discovered in connection with chemical control programs that ceased working.

Synthetic organic insecticides owe their toxicity to selective action on particular target tissues or organs. This concept is by no means new; as long ago as 1898, Paul Ehrlich suggested in a letter to Carl Weigert (Pascoe, 1983) that cells possess "side chains" or particular macromolecules in membranes or cytoplasm that can combine with foreign substances such as drugs or poisons. Binding with a foreign substance effects a change in the cell; if the change is of sufficient magnitude, toxicity ensues. Ehrlich reasoned that different cells must have different "receptors" which respond only to certain complementary foreign substances; receptor selectivity thus gives rise to the observed specificity of action of many drugs or poisons. Essentially the same concept applies today, and since the turn of the century considerable effort has gone into identifying and characterizing these "receptors" or target sites.

2. TSI AND SYNTHETIC INSECTICIDES

2.1. The Sodium Gate

In 1951, Busvine noticed several differences in the kind of DDT resistance displayed by several strains of the house fly, *Musca domestica* L.

(Diptera: Muscidae). Flies of the 'normal' "Rome" strain were exceedingly susceptible to the paralytic action of topically applied DDT, so susceptible, in fact, that they were literally "knocked down" in mid-flight. An "Italian" strain, however, was almost completely immune to the knock-down effect of DDT. Even more surprising was the fact that these flies, selected only for resistance to DDT in the laboratory, were also resistant to pyrethrins. Other forms of resistance to DDT (e.g., as manifested by the "Sardinian" strain) were associated with resistance to other chlorinated hydrocarbons, rapid dehydrochlorination apparently sufficing as a common mode of detoxification. The so-called kdr ("knock-down resistance") flies, however, dehydrochlorinated DDT very slowly; their resistance could not, therefore, be attributed to more efficient dehydrochlorination of DDT. Busvine concluded with the statement that "one is driven to postulate an additional defense mechanism in the Italian strain".

The additional defense mechanism postulated by Busvine was subsequently revealed to be a form of target site insensitivity due to an alteration in the target site. The reason that the kdr flies consistently show cross resistance to DDT and pyrethroids is that the two insecticides share a common target site; both interfere with the function of sodium channels in nerve cells. Normally, nerve cells conduct signals (action potentials) by changes in permeability to certain ions. As an action potential moves along the nerve cell, electrical currents reduce the potential across the membrane in advance of the action potential. This decrease in potential causes a rapid increase in permeability of the membrane to positively charged sodium ions, abundant outside the cell. Sodium ions then begin to rush in, reducing the membrane potential still further and allowing even more sodium to enter. The increase in sodium permeability is brought about by the opening of sodium "channels" or "gates" (glycoproteins spanning the membrane). When the voltage inside the cell changes from negative to positive during the action potential, the sodium channels quickly close and the influx of sodium is halted. The decrease in cell membrane potential in front of the action potential, however, at the same time increases the permeability of the membrane to potassium ions, which are abundant inside the cell. This increase in potassium permeability occurs so slowly that appreciable amounts of potassium do not begin to leave the cell until the sodium channels have closed. This efflux of positively-charged potassium ions reestablishes the normal resting potential.

The sodium channel has been identified as a glycoprotein with a hydrophilic core that provides a channel for sodium ions to enter the axon (Stevens, 1984). Pyrethrins and DDT bind to the membrane in such a way that the sodium channels are maintained in an open configuration. The continued influx of sodium then causes a depolarizing afterpotential following the action potential, with repetitive firing _ad infinitum_. This repetitive firing causes the tremor, convulsions and general incoordination associated with DDT and pyrethroid toxicity. Kdr-based pyrethroid resistance in _Musca domestica_ may be attributable to a reduction in the number of insecticide binding sites at or near sodium channels (Chang and Plapp, 1983); other possible mechanisms include changes in the channel protein itself (Osborne and Smallcombe, 1983) or changes in nerve membrane fluidity due to altered phospholipids (Chialiang and Devonshire, 1982). While most examples of kdr are found in species important in medical or veterinary science, there is a report (Gammon, 1980) of resistance to permethrin, a synthetic pyrethroid, in a herbivorous insect. The Egyptian armyworm, _Spodoptera littoralis_ (Boisduval) (Lepidoptera: Noctuidae), is a polyphagous plant feeder in which resistance to pyrethroids shows many similarities to kdr resistance. Thus, the capacity to develop TSI to pyrethroid insecticides may exist in foliage-feeding insects. To date, no studies have been done specifically examining adapted insect associates of

Tanacetum (=Chrysanthemum) cinerariaefolium, the plant source of pyrethrins, for kdr resistance; such studies should prove informative about the evolutionary significance of TSI as a resistance mechanism.

2.2. The Synapse

The nicotinic receptor for acetylcholine is another target site for insecticide action. This receptor, the first identified for a neurotransmitter, was postulated to exist as long ago as 1906 by Langely, who asserted that "a receptive substance . . . combines with nicotine and curari [sic] [in chickens]" (Changeux et al., 1984). Acetylcholine (ACh), the neurotransmitter, is formed in the presynaptic cell and released via exocytosis into the synaptic cleft. It then binds to receptors on the postsynaptic side, creating a post-synaptic potential (a local change in the cell's membrane potential), that may, in turn, cause generation of an action potential. The magnitude of this post-synaptic potential, and therefore the probability that it will initiate an action potential, is proportional to the number of ACh molecules that bind with the post-synaptic membrane receptor proteins. Finally, an enzyme, acetylcholinesterase (AChE), releases the acetylcholine from the receptor by hydrolysis to acetate and free choline, at which point the membrane returns to resting potential.

Many insecticides, natural and otherwise, interfere with the action of AChE and are collectively known as acetylcholinesterase inhibitors, or anticholinesterases. Organophosphate (OP) insecticides, for example, bind to the enzyme in place of ACh. While the acetylated enzyme resulting from the reversible complex with ACh breaks down in a few milliseconds to reactivate the enzyme and free the receptor for the next acetylcholine molecules, the phosphorylated enzyme resulting from an organophosphate - enzyme complex breaks down very slowly, in minutes to hours. While it is phosphorylated, the enzyme is inactivated and acetylcholine remains bound to the receptor, continuing to depolarize the post-synaptic cell and thus producing tremors, convulsions and other physiological disturbances. Carbamate insecticides share a similar mode of action, carbamylating the enzyme instead of phosphorylating it, but with much the same result. Insensitivity to organophosphate and carbamate insecticides may result from a change in the structure of the enzyme active site which reduces affinity of acetylcholinesterases for these compounds. This reduced affinity is reflected by the increased Km values of these compounds (Wilkinson 1976). Not only do these TSI resistant insects display a reduced affinity for OPs and carbamates, they often show a reduced affinity for the normal substrate as well. In the two-spotted spider mite, Tetranychus urticae Koch (Acari: Tetranychidae), for example, AChE in resistant populations was only one third as reactive toward ACh as was AChE from susceptible populations; multiple forms of the AChE enzyme may be involved in this overall reduced sensitivity (Smissaert, 1964). The green rice leafhopper, Nephotettix cincticeps Uhleri (Homoptera: Cicadellidae), has an altered AChE sensitivity to carbamates, and, as in most cases in which carbamate insensitivity is reported, it was found that OP sensitivity had developed earlier (Hama, 1983). In addition to reduced sensitivity of AChE to inhibition by OP and carbamates, the green rice leafhopper possesses the capacity to detoxify these insecticides (Hama, 1983).

3. TSI IN NATURAL SYSTEMS

Brattsten (1984) pointed out that, in three known cases, TSI follows metabolic detoxification over time, Table 1. Why TSI comes about and why it appears to do so later than metabolic resistance are subject to speculation. Plapp (1974) suggested that target site insensitivity was the

Table 1. Dates of reported acquisition of metabolic or target site resistance to various insecticides.

Compound	Introduced	Metabolic Resistance	Target site Resistance
DDT	1942	1949	1951
Me-parathion	1944	1958	1967
Carbaryl	1956	1961	1971

(from Brattsten, 1984)

result of an incapacity of metabolic detoxification to keep up with intense insecticide selection. One might then argue that TSI is exclusively associated with synthetic organic insecticide resistance, since the nature of the selective pressures exerted by pesticide applicators is fundamentally different from the selective pressures exerted by naturally occurring allelochemicals. A major difference is that most synthetic organic insecticides are acutely toxic at very low doses and fast-acting whereas the majority of natural compounds at equivalent concentrations have more subtle effects, such as feeding deterrency, developmental aberrations and decreased fertility. Moreover, the majority of synthetic organic insecticides, at least historically, have been nerve poisons that owe their efficacy at least in part to the fact that they are contact or topical insecticides and, by cuticular penetration, largely circumvent metabolic detoxification. In contrast, the vast majority of known natural toxicants in plants are ingested by herbivorous insects. Even many of the botanical products preempted for insecticide use are much less effective as stomach poisons than as contact insecticides (Crosby and Jacobson, 1969).

It is not altogether unreasonable, however, to expect target site insensitivity to occur in natural situations. There are some natural toxicants that are acutely toxic and fast-acting; there is even a natural carbamate, physostigmine, occurring in the calabar or ordeal bean <u>Physostigma venenosum</u> (Leguminosae) (Metcalf and March, 1950; Windholz et al., 1983). There are also many natural products that act as contact toxicants, including precocenes (Bowers, 1980), other chromenes (Proksch et al. 1984), tridecanone (Dimock et al., 1981), and furanocoumarins (Kagan and Chan, 1984).

Most importantly, selective pressure by plants on insects is in all likelihood intense enough even in natural situations to overcome the limitations of metabolic detoxification. An allelochemical effect such as feeding deterrency can be an effective plant defense if the insect has other options - for example, if it can move off the offending plant onto another one nearby. Such options may not, however, always be available. Extremely specialized insects are more or less committed to a handful of genera or even species and may not be able to switch hosts freely due to behavioral restrictions, e.g., dependency on the presence of a feeding stimulant or physiological limitations. Extremely sedentary species, particularly sedentary specialists, are limited as to where they can go to avoid the effects of overloading their metabolic systems. Target site insensitivity, then, could be expected to arise in insects that are highly dependent upon a particular plant or group of plants for survival with few or no options for host-switching, such as oligophages, particularly soft-bodied concealed feeders such as armored scales, seed borers, leaf miners, and the like. Highly mobile insects such as grasshoppers probably exploit the 'eat and move on' approach, as do mammalian herbivores, which would obviate a metabolic overload (Freeland and Janzen, 1974).

There are several situations, then, that could in theory select for target site insensitivity. TSI would most likely arise when natural selection pressures resemble selection pressures exerted by insecticide application. This would be the case:

- in an oligophage with restricted movement;
- in the presence of an allelochemical found in all edible plant parts, thereby ruling out selective feeding and behavioral resistance;
- with year-to-year predictability of allelochemical presence;
- in the absence of refuges such as alternate hosts without the allelochemical;
- when allelochemical selection is exerted against both larvae and adults;
- when selection occurs every generation;
- when toxicity is sufficiently high to kill a substantial fraction of the population at the time the allelochemical is initially encountered (see Georghiou and Taylor, 1976).

4. EXAMPLES OF TSI IN INSECT-PLANT INTERACTIONS

4.1. Cardenolide insensitivity

One excellent example of target site insensitivity in a natural situation involves insect associates of Asclepiadaceae (the milkweeds) containing cardiac glycosides. These compounds, also called cardenolides, are C-23 steroids characterized by a C17 lactone, a C14 beta hydroxyl group, and a C3 ether linkage to one or more sugars. When ingested by vertebrates, they induce cardiac arrhythmias, visual disturbance, delirium and emesis (Windholz et al., 1983). A possible site of action for the cardenolides is the enzyme that catalyzes metal cation transport across membranes, Na^+,K^+ ATPase (sodium- and potassium-dependent adenosine triphosphate phosphohydrolase), which is essential for nerve cell function. The cardenolide ouabain (G-strophanthin), derived from *Strophanthus gratus* (Apocynaceae), specifically inactivates this enzyme. The fauna of both Apocynaceae and Asclepiadaceae consists almost exclusively of highly specialized insects, among the more conspicuous of which are species in the genus *Danaus* (Lepidoptera: Danaiidae). Vaughan and Jungreis (1977) demonstrated that neuronal tissue from larvae of the monarch butterfly, *D. plexippus* (L.), is 300 times less sensitive *in vitro* to inactivation by ouabain than comparable tissue from two non-asclepiad feeders, *Manduca sexta* (L.) (Lepidoptera: Sphingidae) and *Hyalophora cecropia* (L.) (Lepidoptera: Saturniidae). All three species derive a measure of protection in vivo by high K^+ concentration in the hemolymph. While 30 mM K^+ totally protects the *D. plexippus* enzyme from 7.5 mM ouabain, it permits only 65% activity of the enzyme from the other two species (Vaughan and Jungreis 1977).

A similar mechanism operates in *Oncopeltus fasciatus* (Dallas) (Homoptera: Lygaeidae), the large milkweed bug (Moore and Scudder, 1985). Its gut is highly permeable to cardenolides and like the monarch, it sequesters the intact plant allelochemicals (Scudder and Meredith, 1982). *O. fasciatus* was able to tolerate injected ouabain at concentrations 1954 and 7288 times the LD_{50} for a grasshopper and a cockroach, respectively. Nerve tissue lyophilates displayed up to 200 fold less sensitivity to inhibition of Na^+,K^+ ATPase by cardenolides. Systemic cardenolide tolerance in *O. fasciatus* can thus be attributable to a cardenolide-insensitive Na^+,K^+ ATPase (Moore and Scudder 1985). In other organisms, enzyme resistance is thought to derive either from a structural change in the enzyme, destabilizing enzyme - cardenolide complexes, from a decrease in the number of ouabain binding sites, from an altered response to bound ouabain, or from some combination of these factors. Over and above target enzyme insensitivity, however, ouabain and digitoxin can be metabolized by *O. fasciatus* (Meredith et al., 1984; Scudder et al., 1985 and references therein).

4.2 Cyanide insensitivity

Another major source of TSI is the elimination of the target altogether. Some forms of resistance to cyanide are believed to involve such a mechanism. Cyanide is poisonous to most organisms because it interferes with mitochondrial electron transport and hence with cellular respiration (Conn, 1979). CN^-, like CO, NO and N_3^-, binds directly to the heme moiety of the terminal cytochrome oxidase to which molecular oxygen normally binds. California red scale, *Aonidiella aurantii* (Maskell) (Homoptera: Diaspididae), developed resistance to cyanide shortly after its introduction as a chemical control agent (Quay, 1916). Since oxidative respiration appears to remain a major pathway for energy production and there is no evidence of anoxia tolerance or anaerobic hydrogen acceptance to maintain a hydrogen "store", the resistance is presumed due to the existence of metal-free respiratory enzymes, such as an autooxidizable flavoprotein (Yust and Shelden 1952).

While in the case of the California red scale resistance was generated by insecticide pressure, similar resistance may occur naturally. Millipedes that produce their own cyanogenic glycosides endogenously for defense show a similar resistance to HCN distinct from low oxygen tolerance (Hall et al., 1971), as do cyanogenic zygaenid moths (Levinson et al., 1973). Conceivably, zygaenid moths, heliconian butterflies and other aposematic insects that sequester cyanogenic glycosides from plants or synthesize them endogenously (Wray et al., 1983; Nahrstedt and Davis, 1981) may possess some form of cyanide-insensitive respiration (see also Ahmad et al., Chapter 3 this text).

4.3. Canavanine insensitivity

Some forms of resistance involve sensitivity on the part of carrier molecules that transport xenobiotics. A nonprotein amino acid, l-canavanine, is produced in large concentrations in the seeds of several tropical legumes; *Canavalia* and *Dioclea* spp. contain up to five to ten percent of the dry seed weight in the form of l-canavanine (Rosenthal et al., 1976). It is a structural analogue of l-arginine; in some animals, arginyl-tRNA synthetase esterifies l-canavanine to arginyl-tRNA, allowing it to be incorporated into proteins which then do not function properly. This incorporation results in a variety of attendant toxic effects (Rosenthal and Dahlman, 1975). The beetle *Caryedes brasiliensis* Thunberg (Coleoptera: Bruchidae) lives inside the seeds of *Canavalia* and has little choice but to ingest l-canavanine. It has a modified arginyl-tRNA synthetase that does not bind l-canavanine. After injection of radiolabelled l-canavanine, no label appears in bruchid larval proteins. In contrast, tobacco hornworm larval proteins show a large amount of radiolabel and upon hydrolysis yield at least 3.5 percent of the injected l-canavanine as labelled CO_2 (Rosenthal et al., 1976). *Caryedes* is an oligophage with restricted movement, surrounded by the allelochemical in question (up to 80 percent of the amino acid nitrogen can be in the form of l-canavanine), and is subject to selection every generation at both larval and adult stages; in other words, it is a logical candidate for the evolution of target site insensitivity.

4.4. Nicotine insensitivity

Nicotine resistance in *Manduca sexta* is another good example in which resistance involves a failure of the allelochemical to reach a target site. Nicotine, a major alkaloidal component of the foliage of tobacco, is a structural analogue of acetylcholine and mimics the action of acetylcholine by binding to ACh receptors. It is highly toxic to many insects, as well as to mammals, birds, and many other organisms (Matsumura, 1976). While it is the ionized form of nicotine that mimics acetylcholine and binds to the receptor, only the free base is an effective insecticide, since the insect central nervous system is protected by an ion-impermeable sheath. In the hemolymph of *Manduca sexta*, with its pH of 6.6, over 98 percent of absorbed nicotine is present in the form of the nicotinium ion. In its charged form, the nicotinium ion cannot cross the ion-impermeable barrier and is excreted with no ill effect (Self et al., 1964). Target site insensitivity *sensu stricto* may also be involved in the resistance of *M. sexta* to nicotine. Intact dorsal nerve cord in *M. sexta* is 100-fold less sensitive to nicotine than that of *Periplaneta americana* L. (Orthoptera: Blattidae). Desheathing increases the sensitivity of *M. sexta* nervous tissue to nicotine, but it is still substantially less sensitive than that of *P. americana* (Morris, 1984).

5. COEVOLUTIONARY CONSEQUENCES OF TSI

5.1. Consequences for the insects

Target site insensitivity has very different consequences for plant and insect as far as coevolution is concerned. With target site insensitivity, there may be no penalty to the insect if metabolism is very slow or nonexistent. If the target site is unreactive or unreachable, the allelochemical can circulate unchanged in insect hemolymph without toxic effects. This opens up the possibility that the insect can make use of the plant toxicant for its own defense. Black swallowtails, *Papilio polyxenes* Fabr. (Lepidoptera: Papilionidae), for example, metabolize up to 98 percent of ingested furanocoumarins within 90 minutes and thus contain little if any intact furanocoumarin in the body (Ivie et al., 1983; Bull et al., 1984). They thereby eliminate the possibility of sequestering an effective DNA inhibitor and feeding deterrent for defense against predators. In contrast, monarch caterpillars, insensitive to the effects of asclepiadaceous toxicants, store cardenolides from their hostplants for the duration of their development and acquire the emetic characteristics of their hostplants. Many known cases of sequestration of intact or unmetabolized plant chemicals involve compounds that have been associated with target site insensitivity, notably, cardenolides in monarchs and milkweed bugs, and cyanogenic glycosides in zygaenids.

Simply because a plant chemical is sequestered does not necessarily mean that there is no metabolic detoxification. This distinction was realized early by pesticide toxicologists. In his study of OP resistance in the two-spotted spider mite, Smissaert (1964) remarked, "The differences in rates of inhibition [of acetylcholinesterase] would only delay death, unless the inhibitor was detoxified or counteracted in some other way. It is assumed, therefore, that some mechanism which both the S and R strains of mites must have in common, has sufficient time to eliminate the poison in the resistant but not in the susceptible mites." This is probably equally true in situations involving natural plant products. Many cyanide-producing arthropods possess enzymatic systems capable of detoxifying cyanide (see Chapters by Brattsten and Ahmad et al., this text). The monarch butterfly can metabolize cardenolides with a non-microsomal NADPH-dependent reductase in the soluble fraction of midgut cells (Marty and Krieger, 1984), and the tobacco hornworm is capable of converting nicotine to cotinine enzymatically (Barbosa and Saunders, 1985).

With synthetic organic insecticides, the discovery of resistance due to metabolic detoxification invariably preceded the discovery of target site insensitivity, often by many years (Brattsten, 1984). The possibility exists that TSI is more difficult to detect experimentally and that the time lag in discovery is artifactual. However, the failure of increasingly sophisticated technology to uncover widespread TSI in many agricultural systems suggests that it really is not common. The same is probably the case in natural situations; TSI may follow the developed capacity for and subsequent overload of metabolic detoxification.

At present, virtually nothing is known of genetic variability and mutational rates involved in TSI systems, so it is difficult even to speculate on evolutionary rates of change in the establishment of target site insensitivity. Evidence that suggests that TSI genes may be relatively rare in populations can be found in the data of Decker and Bruce (1952), who used a number of house fly strains to select for resistance to a variety of insecticides via topical application over many generations. Resistance was slowest to develop to any great extent to paraoxon, an organophosphate, and to a 10:1 mixture of piperonyl butoxide (a cytochrome P-450 inhibitor and synergist) and pyrethrins. In the latter case, after 30 generations of inbreeding and selection, only a 20-fold increase in tolerance was achieved. In contrast, DDT resistance in at least two strains increased almost 1000-fold in less than 20 generations. Resistance to synergized pyrethrins is consistent with, although may not be identical to TSI; if these flies did indeed develop TSI resistance to pyrethroids, then its acquisition is slow relative to other forms of resistance. The slow establishment of TSI resistance is consistent with selection for genes that are initially rare (see Crow, 1952).

No doubt, there are evolutionary consequences of abandoning what Plapp (1974) called a "biologically preferred or more available . . . mechanism for protection against xenobiotics", i.e., metabolic detoxification. Eliminating the requirement for *de novo* protein synthesis in response to the presence of an allelochemical in the diet may affect the metabolic costs involved in consuming plants with toxicants; target site insensitivity is essentially "built in" and effective at all times. Jonathan Neal (in preparation) compared larval development time and prepupal weights in *Heliothis zea* Boddie (Lepidoptera: Noctuidae) on a standard artificial diet and on the same diet to which was added pentamethylbenzene (PMB), a xenobiotic known to induce detoxifying enzymes (Brattsten, 1979). PMB in the diet effects a reduction in the approximate digestibility (AD) and the prepupal weight in comparison to larvae reared on control diet alone; in two of three replicates, the efficiency of conversion of ingested food (ECI) was also reduced, Table 2. While this sort of study can determine only the overall effects of PMB ingestion *including* toxicity, it is conceivable that the observed reductions in food utilization indices result from metabolic expenditure due to enzyme induction. Assuming there are no other effects on viability associated with TSI, an assumption that appears to be true in many cases for insecticide-associated TSI, then the reduced dependence on metabolic detoxification may translate into an intraspecific competitive edge in the form of more efficient food utilization and faster growth in the TSI insect in the presence of a host plant toxicant.

Table 2. Growth and development of *Heliothis zea* in the presence and absence of a cytochrome P-450 inducer (PMB) [a,b]

	AD (%)	ECI (%)	ECD (%)	Prepual weight (g)
Control	41.5	19.9	49.5	154.7
PMB (0.2%)	38.8	18.9	50.0	149.3
t-test: $p <$	0.05	0.40	0.50	0.05

a Values are averages of three replicate paired experiments.
b See text for key to abbreviations.

5.2. Consequences for the plants

The evolution of TSI in herbivorous insects can be expected to be accompanied by a radically different set of selection pressures for plants. One implication of target site insensitivity, already apparent to insecticide toxicologists, is that synergists fail to enhance the toxicity of a xenobiotic. For example, natural pyrethrin insecticides, expensive to produce, are often synergized by methylenedioxyphenyl compounds that act as competitive inhibitors by binding to cytochrome P-450 and preventing the insecticide from being oxidized (Casida, 1970). Since kdr resistance does not involve metabolism per se, piperonyl butoxide and other commercial synergists do not enhance natural pyrethroid toxicity in kdr insects. Many allelochemicals may act as synergists in the plants producing them. Myristicin, for example, is a methylenedioxyphenyl phenylpropene component of the essential oil that co-occurs widely with the furanocoumarins in the Umbelliferae. Myristicin increases the toxicity of xanthotoxin, a furanocoumarin, to the generalist *Heliothis zea* almost five-fold. The LC_{50} of unsynergized xanthotoxin is 9.6 mg per g but in the presence of 0.10 percent myristicin, it is 1.9 mg per g. Thus, every mg of 11-carbon myristicin can "save" the plant 77 mg of 12-carbon furanocoumarin, quite possibly a significant saving in the energy expenditure involved in biosynthesis (Berenbaum and Neal, 1985). Many other plant products which share structural features with known insecticide synergists and may also function in synergizing co-occurring plant toxicants (Berenbaum, 1985) remain unexplored, largely due to operational difficulties in designing bioassays to detect synergism. This does not, however, diminish their potential importance in natural systems. Synergists should have less toxicological impact in systems in which TSI is the mechanism of resistance; they may in fact be important selective agents *for* the evolution of TSI in insects.

Just how TSI selective pressure on plant chemistry, in the sense of Janzen's 1980 definition of coevolution, might differ from metabolic and other types of resistance is open to speculation. TSI might give rise to the paradoxical situation in which herbivory selects for *reduced* amounts of allelochemicals or to alterations in chemical structure to *reduce* toxicity. If, for example, the TSI insect depends on unmetabolized sequestered plant products for its own defense against predators, individuals feeding on plants with lower quantities of or less toxic allelochemicals will contain less effective sequestered defenses, and may suffer an increased probability of being picked off by a predator. *Asclepias syriaca*, the principal milkweed host plant of *Danaus plexippus* throughout the northeastern U.S., varies considerably in its cardenolide content. This variation is reflected in the cardenolide content and attendant toxicity of *D.*

plexippus raised on this plant (Brower et al., 1972; Brower et al., 1982), leaving some individuals in the population as unprotected "mimics" of their conspecifics. TSI might also act to reduce the chemical diversity of a plant population or species in the sense that metabolic synergism is not a viable plant option for enhancing toxicity against the TSI insect.

On the other hand, TSI might select for increased chemical diversity in which different chemical classes act on different target sites, or might select for chemicals with multiple target sites. Furanocoumarins, phototoxic chemicals in the Umbelliferae and Rutaceae (Murray et al., 1982), form cycloadducts with pyrimidine bases in DNA and irreversibly interfere with transcription. Their phototoxicity has been linked directly to this binding capacity (Perlman et al., 1985). There is evidence, however, that furanocoumarins are also capable of photobinding irreversibly with proteins (Granger and Helene, 1983) and lipids. Moreover, several furanocoumarins when irradiated can generate singlet oxygen and free oxygen radicals (Poppe and Grossweiner, 1975; Joshi and Pathak, 1983). These toxic oxygen species can cause oxidative damage to both nuclear DNA and cell membranes. Thus plants in the Umbelliferae produce a single class of phototoxins with potentially multiple target sites. Moreover, many umbellifer species also produce polyacetylenes or saponins which damage membranes (Garrod, 1978; Applebaum and Birk, 1979), alkaloids which interfere with nerve function (Fairbairn, 1971), and cytotoxic terpenoid derivatives (Mabry and Gill, 1979). A single umbelliferous species, for instance, Pastinaca sativa, can contain dozens of potential toxicants with as many different target sites (Berenbaum, 1985).

Simultaneous evolution of TSI at multiple sites may be evolutionarily slow or even altogether improbable. Whether or not the allelochemicals ineffective against TSI insects are out of the genome depends on a number of factors, not the least of which is the fact that plants are generally subject to herbivore selective pressure from a wide variety of species, many or perhaps most of which may be unprotected by TSI.

All in all, like Busvine in 1951, "one is driven" to speculate with regard to the significance of target site insensitivity in natural coevolved systems: its taxonomic distribution, its ecological correlates, and its potential evolutionary significance for the chemical composition of plants. Speculation must suffice, as well, until the mode of action and target site of considerably more plant allelochemicals in ecologically appropriate herbivores have been identified.

6. SUMMARY

Among the ecologically lesser known or understood forms of resistance to xenobiotics is target-site insensitivity (TSI), that is, failure of a toxicant to bind to the site of action or target site due to an alteration in the structure or accessibility of that target site. TSI was first recognized and is best known in interactions involving synthetic insecticides. TSI as a basis for resistance to DDT and pyrethroids is thought to involve a change or changes of the sodium channel in nerve cells, whereas TSI in some forms of resistance to organophosphate and carbamate insecticides may derive from changes in the structure of the enzyme active site, resulting in reduced affinity of acetylcholinesterases for compounds which otherwise impair nerve function by inhibiting them.

In natural systems, TSI may be restricted to ecological situations in which selection pressures resemble in magnitude the selection pressures in systems involving human applications of synthetic insecticides. TSI may arise when, due to intense selection, metabolic detoxication has been overcome or circumvented. Extremely specialized sedentary herbivores, subject to continual selection at most or all life stages, may be particularly prone to developing TSI. Insensitivity to cardenolides via insensitivity of Na^+,K^+ ATPase is characteristic of many oligophagous associates of cardenolide-producing Asclepiadaceae; cyanide insensitivity, due to the presumed existence of metal-free respiratory enzymes, is characteristic of insects which either sequester or synthesize cyanogenic glycosides. Insensitivity in oligophagous bruchids to toxic amino acids in leguminous plants and insensitivity to nicotine toxicity in Solanaceae-feeders can also be considered examples of TSI in coevolved systems.

The evolutionary consequences of TSI differ considerably from the consequences of metabolic detoxification to both herbivore and host plant. For insects, TSI may reduce their metabolic costs of feeding and may facilitate sequestration of unmetabolized allelochemicals for their own defense. For the plant many scenarios can be devised; among these, TSI may select for multiple target sites or multiple chemical defense systems and, thus, may well be a factor underlying the impressive biochemical diversity of angiosperm plants.

7. ACKNOWLEDGEMENT

I thank L. B. Brattsten for editorial and bibliographic assistance with this paper, and my colleagues, students, and friends in the Department of Entomology at the University of Illinois at Urbana-Champaign, in particular, F. Delcomyn, D. Seigler, E. Heininger, J. Neal, J. Nitao, and A. Zangerl for comments on the manuscript and for ensuring that the speculation therein fell somewhere short of arrant.

8. REFERENCES

Anstee, J. and K. Bowler, 1976. Ouabain-sensitivity of insect epithelial tissues, Comp. Biochem. Physiol., 62A:763-769.

Applebaum, S. W. and Y. Birk, 1979. Saponins, in: "Herbivores, Their Interaction with Secondary Plant Metabolites", G. A. Rosenthal and D. H. Janzen, eds., pp. 539-566, Academic Press, New York.

Barbosa, P. and J. Saunders, 1985. Plant allelochemicals: linkages between herbivores and their natural enemies. Rec. Adv. Phytochem., 19:107-137.

Beesley, S., S. G. Compton, and D. Jones, 1985. Rhodanese in insects, J. Chem. Ecol., 11:45-50.

Berenbaum, M., 1985. Brementown revisited: interactions among allelochemicals in plants, Rec. Adv. Phytochem., 19:139-169.

Berenbaum, M. and J. J. Neal, 1985. Synergism between myristicin and xanthotoxin, a naturally co-occurring plant toxicant, J. Chem. Ecol., 11:1349-1358.

Bowers, W. S., 1983. Phytochemical action on insect morphogenesis, reproduction and behavior, in: "Natural Products for Innovative Pest Management", D. Whitehead and W. S. Bowers, eds., pp. 313-321, Pergamon Press, New York.

Brattsten, L. B., 1979. Biochemical defense mechanism in herbivores against plant allelochemicals, in: "Herbivores, Their Interaction with Secondary Plant Metabolites", pp. 199-271, Academic Press, New York.

Brattsten, L. B., 1984. Presentation at the 17th International Congress of Entomology, Hamburg, Germany. August 1984.

Brooks, G. T., 1976. Penetration and distribution of insecticides, in: "Insecticide Biochemistry and Physiology", C. F. Wilkinson, ed., pp. 3-60, Plenum Publ. Corp., New York.

Brower, L. P., P. McEvoy, K. Williamson, and M. Flannery, 1972. Variation in cardiac glycoside content of monarch butterflies from natural populations in eastern North America, Science, 177:426-429.

Brower, L. P., J. Seiber, C. Nelson, S. Lynch and P. Tuskes, 1982. Plant-determined variation in the cardenolide content, thin-layer chromatography profiles, and emetic potency of monarch butterflies, Danaus plexippus, reared on the milkweed, Asclepias eriocarpa in California, J. Chem. Ecol., 8:579-633.

Bull, D. L., G. W. Ivie, R. Beier, N. Prior and E. Oertli, 1984. Fate of photosensitizing furanocoumarins in tolerant and sensitive insects, J. Chem. Ecol., 10:893-911.

Busvine, J. R., 1951. Mechanism of resistance to insecticide in houseflies, Nature, 168:193-195.

Casida, J. E., 1970. Mixed-function oxidase involvement in the biochemistry of insecticide synergists, J. Agr. Food Chem., 18:753-760.

Chang, C. P. and F. W. Plapp, 1983. DDT and pyrethroids: receptor binding in relation to knockdown resistance (kdr) in the house fly, Pestic. Biochem. Physiol., 20:86-91.

Chang, C. P. and F. W. Plapp, 1983. DDT and pyrethroids: receptor binding and mode of action in the house fly, Pestic. Biochem. Physiol., 20:76-85.

Changeux, J. P., A. Devillers-Thiery, and P. Chemouilli, 1984. Acetylcholine receptor: an allosteric protein, Science, 225:1335-1345.

Chialiang, C. and A. L. Devonshire, 1982. Changes in membrane phospholipids, identified by Arrhenius plots of acetylcholinesterase and associated with pyrethroid resistance (kdr) in house flies (Musca domestica), Pestic. Sci., 13:156-160.

Conn, E. E., 1979. Cyanide and cyanogenic glycosides, in: "Herbivores, Their Interaction with Secondary Plant Metabolites", G. A. Rosenthal and D. H. Janzen, eds., pp. 387-412, Academic Press, New York.

Crosby, D. G. and M. Jacobson, 1971. "Naturally Occurring Insecticides", Marcel Dekker Press, New York.

Crow, J. F., 1952. Some genetic aspects of selection for resistance, in: "Conference on Insecticide Resistance and Insect Physiology", Publ. No. 219, pp. 72-78, National Academy of Sciences National Research Council, Washington.

Decker, G. C. and W. Bruce, 1952. Illinois Natural History Survey Research on House fly resistance to chemicals, in: "Conference on Insecticide Resistance and Insect Physiology", Publ. No. 219, pp. 25-33, National Academy of Sciences National Research Council, Washington.

Dimock, M. B., G. Kennedy, and W. Williams, 1982. Toxicity studies of analogs of 2-tridecanone, a naturally occurring toxicant from a wild tomato, J. Chem. Ecol., 8:837-842.

Devonshire, A. L. and G. Moore, 1984. Different forms of insensitive acetylcholinesterase in insecticide-resistant house flies (Musca domestica), Pestic. Biochem. Physiol., 21:336 340.

Duffey, S. S., 1980. Sequestration of plant natural products by insects. Annu. Rev. Entomol., 25:447-477.

Fairbairn, J. W., 1971. The alkaloids of hemlock (Conium maculatum L. or Conium maculatum L.: The odd man out), in: "The Biology and Chemistry of the Umbelliferae", V. Heywood, ed., Bot. J. Linn. Soc. 64 Suppl. 1:361-368.

Freeland, W. and D. H. Janzen, 1974. Strategies in herbivory by mammals: the role of plant secondary compounds, Am. Nat., 108:269-289.

Gammon, W., 1980. Pyrethroid resistance in a strain of Spodoptera littoralis is correlated with decreased sensitivity of the CNS in vitro, Pestic. Biochem. Physiol., 13:53-57.

Georghiou, G. P., 1972. The evolution of resistance to pesticides, Annu. Rev. Ecol. Syst., 3:133-168.

Georghiou, G. P., 1983. Management of resistance in arthropods, in: "Pest Resistance to Pesticides", G. P. Georghiou and T. Saito, eds., pp. 769-792, Plenum Publ. Corp., New York.

Georghiou, G. P. and R. Mellon, 1983. Pesticide resistance in time and space, in: "Pest Resistance to Pesticides", G. P. Georghiou and T. Saito, eds., pp. 1-46, Plenum Publ. Corp., New York.

Georghiou, G. P. and T. Saito, eds., 1983. "Pest Resistance to Pesticides", Plenum Publ. Corp., New York.

Georghiou, G. P. and C. E. Taylor, 1976. Pesticide resistance as an evolutionary phenomenon, Proc. XV Int. Cong. Ent. 1976. 759-785.

Granger, M. and C. Helene, 1983. Photoaddition of 8-methoxypsoralen to E. coli DNA polymerase. I. Role of psoralen photoadducts in the photosensitized alterations of pol I enzymatic activities, Photochem. Photobiol., 38:563-568.

Hall, F. R., R. Hollingsworth and D. Shankland, 1971. Cyanide tolerance in millipedes: the biochemical basis, Comp. Biochem. Physiol., 38B:723-737.

Hama, H., 1983. Resistance to insecticides due to reduced sensitivity of acetylcholinesterase, in: "Pest Resistance to Pesticides", G. P. Georghiou and T. Saito, eds., pp. 299-332, Plenum Publ. Corp., New York.

Ivie, G. W., D. Bull, R. Beier, N. Pryor, and E. Oertli, 1983. Metabolic detoxification: mechanism of insect resistance to plant psoralens, Science, 221:374-376.

Jackson, F. R., S. D. Wilson, and L. M. Hall, 1984. Two types of mutants affecting voltage-sensitive sodium channels in Drosophila melanogaster, Nature, 308:189-191.

Jacobson, M. and D. Crosby, 1971. "Naturally Occurring Insecticides", Marcel Dekker, New York, 585 pp.

Janzen, D. H., 1980. When is it coevolution? Evolution, 34:611-612.

Joshi, P. C. and M. Pathak, 1983. Production of singlet oxygen and superoxide radicals by psoralens, Biochem. Biophys. Res. Commun., 112:638-646.

Kagan, J. and G. Chan, 1983. The photoovicidal activity of plant components towards Drosophila melanogaster. Experientia, 39:402-403.

Levinson, H. Z., K. E. Kaissling and A. R. Levinson, 1973. Olfaction and cyanide sensitivity in the six spot burnet moth Zygaena filipendulae and the silk moth Bombyx mori, J. Comp. Physiol., 86:209-214.

Long, K. Y. and L. B. Brattsten, 1982. Is rhodanese important in the detoxification of cyanide in southern armyworm (Spodoptera eridania Cramer) larvae? Insect Biochem., 12:367-375.

Lund, A. E., 1985. Insecticides: effects on the nervous system, in: "Comprehensive Insect Physiology, Biochemistry and Pharmacology", G. A. Kerkut and L. I. Gilbert, eds., Vol. 12, pp. 9-56, Pergamon Press, New York.

Mabry, T. and J. Gill, 1979. Sesquiterpene lactones and other terpenoids, in: "Herbivores, Their Interaction with Secondary Plant Metabolites", G. A. Rosenthal and D. H. Janzen, eds., pp. 501-537, Academic Press, New York.

Marty, M. and R. I. Krieger, 1984. Metabolism of uscharidin, a milkweed cardenolide, by tissue homogenates of monarch butterfly larvae, Danaus plexippus, J. Chem. Ecol., 10:945-956.

Matsumura, F., 1976. "Toxicology of Insecticides", Plenum Publ. Corp., New York, 503 pp.

Matsumura, F., 1983. Penetration, binding and target insensitivity as causes of resistance to chlorinated hydrocarbon insecticides, in: "Pest Resistance to Pesticides", G. P. Georghiou and T. Saito, eds., pp. 376-386, Plenum Publ. Corp., New York.

Meredith, J., L. Moore, and G. G. E. Scudder, 1984. The excretion of ouabain by the Malpighian tubules of O. fasciatus. Am. J. Physiol. 246 (Regulatory Integrative Comp. Physiol. 15):R705-R715.

Metcalf, R. L. and R. B. March, 1950. Properties of acetylcholine esterases from the bee, the fly, and the mouse and their relation to insecticide action, J. Econ. Entomol., 43:670-677.

Miller, T. A., J. Kennedy and C. Collins, 1979. CNS insensitivity to pyrethroids in the resistant kdr strain of house flies, Pestic. Biochem. Physiol., 12:224-230.

Miller, T. A., V. L. Salgado and S. Irving, 1983. The kdr factor in pyrethroid resistance, in: "Pest Resistance to Pesticides", G. P. Georghiou and T. Saito, eds., pp. 376-386, Plenum Publ. Corp., New York.

Moore, L. V. and G. G. E. Scudder, 1986. Ouabain resistant Na, K-ATPases and cardenolide tolerance in the large milkweed bug, Oncopeltus fasciatus. J. Insect Physiol., 32:27-33.

Morris, C. E., 1984. Electrophysiological effects of cholinergic agents in the central nervous system of a nicotine-resistant insect, the tobacco hornworm (Manduca sexta), J. Exp. Zool., 229:361-374.

Murray, R. D. H., J. Mendez and S. A. Brown, 1982. "The Natural Coumarins", J. Wiley and Sons, Ltd., Chichester.

Nahrstedt, A. and R. H. Daves, 1981. Occurrence of the cyanoglucosides, linamarin and lotaustralin, in Acrea and Heliconius butterflies, Comp. Biochem. Physiol., 68(B):575-578.

Narahashi, T., 1983. Resistance to insecticides due to reduced sensitivity of the nervous system, in: "Pest Resistance to Pesticides", G. P. Georghiou and T. Saito, eds., pp. 376-386, Plenum Publ. Corp., New York.

Nelson, C. J., J. Seiber, and L. P. Brower, 1981. Seasonal and intraplant variation of cardenolide content in the California milkweed, Asclepias eriocarpa, and implications for plant defense, J. Chem. Ecol., 7:981-1010.

Osborne, M. P. and A. Smallcombe, 1983. Site of action of pyrethroid insecticides in neuronal membranes as revealed by the kdr resistance factor. in: "Mode of Action, Metabolism and Toxicology", S. Matsunaka, D. Hutson and S. Murphy, eds., Vol. 3 of Pesticide Chemistry: Human Welfare and the Environment, pp. 103-107, Pergamon Press, New York.

Pascoe, D., 1983. "Toxicology", E. Arnold and Co., London.

Pearlman, D. A., S. R. Holbrook, D. Pirkle and S. H. Kim, 1985. Molecular models for DNA damage by photoreaction. Science, 227:1304-1308.

Plapp, F. W., 1974. Biochemical genetics of insecticide resistance, Annu. Rev. Entomol., 21:179-197.

Plapp, F. W. and T. C. Wang, 1983. Genetic origins of insecticide resistance, in: "Pest Resistance to Pesticides", G. P. Georghiou and T. Saito, eds., pp. 376-386, Plenum Publ. Corp., New York.

Poppe, W. and L. Grossweiner, 1975. Photodynamic sensitization by 8-methoxypsoralen via the singlet oxygen mechanism, Photochem. Photobiol., 22:217-222.

Proksch, P., M. Proksch, G. H. N. Towers, and E. Rodriguez, 1983. Phototoxic and insecticidal activities of chromenes and benzofurans from Encelia, J. Nat. Prod., 46:331-334.

Quay, G. H., 1916. Are scales becoming resistant to fumigation? Cal. Univ. J. Agr., 3:333-334, 358.

Rosenthal, G. A. and D. L. Dahlman, 1975. Non-protein amino acid-insect interactions II. Effects of canaline on growth and development of the tobacco hornworm, Manduca sexta L. (Sphingidae), Comp. Biochem. Physiol., 52A:105-108.

Rosenthal, G. A., D. L. Dahlman, and D. H. Janzen, 1976. A novel means for dealing with L-canavanine, a toxic metabolite, Science, 192:256-258.

Scudder, G. G. E., L. Moore and M. B. Isman, 1986. Sequestration of cardenolides in Oncopeltus fasciatus: morphological and physiological adaptations, J. Chem. Ecol., 13: in press.

Scudder, G. E. and J. Meredith, 1982. The permeability of the midgut of three insects to glycosides, J. Insect Physiol., 28:689-694.

Self, L. S., F. Guthrie and E. Hodgson, 1964. Adaptation of tobacco hornworms to the ingestion of nicotine, J. Insect Physiol., 12:224-230.

Smissaert, H. R., 1964. Cholinesterase inhibition in spider mites susceptible and resistant to organophosphate, Science, 143:129-131.

Soderlund, D. M., S. M. Guiasuddin, and D. W. Helmuth, 1983. Receptor-like stereospecific binding of a pyrethroid insecticide to mouse brain membranes, Life Sci., 33:261-267.

Stevens, C. F., 1984. Biophysical studies of ion channels, Science, 225:1346-1350.

Vaughan, G. L. and A. M. Jungreis, 1977. Insensitivty of lepidopteran tissues to ouabain: physiological mechanisms for protection from cardiac glycosides, J. Insect Physiol., 23:585-589.

Wilkinson, C. F., ed., 1976. "Insecticide Biochemistry and Physiology", Plenum Publ. Corp., New York.

Windholz, M., S. Budavari, R. Blumetti, and E. Otterbein, 1983. "The Merck Index", Merck and Co., Inc. Rahway.

Wray, Y., R. H. Davis and A. Nahrstedt. 1983. Biosynthesis of cyanogenic glycosides in butterflies and moths: Incorporation of valine and isoleucine into linamarin and lotaustralin by Zygaena and Heliconius species (Lepidoptera), Z. Naturforsch., Sect. C Biosci., 38:583-588.

Yamamoto, I., Y. Takahashi and N. Kyomura, 1983. Suppression of altered acetylcholinesterase of the green rice leafhopper by N-propyl and N-methyl carbamate combinations, in: "Pest Resistance to Pesticides", G. P. Georghiou and T. Saito, eds., pp. 376-386, Plenum Publ. Corp., New York.

Yust, H. R., and F. Shelden, 1952. A study of the physiology of resistance to hydrocyanic acid in the California red scale, Ann. Ent. Soc. Amer., 45:220-228.

BEHAVIORAL ADAPTATIONS IN INSECTS TO PLANT ALLELOCHEMICALS

Douglas W. Tallamy

University of Delaware
Department of Entomology and Applied Ecology
Newark, Delaware 19717-1303

1. INTRODUCTION

Insects which obtain their nutrition from plants encounter an arsenal of physical and chemical defenses protecting plant tissues (Levin, 1976; Rhoades and Cates, 1976; Wallace and Mansell, 1976; Rhoades, 1979; Rosenthal and Janzen, 1979; Futuyma, 1983). Plants deploy their chemical defenses in two ways (Levin, 1976). A base-line defense is maintained at all times to deter herbivory through direct toxicity or by reducing the digestibility of plant tissues. A second line of defense behaves facultatively: plants mobilize or produce allelochemicals de novo in response to tissue damage from herbivores (Rhoades, 1979, 1983; Ryan, 1983). In combination, the constitutive and inducible (facultative) defenses of a particular plant are comprised of a species-specific set of chemicals that deters all herbivores except those few with an appropriate suite of counter-adaptations (Ehrlich and Raven, 1965; Feeny, 1975, 1976; Rhoades and Cates, 1976; Rhoades, 1985).

Physiological and biochemical adaptations which enable insect herbivores to detoxify, metabolize, store, or excrete plant allelochemicals are many and varied (Rhoades, 1977; Millburn, 1978; Brattsten, 1979; Brattsten, Chapter 6 in this text; Berenbaum, 1980; Berenbaum, Chapter 7 in this text; Dowd et al., 1983; Jaenike, et al 1983). But physiological adaptations are unnecessary if herbivores simply avoid ingesting deleterious plant chemicals (Gould, 1984). Though this aspect of plant-insect interactions has received relatively little attention (see Rhoades 1983, 1985), there is growing evidence that phytophagous insects exhibit an array of avoidance behaviors which are derived from slight extensions of various foraging, feeding and diapause behaviors already in place. In this chapter I review the major evidence that most if not all insect herbivores are behaviorally adept at minimizing direct exposure to plant allelochemicals. Adaptations discussed include those which enable insects to:

- avoid constitutive and inducible defenses in time;
- avoid constitutive and inducible defenses in space;
- avoid triggering inducible defenses;
- overwhelm the chemical defenses of their host;
- block inducible defenses that have already been triggered.

I also suggest that these behavioral adaptations are not exercised without

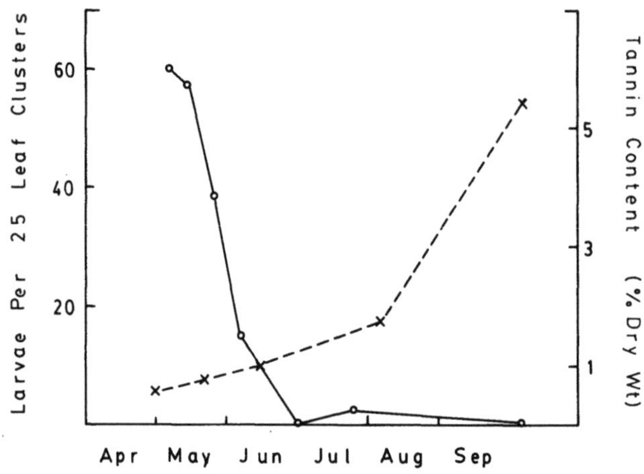

Fig. 1. Seasonal variation of number of oak feeding lepidopterous caterpillars per 25 leaf clusters (solid line) and total tannin content (hydrolyzable and condensed) of upper sun leaves of Quercus robur (dotted line). Modified from Feeny 1970.

considerable ecological costs: costs which may help balance the interactions between insects and their host plants.

2. AVOIDING DEFENSES IN TIME

In the past, plant allelochemicals were considered static in their effects on herbivores (Dixon, 1970; Dawkins and Krebs, 1979; Rhoades, 1985). Inducible defenses were not yet recognized and standing defenses were thought to be omnipresent and invariant. We now realize this is not the case. There is substantial heterogeneity in the temporal and spatial distribution of several components of leaf quality including water content, toughness, nitrogen and the concentration of defensive chemicals (Whitham, 1981; Schultz, 1983b). If standing plant defenses vary through time, herbivores have the evolutionary option of adjusting consumer phases of their life histories to periods when host defenses are minimal. One of the classic studies which served as a catalyst for serious investigations of plant-insect interactions concerned an insect guild that avoids deteriorating leaf resources in time (Feeny, 1966, 1968, 1969, 1976; Feeny and Bostock, 1968). Feeny examined seasonal changes in the toughness, tannin content, and nutrients of English oak, Quercus robur L. (Fagaceae) in relation to herbivory patterns of oak-feeding lepidopterans. He noted that the density of external folivores declined dramatically as the season progressed, was negatively correlated with the total tannin content of oak leaves, Fig. 1, and was positively correlated with leaf nitrogen content. By adding oak tannin to an artificial diet, Feeny was able to reduce significantly larval growth rate and pupal weight of the winter moth, Operophtera brumata L. (Lepidoptera:Geometridae), an early spring oak herbivore. He concluded that the high quality of early season leaves was responsible for the skewed seasonality of most oak lepidopterans.

Other authors also argue that tannins may deter herbivory by decreasing palatability, binding with available proteins, or acting as a direct toxic agent (Pridham, 1963; Haslam, 1966; Goldstein and Swain, 1965; Rhoades and Cates, 1976; Swain, 1979). Tannins and other polyphenols as well as lignins, resins, and silica are examples of "quantitative" defenses (Feeny 1975, 1976), so named because they impact herbivores in a dosage-dependent manner. Until recently it was thought that even specialists physiologically adapted to toxic "qualitative" defenses such as enzyme inhibitors, alkaloids and cyanogens are deterred by increasing concentrations of quantitative defenses. It is now clear, however, that this generalization does not always hold, for at least some insects are physiologically resistant to hydrolyzable and/or condensed tannins (Fox and McCauley, 1977; Berenbaum, 1980; Bernays, 1981; Zucker, 1983; Martin and Martin, 1984; Martin et al., 1985), if only during periods before concentrations have reached seasonal highs (Schultz, 1983a). Nevertheless, quantitative defenses generally accumulate slowly in plant tissues: as in oaks it may be weeks before concentrations reach deleterious levels (Thielges, 1968; Baltensweiler et al., 1977; Benz, 1977; Higgins et al., 1977; Niemala et al., 1979; Rhoades, 1977, 1983; Wallner and Walton, 1979; Davies and Schuster, 1981; Schultz and Baldwin, 1982; McNaughton and Tarrants, 1983). It is during this period of defense build-up that susceptible herbivores can successfully exploit tissues soon to become unavailable.

Selection for rapid early season development can be so strong that many insects forsake months of environmentally favorable conditions. Trirhabda baccharidis (Weber) (Coeloptera:Chrysomelidae), a specialist on the groundsel tree Baccharis halimifolia L. (Compositae) (Kraft and Denno, 1982), the fall cankerworm, Alsophila pometaria (Harris) (Lepidoptera: Geometridae) (Jones and Schaffner, 1959), the imported willow leaf beetle, Plagiodera versicolora Liach. (Coleoptera:Chrysomelidae) (Raupp pers. comm.), Malacosoma tent caterpillars (Lepidoptera:Lasiocampidae) (Becker, 1938) and the gypsy moth, Lymantria dispar L. (Lepidoptera:Lymantriidae) (Nichols, 1961) are but a few species which avoid summer defenses of their hosts by developing on spring foliage, then passing the remainder of the season in reproductive diapause. I do not imply that behavioral avoidance of plant defenses is the sole cause for spring specialization among insects. Certainly the high nitrogen content of early season foliage is a primary factor encouraging herbivores to exploit these tissues (Feeny, 1970; Scriber and Slansky, 1981; Kraft and Denno, 1982). But plant allelochemicals, particularly those that increase in concentration and effectiveness over time, have likely played an important role as well.

Inducible defenses can also be avoided in time by herbivores. The inducible defense system is a dynamic one, often responding to tissue damage with toxic qualitative metabolites (Loper, 1968; Green and Ryan, 1972; Puritch and Nijholt, 1974; Benz, 1977; Rhoades, 1979; Carroll and Hoffman, 1980). The rate at which induced compounds become effective is poorly understood for most plants, but even rapid induction rates provide herbivores with some opportunity to feed before defenses defile the surrounding tissues. For example, Carroll and Hoffman (1980) report that chemical defenses in the Cucurbitaceae can be mobilized within 40 minutes of simulated insect damage. Though this is the fastest known induction rate, it leaves ample time for a quick "snack" by cucurbit herbivores. By adopting what Rhoades (1985) has called a "surprise" feeding pattern, that is, by restricting feeding to naive tissues in which defensive chemicals have not yet accumulated, insects can consistently consume high quality tissues.

Nearly every plant bears circumstantial evidence that insects avoid inducible defenses to some extent. A visual survey of plant leaves quickly demonstrates that most leaves are only partially consumed by herbivores, Fig. 2. In fact, usually only small areas of a leaf are destroyed, convincing testimony that many insects feed for only a short time on one leaf before abandoning it for another leaf. Such ephemeral feeding patterns are thought to prevent predators from forming search images based on feeding damage (Heinrich, 1979; Heinrich and Collins, 1983). This behavior may also enable insects to beat the defenses of naive host plants by departing

Fig. 2. Characteristic sporadic feeding patterns of phytophagous insects on Fagus grandifolia Ehrhart. Photograph by the author.

before the defensive response of the injured plant can become effective (Hargrove and Crossley, 1985). Rhoades (1985) postulates that the surprise feeding strategy may form the evolutionary basis for migration and dispersal behavior in insects such as locusts (Orthoptera:Acrididae), spruce budworm, Choristoneura fumiferana (Clemens) (Lepidoptera: Tortricidae), larch bud moth Zeiraphera griseana (Hubner) (Ibid.), tent caterpillars (Lepidoptera:Lasiocampidae) and armyworms (Lepidoptera: Noctuidae). By immigrating into populations of naive plants and then emigrating before allelochemicals are fully mobilized, these insects may avoid chemical defenses of their hosts. Studies of migration in neotropical Urania species (Lepidoptera:Uraniidae) support this hypothesis (Smith, 1982, 1983). Urania moths are specialists on Omphalea (Euphorbiaceae) lianas. When larvae are forced to feed on vines which have been grazed by previous generations, Urania fitness declines significantly. On four or eight year cycles, Urania stages huge unidirectional flights from Omphalea patches throughout the tropical Americas. The periodicity of Urania emigrations may reflect the timing of biochemical changes which they have induced within their host plants.

3. AVOIDING DEFENSES IN SPACE

We have seen that standing defenses can increase to formidable concentrations over the course of a season, and inducible defenses can significantly alter leaf quality within minutes of tissue trauma. Some insects have capitalized on these temporal changes in allelochemical concentrations by restricting feeding bouts to periods when host defenses are low. Yet it is spatial variation in plant allelochemicals that provides herbivores with their best chance of minimizing exposure to deleterious defenses. Substantial qualitative and quantitative diversity in the secondary metabolites of different plant species has been demonstrated repeatedly (Levin, 1976; Chew and Rodman, 1979), but we have only recently recognized the degree to which chemical defenses vary within species, populations, and particularly individuals (Dolinger et al., 1973; Hare and Futuyma, 1978; Moore, 1978; McKey, 1979; Rodman and Chew, 1980; Schultz et al., 1981; Whitham, 1981; Schultz et al., 1982; Louda and Rodman, 1983a, 1983b; Schultz, 1983a).

Whitham (1981) has reviewed the principal sources of within and between plant variation. An individual plant can be considered the genetic equivalent of a population of thousands of self-replicating buds (Harper, 1979). Somatic mutations occurring at any point in these replications are heritable and can be preserved by sexual or asexual reproduction (Hartmann and Kester, 1975; Stewart and Dermen, 1979). Developmental patterns such as the distribution of juvenile and adult phases and the differential allocation of toxins and/or digestibility-reducing phenolics to leaves of various age classes contribute to the heterogeneity of single plants. Environmental factors can also be a principal source of variation within plants. The root system of a tree, for example, can encounter differences in soil type, nutrients, water, etc. The branches that a single root supplies may reflect these differences (Richardson, 1958; Zimmerman and Brown, 1971). Furthermore, irregular feeding patterns by herbivores may induce chemical defenses irregularly, increasing the heterogeneity of a single plant. Intraspecific variation in plant defenses can be genetically based (Edmunds and Alstad, 1978, 1981; Rice, 1983) or result from effects that prove stressful to individual plants such as water deficit or surplus, various climatic conditions, nutrient deficiences, pollution, herbivory, etc. (Rhoades, 1983).

Because insects are confronted with a spatially complex mosaic of food quality, they have developed foraging behavior patterns which enable them to discriminate between high and low quality resources (McGugan, 1954; Knerer and Atwood, 1973; Bryant and Raske, 1975; Sharik and Barnes, 1976; Cates and Rhoades, 1977; Hassell and Southwood, 1978; Larsson and Tenow, 1979; Schultz 1983a). Schultz (1983b) provides a detailed account of the foraging behavior of fifth instar larvae of the saddled prominent, Heterocampa guttivitta Walker (Lepidoptera:Notodontidae). During a five hour period six larvae moved 970 cm on their sugar maple host, passed 776 leaves, tasted 98 leaves, but fed on only 30 leaves. Leaves that were rejected by one larva were also rejected by the other larvae. In the Eastern tent caterpillar, Malacosoma americanum (F.), foraging is a social activity (Fitzgerald, 1976; Fitzgerald and Edgerly, 1979; Edgerly and Fitzgerald, 1982; Fitzgerald and Peterson, 1983). Colonies of siblings rest between feeding bouts within a silk tent constructed on their cherry and apple host trees. A communal feeding bout is initiated when a few "scout" larvae leave the tent and forage for young leaves. When a suitable leaf source is located scouts return to the tent depositing a recruitment trail pheromone as they travel. The recruitment pheromone excites the remainder of the colony and guides them to the appropriate leaf patch.

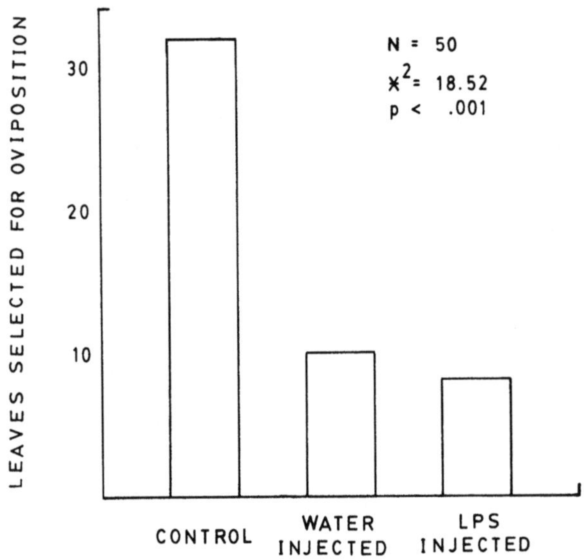

Fig. 3. Epilachna varivestis oviposition preference on control, water-injected, and lipopolysaccharide-injected (LPS) first trifoliate Phaseolus vulgaris L. leaflets. From Schroeder 1984.

In view of the physiological costs associated with the production and maintenance of allelochemicals (Soloman and Crane, 1970; Levin, 1976), defense systems may be seriously compromised when plants are stressed, and thus plants under stress should be preferentially attacked by herbivores (Rhoades, 1979, 1983, 1985; Wright et al., 1979; Waring and Pitman, 1983). For example, grand firs, Abies grandis (Douglas) (Pinaceae), that have been defoliated by folivores can no longer maintain adequate monoterpene production. The fir engraver beetle, Scolytus ventralis Le Conte (Coleoptera:Scolytidae) quickly detects weakened trees and attacks en masse (Wright et al., 1979). Similarly, Louda and Rodman (1983a) found that sun-stressed Cardamine cordifolia (Cruciferae) produced lower concentrations of glucosinolates and were significantly more damaged by insects than were shaded plants.

There is some evidence that host plant defenses can also influence the selection of oviposition sites. If gravid females discriminate between high and low quality leaves before ovipositing, emerging larvae which may be too small to efficiently forage for themselves can immediately consume suitable resources. Schroeder (1984) has shown that gravid females of the Mexican bean beetle, Epilachna varivestis Muls. (Coleoptera:Coccinellidae), taste bean leaves before ovipositing, possibly to determine their quality. When given a choice between undamaged leaves, leaves damaged by water injection, and leaves damaged by lipopolysaccharide injection, females significantly preferred undamaged leaves for oviposition sites, Fig. 3. Apparently, wounds from injection induce a defensive response in bean leaves strong enough to be detected and avoided by bean beetles. Boll weevil, Anthonomus grandis Boheman (Coleoptera: Curculionidae), females reportedly exhibit a similar behavior (Benedict et al., 1979).

The proximate causes of foraging "decisions" such as those described above are largely a consequence of behavioral responses to leaf position or to a variety of chemical attractants and repellants (Dethier, 1954, 1970, 1976, 1979; Schoonhoven 1968, 1972; Mitchell, 1981; Schultz, 1983b).

Ultimately, however, the foraging patterns of insect herbivores represent an evolutionary compromise between the need to maximize resource quality while minimizing exposure to predators and the energetic costs associated with movement (Schultz, 1983b, See also Section 7).

One may speculate that the specialized feeding behavior of leaf miners and pith tunnelers can be viewed as evolutionary attempts to spatially avoid quantitative plant defenses (Feeny, 1970). Some 26 percent of the late season lepidopterous herbivores of English oak are leaf miners. By restricting their consumption to the spongy parenchymal tissues within leaves, these tiny larvae may avoid most of the leaf phenolics which are concentrated in the palisade cells that line the epidermis.

4. MECHANISMS TO AVOID TRIGGERING DEFENSES

Rhoades (1985) has postulated that herbivores exhibit two alternative attack strategies which ultimately determine their population dynamics. Species with small, relatively stable populations exploit their host plants with "stealth". Less numerous but highly visible species with exceptionally variable populations attack plants in "opportunistic" ways (See Section 5). Rhoades suggests that populations of stealthy herbivores do not fluctuate dramatically because such species display behavioral adaptations which minimize their impact on host fitness. By confining their attack to tissues of relatively low value to their host, and by avoiding conspecifics, stealthy herbivores can exploit plants without triggering strong responses from inducible defenses. Thus stealthy herbivores should be characterized by traits such as territorial defense of the resource and/or solitary behavior.

At this writing the concepts of stealth and opportunism as offensive alternatives for herbivores are young and have yet to be scrutinized by the scientific community. Nevertheless, there is clearly a dichotomy in the population dynamics of insect herbivores: under natural conditions most species (stealthy herbivores) do not exhibit cyclic or even occasional population surges. Those that do, however, (opportunistic herbivores) such as the spruce budworm (Morris et al., 1958), bark beetles (Berryman, 1978) and locusts (Rhoades, 1985) can comprise the most destructive and destablizing biotic forces in nature. The extent to which these differences are the result of particular physiological and behavioral adaptations for exploiting plants remains to be seen, but there is mounting evidence that some herbivores may, indeed, be able to suppress a plant's ability to recognize and/or respond to their feeding damage.

It has been proposed that systemic plant responses to insect attacks are triggered by signals released from wounded tissues (Thielges, 1968; Green and Ryan, 1972; Haukioja and Niemela, 1977; Niemela et al., 1979; Rhoades, 1979; McIntyre, 1980; Ryan, 1983; Tallamy, 1985). Recent studies of insect-induced proteinase inhibitors in tomato plants suggest that it is cell wall fragments from badly injured tissues that serve as inter- and intratissue messengers during insect attack (Ryan, 1978; Bishop et al., 1981; Ryan et al., 1982; Ryan, 1983). In view of the tissue trauma required to stimulate a systemic response in plants, insects may be able to avoid inducible systemics by minimizing the amount of mechanical tissue damage they inflict on their host. For instance, aphids (Homoptera: Aphididae) in search of a vascular tube meticulously direct their stylets through intercellular spaces until they reach the phloem (Zweigelt, 1915; Bradley, 1962; Kloft and Ehrhardt, 1962; Miles, 1968; McLean and Kinsey, 1965; Mullin, Chapter 5 in this text). In this way aphids not only avoid vacuoles in which accumulated defenses are concentrated (Campbell et al., 1982; Dreyer and Campbell, 1984), but they may also avoid communicating their presence to the plant by minimizing cellular disruption (Miles, 1968). In fact by reducing structural trauma to plant tissues, the piercing-sucking habits of all phytophagous Hemiptera, Homoptera, and Thysanoptera may not trigger plant defenses as rapidly or as often as the "rip and tear" damage inflicted by mandibulate herbivores. Avoidance of inducible plant defenses may have been an important factor favoring the initial divergence of piercing mouthparts from the ancestral mandibulate condition.

There are few supportive data, but herbivorous insects may also reduce or avoid inducible systemics by interfering with intertissue communication through salivary secretions. Saliva that in some way inhibits the release or transport of wound messengers would allow insects to feed sequentially on several parts of the same plant without triggering a systemic defense throughout the entire plant (Rhoades, 1985).

Plants which rely on the phototoxicity of allelochemicals for defense against insect herbivores (Berenbaum, 1978; Towers, 1980; Arnason et al., 1983; Philogene et al., 1984) are susceptible to species that can avoid or prevent the photoactivation of those compounds. Shielded by plant tissues from ultraviolet radiation, leaf rollers, tiers, folders, and webbers all share behavioral preadaptations which limit the activation of phototoxic defenses. Perhaps the best evidence has resulted from Berenbaum's (1978) work with phototoxic linear furanocoumarins common in the Umbelliferae and Rutaceae. As with other phototoxic compounds, furanocoumarin toxicity is greatly enhanced by UV radiation. Since UV light does not penetrate a rolled or folded leaf (Berenbaum, 1978), insects feeding within such leaves may behaviorally prevent the activation of an otherwise potent chemical defense. It is noteworthy that ten of the 14 common folivores of umbelliferous plants in Tompkins County, New York, are microlepidopterans which feed only within rolled or webbed leaves. In a similar way nocturnal feeding may also enable insects to avoid phototoxic plant defenses.

5. OVERWHELMING PLANT DEFENSES

Most, if not all, plant chemicals which function as antiherbivore defenses are synthesized from smaller molecules at some energetic cost (Soloman and Crane, 1970; Levin, 1976). These costs are often reflected by the inverse relationship between the quantity of defensive compounds in plants and plant fitness (Mothes, 1960, 1976; Hanover, 1966; Foulds and Grime, 1972; Pimentel, 1976; Tester, 1977). When some plants become stressed, they can no longer afford defensive expenditures, and concentrations of allelochemicals decrease (Madden, 1977; Wright et al., 1979; Louda and Rodman, 1983a; Raffa and Berryman, 1982a, 1982b, 1983a, 1983b; Raffa, Chapter 9 in this text; Kimmerer, pers. comm.). Thus, by preferentially exploiting plants under stress, herbivores may reduce their exposure to chemical defenses (Lewis, 1979; Rhoades, 1979, 1983; McLean and Kinsey, 1968; Van Emden, 1972; Wood, 1982; Mares et al., 1977).

The advantages of feeding on plants with lowered defenses are so great that many insects do not wait for environmental conditions to induce stress in their hosts: instead they have assumed an active role by inducing stress through effective behavioral adaptations. Rhoades (1985) has labeled such species "opportunists" and predicts that they should maximize their impact on plant fitness by attacking tissues of high value to the plant and by cooperating in mass attacks to overload plant defense mechanisms. Opportunistic adaptations are particularly obvious in wood boring insects which develop within conifers (Coutts and Dolezal, 1969; Madden, 1977; Christiansen and Horntvedt, 1983; Raffa and Berryman, 1983b; Raffa, Chapter 9 in this text). When healthy lodgepole pines, Pinus contorta Douglas (Pinaceae), for example, are attacked by the mountain pine beetle, Dendroctonus ponderosae Hopkins (Coleoptera:Scolytidae), a series of biochemical alterations including the accumulation of monoterpenes and resin at the site of injury confine the beetle and its fungal symbionts within small necrotic lesions (Reid et al., 1967; Berryman, 1972; Raffa and Berryman, 1982a). The degree to which a particular tree can resist beetle attack changes inversely with the beetle's ability to overwhelm host chemical defenses through rapid aggregation (Thalenhorst, 1958; Berryman, 1972, 1976; Vité et al., 1972; Safranyik et al., 1975; Kravielitzki et al., 1983; Raffa and Berryman, 1983a, 1983b; Raffa, Chapter 9 in this text). As pioneer beetles bore into the subcortical region of a tree, they release aggregating pheromones very attractive to mates and additional colonizers. Each colonizer inoculates the tree with Europhium clavigerum Robinson and Davidson, a pathogenic fungus that contributes to beetle success in two ways. As the fungus is disseminated within the pine, it rapidly invades the ray cells disrupting transpiration and causing desiccation. Once phloem and xylem cells have been killed, the tree is no longer capable of repelling its attackers. The fungus also provides an important source of nutrition for the beetles (Wilson, 1971).

A similar mechanism of overwhelming host defenses has arisen independently in the wood wasp, Sirex noctilio Fab. (Hymenoptera: Siricidae). S. noctilio develop as larvae within their Australian host, Pinus radiata D. Don (Pinaceae). When ovipositing in the tree, females drill from one to three tunnels through the bark with their ovipositor, depending upon the physiological condition of their host (Coutts and Dolezal, 1969; Coutts, 1968; Madden, 1977). On healthy trees primarily single tunnels are drilled. These are injected with a toxic mucus and arthrospores of the symbiotic fungus, Amylostereum areolatum (Fries), but no eggs. The toxin blocks transport of photosynthate from leaves and reduces the synthesis of polyphenols and resins near tunnels. This promotes rapid fungal invasion of the conductive tissues, inducing transpirational stress and reducing exudation of resins that would otherwise kill Sirex eggs. Thus female wasps quickly transform healthy trees into future oviposition sites, either for themselves or their offspring. In less vigorous trees with weak polyphenol and resin defenses females immediately drill multiple tunnels. In one tunnel they deposit an egg; in the others they inject arthrospores and toxins as before. The weakened tree cannot flood the oviposition tunnel with resins, and by egg hatch the fungal invasion has created a suitable resource for larval development.

These exceptional adaptations enable insects to undermine powerful chemical defenses in their hosts, thereby creating superior resources from those that were previously unacceptable. Many insects unable to overwhelm host defenses themselves have become adept at locating stressed plants in which defenses have been subdued by other phenomena (Hanson, 1939; Evans, 1966, 1971; Bright and Stark, 1973; Sickerman and Wangberg, 1983). As in S. noctilio, volatiles specific to physiologically stressed plants identify suitable hosts for many insects over short ranges. But the black fire beetle, Melanophila acuminata DeGeer (Coleoptera:Buprestidae) can locate conifers stressed by fire over distances of many kilometers (Evans, 1966, 1971; Evans and Kuster, 1980). The larvae of these beetles develop best within trees recently scorched and weakened by fire. As such resources are frequently patchy, rare, or ephemeral, these beetles have developed thoracic infrared receptors capable of detecting infrared radiation emitted by burning trees. Their guidance system is analogous to that of a heat-seeking missile and apparently, just as effective: both sexes typically arrive at a burnsite while the trees are still in flames. Mating takes place immediately and females oviposit long before the trees can recover defensively.

Feeding in aggregations is a common behavioral alternative among herbivorous insects that has been credited with overcoming plant toughness (Ghent, 1960; Carne, 1966; Ralph 1976), facilitating defense against predators (Eisner and Kafatos, 1962; Tostawaryk, 1972; Wood, 1975, 1977; Morrow et al., 1976; Nault and Montgomery, 1979), creating nutrient "sinks" (Way and Cammell, 1970; Kidd, 1977), and encouraging mutualistic interactions with ants (McEvoy, 1979; Olmstead and Wood, pers. comm.). Rhoades (1985) suggests that gregarious feeding behavior may also nullify the chemical defenses of plants. Several studies circumstantially support this hypothesis but critical analyses have not been published. When first instars of gregarious stink bugs (Hemiptera:Pentatomidae) (Kiritani, 1964; Kiritani and Kimura, 1966), lace bugs (Hemiptera:Tingidae) (Tallamy and Denno, 1981) and sawflies (Hymenoptera:Diprionidae) (Kalin and Knerer, 1977) are isolated, survivorship is reduced, or development time increased over immatures that feed in groups, even though the solitary insects successfully penetrate and feed on their hosts. If insect saliva can biochemically neutralize plant allelochemicals, the effect may be more powerful when the saliva of many individuals is pooled. There is evidence that the saliva of some hemipterans contains polyphenoloxidase enzymes which may neutralize phenolic plant defenses (Miles, 1978).

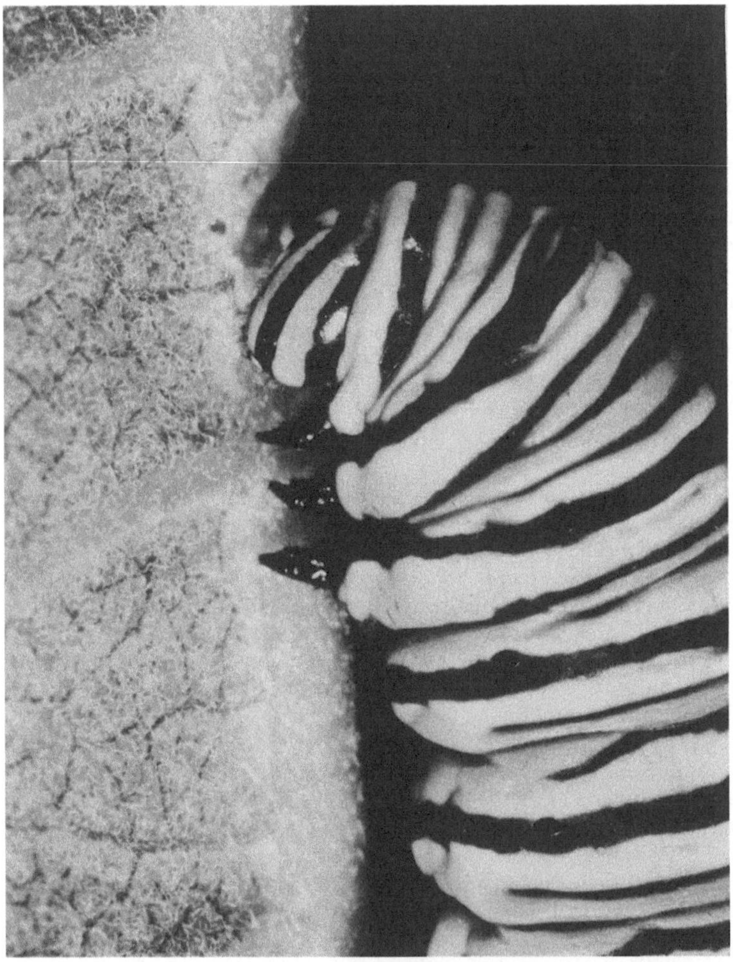

Fig. 4. Danaus plexippus chews through the midrib of Asclepias leaves effectively blocking latex defenses. Photograph by the author.

Fig. 5. Epilachna borealis snips a circular trench in cucurbit leaves before feeding on entrenched tissues. Photograph by R. G. Weber. From Tallamy 1985; reprinted with permission.

6. BLOCKING PLANT DEFENSES

Many plant defenses, particularly inducible defenses mobilized from uninjured areas of the plant, are transported through the vascular system; if vascular bundles are blocked or severed, the palatability and/or quality of the tissues that become isolated is preserved. It is not surprising that a number of mandibulate insects regularly exploit this aspect of inducible defense systems. For instance, the crude and often wasteful feeding behavior of cutworms may be explained as attempts to block host defenses. These caterpillars characteristically chew off young plants near the ground before feeding on the foliage (Metcalf et al., 1962). In so doing, cutworms may shut off the transport of deleterious compounds from stems or roots to the vulnerable leaves. Cutworms are active primarily at night, possibly to avoid avian predators or to reduce the speed with which their severed food supply wilts.

Stem and petiole girdling is a common means by which insects isolate tissues from the vascular system of the plant. Several of the wood boring beetles, especially Oncideres and Agrilus spp. (Coleoptera:Cerambycidae) girdle host twigs before ovipositing in the isolated terminals (Johnson et al., 1976; Mares et al., 1977), while leaf feeding caterpillars, e.g. Heterocampa (Carroll and Hoffman, 1980), Hartigia sawflies (Hymenoptera: Cephidae) (Johnson et al., 1976), and Epilachna beetles (Tallamy, 1985) girdle leaf petioles before feeding on the leaf. Girdling may disrupt inducible plant defense mechanisms in two ways: it may prevent the

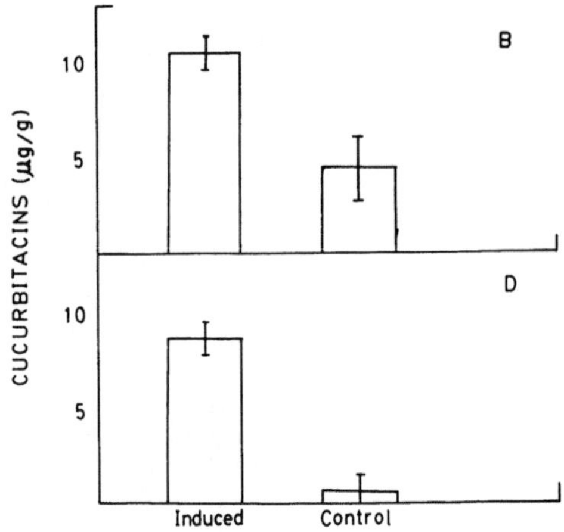

Fig. 6. Comparison of cucurbitacin B and D concentrations (μg/g fresh weight) in Cucurbita pepo "Black" leaves 3 hr after mechanical injury vs concentrations in uninjured leaves. Statistical intervals = standard deviations.

contamination of isolated tissues by defenses transported from distant sources in the plant. It may also interrupt inter-tissue communication by confining potential wound messengers (cell-wall fragments; See Section 4) to the feeding site. If this is the case, the systemic response of the plant is minimized, uninjured tissues remain poorly defended, and the insect attacker is not forced to abandon the plant after one feeding bout.

A related alternative to girdling is to sever the major veins servicing the tissues to be consumed. This behavior is particularly effective against the latex defenses of many plants (Young, 1978; Dillon et al., 1983; Dussourd and Eisner, 1986). Most milkweeds, Asclepias spp. (Asclepiadaceae), for example, are well endowed with a white sticky latex sap that flows freely from wounds on the plant. Although sucking insects can avoid latex by virtue of their mouthparts (See Section 4), mandibulate herbivores are seriously compromised if their mouthparts become smeared with latex; it can "gum up" the mandibles and prevent further feeding. Nevertheless, milkweeds support a number of mandibulate specialists including Danaus spp. (Lepidoptera:Danaidae), Euchaetes egle Drury (Lepidoptera: Arctiidae), Tetraopes spp. (Coleoptera:Cerambycidae) and Labidomera clivicollis (Kirby) (Coleoptera:Chrysomelidae). All have converged on a common feeding strategy: by snipping through the major leaf veins proximal to the feeding site, Fig. 4, these insects can feed unhampered by the gummy latex (Dussourd and Eisner, 1986). Vein snipping is not restricted to insects that encounter latex defenses (MacKay and Wellington, 1977; Dussourd, pers. comm.). It is logical to conclude that such behavior prevents the mobilization of any defensive compound heavily dependent on major veins for its transport.

Adaptations to isolate feeding sites from mobile plant defenses have reached their pinnacle in Epilachna beetles that specialize on cucurbits (Carroll and Hoffman, 1980; Tallamy, 1985). Prior to feeding, larval and adult E. tredecimnotata (Latreille) (Coleoptera:Coccinellidae) and E. borealis Fab. (Ibid.) use enlarged mandibular cusps to cut a circular trench through all but the lower epidermal leaf tissues, Fig. 5. Carroll and Hoffman (1980) were the first to propose that such behavior isolates the feeding area from cucurbitacins, the oxygenated tetracyclic triterpenoids responsible for the bitter taste characteristic of cucurbits. Recent experiments with E. borealis and its zucchini host, Cucurbita pepo L.,

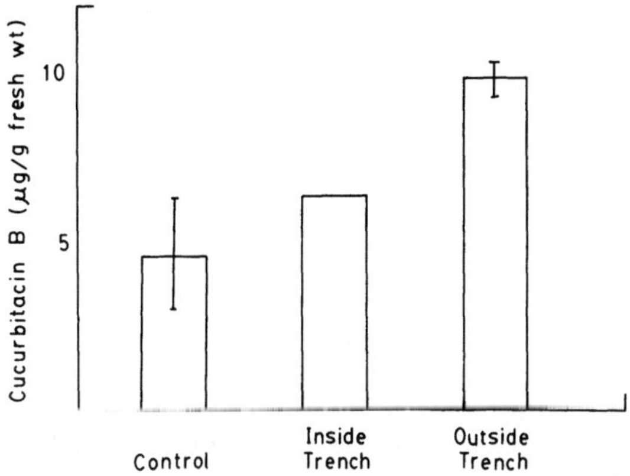

Fig. 7. Comparison of cucurbitacin B concentrations in tissues encircled by an Epilachna borealis feeding trench, in tissue adjacent to but outside of the trench 1 hr post-trenching, and from an undamaged control leaf of Cucurbita pepo "Black" 1 hr post-trenching. Statistical intervals = standard deviations.

support this hypothesis in three ways (Tallamy, 1985). We have shown that mechanically simulating mandibulate feeding damage induces significant increases in the cucurbitacin concentrations of zucchini leaves, Fig. 6. Within three hours of tissue damage cucurbitacins (hereafter called Cucs) B and D more than double in concentration at the site of injury. If beetles are reared on leaves that have been similarly induced, the loss in fitness is striking. Larval survivorship is reduced by a factor of three; adult females are significantly smaller, oviposit significantly later, are fecund for a significantly shorter period, and lay significantly fewer eggs than females reared on control leaves. Furthermore, our work to date suggests that E. borealis feeding trenches do, in fact, block the influx of deleterious chemicals.

If E. borealis females are removed immediately after completing a feeding trench on previously uninjured plants, the concentration of Cucs that subsequently accumulates in tissues outside of the trench can be compared with the Cuc concentration of the tissue inside the trench. Figure 7 depicts this comparison in terms of Cuc B quantified by ultraviolet absorbance one hour post-trenching (Tallamy, 1985). Although the entrenched tissue concentration was obtained from pooled contents of three trenches (UV spectroscopy is not sufficiently sensitive to detect cucurbitacins from smaller samples), the greater Cuc concentration outside of the trench is suggestive and supports the hypothesis that feeding trenches effectively block plant defenses from contaminating the feeding site. Even if trenches are "leaky" and only slow the progress of cucurbit defenses, adult beetles consume entrenched tissues quickly (mean = 21.8 $\pm$ 19 minutes; N = 12) and appear to be well protected while they feed.

It is not unexpected that Epilachna should exhibit behavioral adaptations against plant chemical defenses. Epilachna coccinellids are phytophagous and are thought to comprise a phylogenetically advanced genus within their otherwise predaceous family (Arnett, 1960; Gordon, pers. comm.). If phytophagy is the derived state in this family, predatory Epilachna progenitors probably lacked the physiological preadaptations necessary to detoxify, metabolize, store or excrete the chemical defenses of potential host plants. Those individuals that "discovered" simple mandibular manipulations of plants such as trenching or vein-snipping may have been able to switch directly to plant tissues without ingesting inducible plant defenses. Once avoidance behaviors became genetically established, selection for biochemical or physiological adaptations against inducible allelochemicals may have been reduced or eliminated. This scenario is supported by the extreme sensitivity of E. borealis to the deleterious effects of cucurbitacins within its cucurbit hosts (Tallamy, 1985). Feeding adaptations in this species block exposure to mobile cucurbitacins so completely that the beetle has not developed effective physiological resistance mechanisms.

7. THE COST OF AVOIDING PLANT DEFENSES

It is clear from the above examples that phytophagous insects possess a battery of behavioral adaptations effectively enabling them to reduce or totally avoid exposure to the chemical defenses of their host plants. If we add to these behavioral adaptations a consideration of the many physiological means of rendering allelochemicals harmless (See Chapters 3-7 in this text) one is tempted to view the interaction between plants and insects as remarkably one-sided. This of course is not the case, for as the following examples will illustrate, avoidance adaptations are not without significant physiological and ecological costs which allow plants to indirectly curtail herbivore fitness.

Insects which have adjusted developmental and reproductive phases of their life histories to coincide with seasonal lows in host defenses, considerably slow their genetic contributions to future generations by sacrificing substantial periods of biologically favorable temperature and photoperiod. The forest tent caterpillar (Malacosoma disstria Hubner) (Lepidoptera:Lasiocampidae) for instance, is confined to one brief generation in the spring, presumably to exploit hosts when their defenses are low and nutrition is high. Weather conditions could easily support two additional generations of this moth if resources were equally suitable. The sacrifice of periods favorable to biological processes is particularly restrictive in the temperate zone where insects are necessarily deprived of such conditions up to eight months of the year.

Insects which minimize exposure to allelochemicals by discriminating between spatially variable resources are confounded by three interacting costs (Schultz, 1983a). Any given foraging pattern represents an evolutionary compromise among the metabolic costs of movement between high quality resources, the ecological risks encountered while searching for these resources, and the physiological costs of consuming poor quality resources. One cost can not be minimized without elevating another. Saddled prominent caterpillars (See Section 3), for instance, select only the highest quality leaves that are dispersed among an array of lesser quality leaves (Schultz, 1983a). They are rewarded for discriminating taste by low feeding costs, i.e. minimized reduction in growth rate probably due to minimal leaf toxicity, digestion inhibition, investments in detoxification, or digestive adaptations. Low feeding costs are balanced, however, by elevated costs from excessive movement and exposure to predators. Selective feeding on high quality leaves requires time and energy investments for locating these resources. Metabolic expenditures are high, and so are the risks of being detected by predators attracted by movement. Furthermore, if predators or parasites cue on specific plant tissues such as high quality leaves, caterpillar vulnerability will increase with increases in discriminatory feeding behavior.

The reproductive success of bark beetles that overwhelm conifer defenses by the coordinated attack of many individuals is delicately balanced between the chemical defenses of host trees and their own density (Raffa and Berryman, 1983a). At low beetle densities, healthy conifers

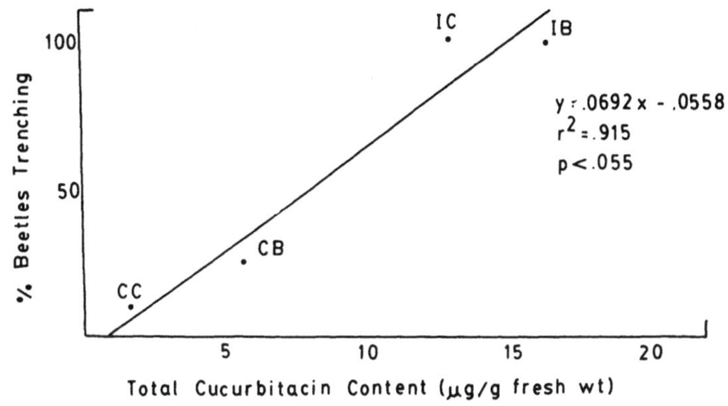

Fig. 8. Relationship between the percentage of Epilachna borealis adults that cut a feeding trench prior to feeding and total cucurbitacin content of induced (IC; N=64) and control (CC; N=55) Early Golden Summer Crookneck leaves and induced (IB; N=85) and control (CB; N=125) Black zucchini leaves.

successfully repel invading beetles. When beetles attack in large numbers through the coordination of aggregation pheromones, conifer defenses are overwhelmed and the beetles become successfully established in the tree. The advantages of coordinated attacks are obvious, but if beetle density becomes too high (more than 62 galleries/m^2 for the mountain pine beetle) intraspecific competition can result in an exponential decline in beetle fitness (Raffa, Chapter 9 in this text).

The benefits of behavioral adaptations that block inducible defenses may also be balanced by costs. If we consider again the trenching behavior of Epilachna squash beetles (See Section 6), we find there are non-trivial time constraints associated with this mode of feeding. There is good evidence that shortly after a beetle completes a feeding trench, tissues of the trenched leaf, Fig. 8, and also those of leaves adjacent to it (Tallamy, 1985), are contaminated with Cucs and are no longer a palatable beetle resource. Thus squash beetles are limited to short feeding bouts before they are forced by induced Cuc defenses to forage some distance for uncontaminated, naive leaves. After consuming the contents within its trench, E. tredecimnotata adults move an average of six meters along host vines before feeding again (Carroll and Hoffman 1980). The time invested in the act of trenching may also handicap squash beetles. E. borealis requires on average 14.4 $\pm$ 11.5 minutes (N=10) to complete a feeding trench. If a beetle cuts only five trenches per day, in one week alone it allocates over eight hours to blocking squash defenses. Such investments may partially explain why the closely related E. varivestis that does not cut trenches before feeding, reaches the age at first reproduction four and a half days sooner at 27°C than E. borealis (Schroeder, 1984).

If blocking defenses is a costly endeavor, insects should only do so when there is no alternative. Facultative vein-snipping in the plantain looper, Autographa precationis Guenee (Lepidoptera:Noctuidae) supports this hypothesis. This polyphagous caterpillar snips the major veins of host leaves that contain latex or resins, but does not exhibit this behavior when feeding on other plants (Dussourd, pers. comm.). Trenching is also expressed facultatively in the squash beetle. If adult E. borealis are provided with excised leaves from cucurbit cultivars that vary in Cuc content, the percentage of beetles that trench is a highly significant positive function of leaf Cuc content, Fig. 9. At high Cuc levels, 100

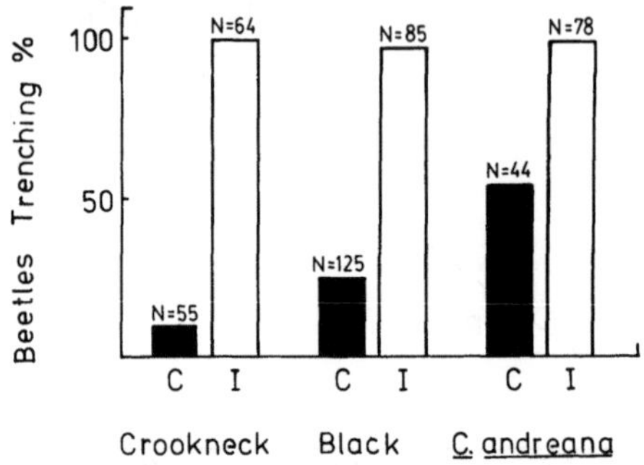

Fig. 9. Propensity of Epilachna borealis to snip feeding 'trenches' in the leaves of cucurbit biotypes varying naturally in cucurbitacin content (C = Control) and varying as a result of induction via prior mechanical tissue damage (I = Induced). Biotypes include Cucurbita pepo 'Crookneck', C. pepo 'Black' and C. andreana.

percent of the beetles cut feeding trenches, but when Cuc content is low, beetles feed without trenching. These data are intriguing, for rarely if ever do beetles naturally encounter leaves with low Cuc concentrations. Even cucurbits with relatively low concentrations in naive leaves quickly exceed the apparent threshold concentration for trenching when they are injured. The expression of trenching behavior has nevertheless remained extremely plastic, suggesting that it is sufficiently demanding in terms of time and energy requirements to justify its disuse at the first opportunity.

Further evidence of the cost of trenching stems from the degree to which squash beetle adults and larvae usurp tissues entrenched by another individual (Tallamy, unpubl.). If more than one larva occupies a leaf simultaneously, the trenching activity of one individual frequently attracts other larvae to the isolated tissues. These "uninvited guests" take advantage of the trencher's labors and consume the encircled tissues.

8. SUMMARY

Phytophagous insects are continuously challenged by deleterious plant chemicals, for most if not all plants are well endowed with such defenses. The power of plant allelochemicals as selective forces shaping the physiological and behavioral attributes of insects can not be overemphasized. As other chapters in this volume suggest, we are accumulating a great deal of information regarding physiological adaptations in insects which permit the exploitation of well defended plants. It is likely, however, that such adaptations are refined only after an insect has behaviorally breached its host's defenses. Physiologically non-adapted herbivores may be unable to utilize defended tissues long enough for physiological adaptations to evolve without first reducing the impact of defensive compounds through some means of behavioral avoidance. Extraordinary temporal and spatial variation in the chemical defenses of plants have provided insects with opportunities to avoid plant defenses which can be easily realized by slight shifts in established behavioral patterns. As the examples above suggest, phytophagous insects can reduce or totally avoid exposure to deleterious host defenses by adjusting their emergence or diapause schedules, by discriminating between resources of greater or lesser toxicity, by actively blocking incoming inducible defenses, by overwhelming host defenses through coordinated attacks with fellow herbivores, or by feeding unobtrusively so as to avoid triggering defenses in the first place. Other 'offensive' behavioral adaptations have been predicted (Rhoades, 1985) but there are currently no data supporting their existence. If there is a unifying conclusion to be drawn from this discussion it is this: in spite of the effectiveness of behavioral adaptations to avoid them, plant chemical defenses succeed in forcing herbivores to compromise, either through life-history trade-offs or increases in energy expenditures, time constraints or risk. Such compromise is a key factor promoting evolutionary stability between the interactions of insects and their host plants.

9. ACKNOWLEDGEMENT

I thank M. J. Raupp, K. F. Raffa, and D. Dussourd for discussions in which many of the ideas presented in this paper arose. I also thank J. C. Schultz, D. Dussourd, P. P. Burbutis, C. B. Keil and R. G. Weber for reviewing the manuscript, R. G. Weber for photographic expertise , and L. A. Horton for technical assistance.

Published with the approval of the Director of the Delaware Agricultural Experiment Station as Miscellaneous Paper No. 1108, Contribution No. 567 of the Department of Entomology and Applied Ecology, University of Delaware, Newark, Delaware.

10. REFERENCES

Arnason, J. T., G. H. N. Towers, B. J. R. Philogene, and J. Lambert, 1983, Naturally occurring photosensitizers and their effects on insects, in: "Mechanisms of Plant Resistance to Insects", P. Hedin, ed., pp. 139-151, Symp. Ser. No. 208, Amer. Chem. Soc., Washington.

Arnett, R. H., 1960, "The Beetles of the United States", The Catholic University of America Press, Washington.

Baltensweiler, W., G. Benz, P. Bovey, and V. Delucchi, 1977, Dynamics of larch bud moth populations, Annu. Rev. Entomol., 22:79-100.

Becker, W. B., 1938, Leaf-feeding insects of shade trees, Mass. Agr. Exp. Sta. Bull., No. 353, 82 pp.

Benedict, J. H., L. S. Bird, and C. Liverman, 1979, Bacterial flora of MAR cottons as a boll weevil resisting character, in: "Proceedings of the Beltwide Cotton Production Research Conferences", J. M. Brown, ed., pp. 228-230, National Cotton Council of America, Memphis.

Benz, G., 1977, Insect induced resistance as a means of self defense in plants, in: "Eucarpia IOBC Working Group Breeding for Resistance to Insects and Mites", SROP, 1977/8, pp. 155-159.

Berenbaum, M., 1978, Toxicity of a furanocoumarin to armyworms: A case of biosynthetic escape from insect herbivores, Science, 201:532-533.

Berenbaum, M., 1980, Adaptive significance of midgut pH in larval Lepidoptera, Am. Nat., 115:138-146.

Bernays, E. A., 1981, Plant tannins and insect herbivores: An appraisal, Ecol. Entomol., 6:353-360.

Berryman, A. A., 1972, Resistance of conifers to invasion by bark beetle -

fungal associations, BioScience, 22:598-602.

Berryman, A. A., 1976, Theoretical explanation of mountain pine beetle dynamics in lodgepole pine forests, Environ. Entomol., 5:1225-1233.

Berryman, A. A., 1978, A synoptic model of the lodgepole pine/mountain pine beetle interaction and its potential application in forest management, in: "Theory and Practice of Mountain Pine Beetle Management in Lodgepole Pine Forests", D. L. Kibbe, A. A. Berryman, G. D. Amman, and R. W. Stark, eds., pp. 98-105, University of Idaho Press, Moscow.

Bishop, P. D., D. J. Makus, G. Pearce, and C. A. Ryan, 1981, Proteinase inhibitor-inducing factor activity in tomato leaves resides in oligosaccharides enzymically released from cell walls, Proc. Natl. Acad. Sci. USA, 78:3536-3540.

Bradley, R. H. E., 1962, Response of the aphid Myzus persicae (Sulz.) to some fluids applied to the mouthparts, Can. Entomol., 94:707-722.

Brattsten, L. B., 1979, Biochemical defense mechanisms in herbivores against plant allelochemicals, in: "Herbivores, Their Interaction with Secondary Plant Metabolites", G. A. Rosenthal and D. H. Janzen, eds., pp. 199-270, Academic Press, New York.

Bright, D. E., Jr., and R. W. Stark, 1973, The bark and ambrosia beetles of California (Coleoptera: Scolytidae and Platypodidae), Bull. Calif. Insect Surv., 16:169.

Bryant, D. G., and A. G. Raske, 1975, Defoliation of white birch by the birch casebearer, Coleophora fuscedinella (Lepidoptera: Coleophorideae), Can. Entomol., 107:217-223.

Campbell, B. C., D. L. McLean, M. G. Kinsey, K. C. Jones, and D. L. Dreyer, 1982, Probing behavior of the greenbug (Schizaphis graminum, biotype C) on resistant and susceptible varieties of sorghum, Ent. Exp. Appl., 31:140-146.

Carne, P. B., 1966, Primitive forms of social behaviour, and their significance in the ecology of gregarious insects, Proc. Ecol. Soc. Australia, 1:75-78.

Carroll, C. R., and C. A. Hoffman, 1980, Chemical feeding deterrent mobilized in response to insect herbivory and counter adaptation by Epilachna tredecimnotata, Science, 209:414-416.

Cates, R. G., and D. F. Rhoades, 1977, Prosopis leaves as a resource for insects, in: "Mesquite: Its Biology in Two Desert Scrub Ecosystems", B. B. Simpson, ed., pp. 61-83, Dowden, Hutchinson, & Ross, Stroudsburg.

Chew, F. S., and J. E. Rodman, 1979, Plant resources for chemical defense, in: "Herbivores, Their Interaction With Secondary Plant Metabolites", G. A. Rosenthal and D. H. Janzen, eds., pp. 271-308, Academic Press, New York.

Christiansen, E., and R. Horntvedt, 1983, Combined Ips/Ceratocystis attack on Norway spruce, and defensive mechanisms of the trees, Z. Angew. Entomol., 96:110-118.

Coutts, M. P., 1968, Rapid physiological change in Pinus radiata following attack by Sirex noctilio and its associated fungus, Amylostereum sp., Aust. J. Sci., 301:275-277.

Coutts, M. P., and J. E. Dolezal, 1969, Emplacement of fungal spores by the woodwasp, Sirex noctilio, during oviposition, For. Sci., 15:412-416.

Davies, E. and A. Schuster, 1981, Intercellular communication in plants: Evidence for a rapidly generated, bidirectionally transmitted wound signal, Proc. Natl. Acad. Sci. U.S.A., 78:2422-2426.

Dawkins, R., and J. R. Krebs, 1979, Arms races between and within species, Proc. R. Entomol. Soc. Lond., Ser. B, 205:489-511.

Dethier, V. G., 1954, Evolution of feeding preferences in phytophagous insects, Evolution, 8:33-54.

Dethier, V. G., 1970, Chemical interactions between plants and insects, in: "Chemical Ecology", E. Sondheimer and J. B. Simeone, eds., pp. 83-102, Academic Press, New York.

Dethier, V. G., 1976, The Hungry Fly, Harvard Univ. Press, Cambridge.

Dethier, V. G., 1979, The importance of stimulus patterns for host-plant recognition and acceptance, Symp. Biol. Hung., 16:67-70.

Dillon, P. M., S. Lowrie, and D. McKey, 1983, Disarming the "evil woman": Petiole constriction by a sphingid larva circumvents mechanical defenses of its host plant, Cnidoscolus urens (Euphorbiaceae), Biotropica, 15:112-116.

Dixon, A. F. G., 1970, Quality and availability of food for a sycamore aphid population, in: "Animal Populations in Relation to Their Food Resources" A. Watson, ed., pp. 271-287, Blackwell, Oxford.

Dolinger, P. M., P. R. Ehrlich, W. L. Fitch, and D. E. Breedlove, 1973, Alkaloid and predation patterns in Colorado lupine populations, Oecologia, 13:191-204.

Dowd, P. F., C. M. Smith, and T. C. Sparks, 1983, Detoxification of plant toxins by insects, Insect Biochem.,.13:453-468.

Dreyer, D. L., and B. C. Campbell, 1984, Association of the degree of methylation of intercellular pectin with plant resistance to aphids and with induction of aphid biotypes, Experientia, 40:224-226.

Dussourd, D., and T. Eisner, 1986. Vein-cutting: insectan counterplay to the latex defense of plants, Science, in press.

Edgerly, J. S., and T. D. Fitzgerald, 1982, An investigation of behavioral variability within colonies of the eastern tent caterpillar Malacosoma americanum (Lepidoptera: Lasiocampidae), J. Kans. Entomol. Soc., 55:145-155.

Edmunds, G. F., Jr., and D. N. Alstad, 1978, Coevolution in insect herbivores and conifers, Science, 199:941-945.

Edmunds, G. F., Jr., and D. N. Alstad, 1981, Responses of black pineleaf scales to host plant variability, in: "Insect Life History Patterns: Habitat and Geographic Variation, R. F. Denno and H. Dingle, eds., pp. 29-38, Springer-Verlag, New York.

Ehrlich, P. R., and P. H. Raven, 1965, Butterflies and plants: a study in coevolution, Evolution, 18:586-608.

Eisner, T., and F. C. Kafatos, 1962, Defense mechanisms of arthropods. X. A pheromone promoting aggregation in an aposematic distasteful insect, Psyche, 69:53-61.

Evans, W. G., 1966, Perception of infrared radiation from forest fires by Melanophila acuminata DeGeer (Buprestidae, Coleoptera), Ecology, 47:1061-1065.

Evans, W. G., 1971, The attraction of insects to forest fires, in: "Proc. Tall Timbers Conference on Ecological Animal Control by Habitat Management", Feb. 1971.

Evans, W. G., and J. E. Kuster, 1980, The infrared receptive fields of Melanophila acuminata (Coleoptera: Buprestidae), Can. Entomol., 112:211-216.

Feeny, P. P., 1966, Some effects on oak-feeding insects of seasonal changes in the nature of their food, Ph.D. thesis, Oxford.

Feeny, P. P., 1968, Effect of oak leaf tannins on larval growth of the winter moth Operophtera brumata, J. Insect Physiol., 14:805-817.

Feeny, P. P., 1969, Inhibitory effect of oak leaf tannins on the hydrolysis of proteins by trypsin, Phytochemistry, 8:2119-2126.

Feeny, P. P., 1970, Seasonal changes in oak leaf tannins and nutrients as a cause of spring feeding by winter moth caterpillars, Ecology, 51:565-581.

Feeny, P. P., 1975, Biochemical coevolution betwen plants and their insect herbivores, in: "Coevolution of Animals and Plants", L. E. Gilbert and P. H. Raven, eds, pp. 3-19, University of Texas Press, Austin.

Feeny, P. P., 1976, Plant apparency and chemical defense, Rec. Adv. Phytochem., 10:1-40.

Feeny, P. P., and H. Bostock, 1968, Seasonal changes in the tannin content of oak leaves, Phytochemistry, 7:871-880.

Fitzgerald, T. D., 1976, Trail marking by larvae of the eastern tent caterpillar, Science, 194:961-963.

Fitzgerald, T. D., and J. S. Edgerly, 1979, Specificity of the trail markers of forest and eastern tent caterpillars, J. Chem. Ecol., 5:565-574.

Fitzgerald, T. D., and S. C. Peterson, 1983, Elective recruitment by the eastern tent caterpillar, Anim. Behav., 31:417-423.

Foulds, W., and J. Grime, 1972, The response of cyanogenic and acyanogenic phenotypes of Trifolium repens to soil moisture supply, Heredity, 28:181-187.

Fox, L. R., and B. J. MaCauley, 1977, Insect grazing on Eucalyptus in response to variation in leaf tannins and nitrogen, Oecologia, 29:145-162.

Futuyma, D. J., 1983, Evolutionary interactions among herbivorous insects and plants, in: "Coevolution", D. J. Futuyma and J. M. Slatkin, eds., pp. 207–231, Sinauer Associates Inc., Sunderland.

Ghent, A. W., 1960, A study of the group-feeding behaviour of larvae of the Jack Pine Sawfly, Neopridion pratti banksianae Roh., Behaviour, 16:110–148.

Goldstein, J. L., and T. Swain, 1965, The inhibition of enzymes by tannins, Phytochemistry, 4:185–192.

Gould, F. 1984, Role of behavior in the evolution of insect adaptation to insecticides and resistant host plants, Bull. Ent. Soc. Am., 30:34–41.

Green, T. R., and C. A. Ryan, 1972, Wound-induced proteinase inhibitor in plant leaves: A possible defense mechanism against insects, Science, 175:776–777.

Hanover, J. W., 1966, Genetics of terpenes. I. Gene control of monoterpene levels in Pinus monticola, Heredity, 21:73–84.

Hanson, H.S., 1939, Ecological notes on the Sirex wood wasps and their parasites, Bull. Entomol. Res., 30:27–65.

Hare, J. D., and D. J. Futuyma, 1978, Different effects of variation in Xanthium strumarium L. (Compositae) on two insect seed predators, Oecologia, 37:109–120.

Hargrove, W. W. and D. A. Crossley, Jr., 1985, Within-leaf feeding site selection by black locust herbivores: implications for plant defense, Bull. Ecol. Soc. Amer., 66:187.

Harper, J. L., 1977, "Population Biology of Plants", Academic Press, London.

Hartmann, H. T., and D. E. Kester, 1975, "Plant Propagation", Prentice-Hall, Englewood Cliffs.

Haslam, E., 1966, "Chemistry of Vegetable Tannins", Academic Press, London.

Hassell, M. P., and T. R. E. Southwood, 1978, Foraging strategies of insects, Annu. Rev. Ecol. Syst., 9:75–78.

Haukioja, E., and P. Niemela, 1977, Retarded growth of a geometrid larva after mechanical damage to leaves of its host tree, Ann. Zoo. Fenn., 14:48–52.

Heinrich, B., 1979, Foraging strategies of caterpillars: leaf damage and possible predator avoidance strategies, Oecologia, 42:325–337.

Heinrich, B., and S. L. Collins, 1983, Caterpillar leaf damage, and the game of hide-and-seek with birds, Ecology, 64:592–602.

Higgins, K. M., J. E. Browns, and B. A. Haws, 1977, The black grass bug (Labops hesperius Uhler): Its effect on several native and introduced grasses, J. Range Manage., 30:380–384.

Jaenike, J., D. A. Grimaldi, A. E. Sluder, and A. L. Greenleaf, 1983, Amanitin tolerance in mycophagous Drosophila, Science, 221:165–166.

Johnson, W. T., and H. H. Lyon, 1976, "Insects that Feed on Trees and Shrubs", Cornell University Press, Ithaca.

Jones, T. H., and J. V. Schaffner, 1959, Cankerworms, USDA Leaflet 183.

Kalin, M., and G. Knerer, 1977, Group and mass effects in diprionid sawflies, Nature, 267:427-429.

Kidd, N. A. C., 1977, Factors influencing aggregation between nymphs of the lime aphid, Eucallipterus tiliae (L.), Ecol. Entomol., 2:273-277.

Kiritani, K., 1964, The effect of colony size upon the survival of larvae of the southern green stink bug., Nezara viridula, Jpn. J. Appl. Entomol. Zool. 8:45-54.

Kiritani, K., and K. Kimura, 1966, A study on the nymphal aggregation of the cabbage stink bug, Eurydema rugosum Motschulsky (Heteroptera: Pentatomidae), Appl. Entomol. Zool., 1:21-28.

Kloft, W., and P. Ehrhardt, 1962, Studies on the assimilation and excretion of labelled phosphate in aphids, in: "Radioisotopes and Radiation in Entomology", pp. 181-190, International Atomic Energy Agency, Vienna.

Knerer, G., and C. E. Atwood, 1973, Diprionid sawflies: Polymorphism and speciation, Science, 179:1090-1099.

Kraft, S. K., and R. F. Denno, 1982, Feeding responses of adapted and non-adapted insects to the defensive properties of Baccharis halimifolia L. (Compositae), Oecologia, 52:156-163.

Kravielitzki, S., J. P. Vité, U. Sturm, W. Francke, 1983, Interactions between resin flow and subcortically feeding Coleoptera, Z. Angew. Entomol., 96:140-146.

Larsson, S., and O. Tenow, 1979, Utilization of dry matter and bioelements in larvae of Neodiprion sertifer Geoffr. (Hym., Diprionidae) feeding on Scots pine (Pinus sylvestris L.), Oecologia, 43:157-172.

Levin, D. A., 1976, The chemical defenses of plants to pathogens and herbivores, Annu. Rev. Ecol. Syst., 7:121-59.

Lewis, A. C., 1979, Feeding preference for diseased and wilted sunflower in the grasshopper Melanopus differentialis, Ent. Exp. Appl., 26:202-207.

Loper, G. M., 1968, Effect of aphid infestation on the coumestrol content of alfalfa varieties differing in aphid resistance, Crop. Sci., 8:104-106.

Louda, S. M., and J. E. Rodman, 1983a, Ecological patterns in the glucosinolate content of native mustard Cardamine cordifolia in the Rocky Mts. (USA), J. Chem. Ecol., 9:397-422.

Louda, S. M., and J. E. Rodman, 1983b, Concentration of glucosinolates in relation to habitat and insect herbivory for the native crucifer Cardamine cordifolia, Biochem. Syst. Ecol., 11:199-207.

Mackay, P. A., and W. G. Wellington, 1977, Notes on the life history and habits of the red-backed sawfly, Eriocampa ovata (Hymenoptera:

Tenthredinidae), Can. Entomol., 109:53-58.

Madden, J. L., 1977, Physiological reactions of Pinus radiata to attack by woodwasp, Sirex noctilio F. (Hymenoptera: Siricidae), Bull. Entomol. Res., 67:405-426.

Mares, M. A., F. A. Enders, J. M. Kingsolver, J. L. Neff, and B. B. Simpson, 1977, Prosopis as a niche component, in: "Mesquite", B. B. Simpson, ed., pp. 123-149, Dowden, Hutchinson and Ross, Stroudsburgh.

Martin, M. M., and J. S. Martin, 1984, Surfactants: Their role in preventing the precipitation of proteins by tannins in insect guts, Oecologia, 61:342-345.

Martin, M. M., D. C. Rockholm, and J. S. Martin, 1985, Effects of surfactants, pH, and certain cations on precipitation of proteins by tannins, J. Chem. Ecol., 11:485-494.

McEvoy, P. B., 1979, Advantages and disadvantages to group living in treehoppers (Homoptera: Membracidae), Misc. Publ. Entomol. Soc. Am. 11:1-13.

McGugan, B. M., 1954, Needle-mining habits and larval instars of the spruce budworm, Can. Entomol., 86:439-454.

McIntyre, J. L., 1980, Defenses triggered by previous invaders: Nematodes and insects, in: "Plant Disease", J. G. Horsfall and E. B. Cowling, eds., Vol 5, pp. 333-343, Academic Press, New York.

McKey, D., 1979, The distribution of secondary compounds within plants. in: "Herbivores, Their Interaction With Secondary Plant Metabolites", G. A. Rosenthal and D. H. Janzen, eds., pp. 56-133, Academic Press, New York.

McLean, D. L., and M. G. Kinsey, 1965, Identification of electrically recorded curve patterns associated with aphid salivation and ingestion, Nature, 205:1130-1131.

McLean, D. L., and M. G. Kinsey, 1968, Probing behavior of the pea aphid, Acyrthosiphon pisum II. Comparisons of salivation and ingestion in host and non-host leaves, Ann. Entomol. Soc. Am., 61:730-739.

McNaughton, S. J., and J. L. Tarrants, 1983, Gross leaf silicification: natural selection for an inducible defense against herbivores, Proc. Natl. Acad. Sci. USA., 80:790-791.

Metcalf, C. L., W. P. Flint, and R. L. Metcalf, 1962, "Destructive and Useful Insects: Their Habits and Control", 4th edition, McGraw Hill Book Company Inc., New York.

Miles, P. W., 1968, Insect secretions in plants, Annu. Rev. Phytopath., 6:137-164.

Miles, P. W., 1978, Redox reactions of hemipterous saliva in plant tissues, Ent. Exp. Appl., 24:534-539.

Millburn, P., 1978, Biotransformation of xenobiotics by animals, in: "Biochemical Aspects of Plant and Animal Coevolution", J. B. Harborne, ed., pp. 35-73, Academic Press, New York.

Mitchell, R., 1981, Insect behavior, resource exploitation, and fitness,

Annu. Rev. Entomol., 26:373-396.

Moore, L. R., 1978, Seed predation in the legume Crotalaria, I. Intensity and variability of seed predation in native and introduced populations of C. pallida Ait., Oecologia, 34:185-202.

Morris, R. F., C. A. Miller, D. O. Greenbank, and D. G. Mott, 1958, The population dynamics of the spruce budworm in eastern Canada, in: "Proc. 10th Int. Cong. Entomol.", 4:137-149.

Morrow, P. A., T. E. Bellas, and T. Eisner, 1976, Eucalyptus oils in the defensive oral discharge of Australian sawfly larvae (Hymenoptera: Pergidae), Oecologia, 24:193-206.

Mothes, K., 1960, Alkaloids in the plant, Alkaloids (N.Y.), 7:1-29.

Mothes, K., 1976, Secondary plant substances as materials for chemical high quality breeding in higher plants, Rec. Adv. Phytochem., 10:385-405.

Nault, L. R., and M. E. Montgomery, 1979, Aphid alarm pheromones, Misc. Publ. Entomol. Soc. Am., 11:23-31.

Nichols, J. O., 1961, The gypsy moth in Pennsylvania, Penn. Dept. Agr. Misc. Bull. 4404, 82 p.

Niemela, P., A. M. Aro, and E. Haukioja, 1979, Birch leaves as a resource for herbivores. Damage-induced increase in leaf phenols with trypsin-inhibiting effects, Rep. Kevo Subarct. Res. Stn., 15:37-40.

Philogene, B. J. R., J. T. Arnason, G. H. N. Towers, Z. Abramowski, F. Campos, D. Champagne, and D. McLachlan, 1984, Berberine: A naturally occurring phototoxic alkaloid, J. Chem. Ecol., 10:115-123.

Pimentel, D., 1976, World food crisis: Energy and pests, Bull. Entomol. Soc. Am., 22:20-26.

Pridham, J. B., 1963, "Enzyme Chemistry of Phenolic Compounds", Pergamon Press, Oxford.

Puritch, G. S., and W. W. Nijholt, 1974, Occurrence of juvabione-related compounds in grand fir and pacific silver fir infested by balsam wooly aphid, Can. J. Bot., 52:585-587.

Raffa, K. F., and A. A. Berryman, 1982a, Physiological differences between lodgepole pines resistant and susceptible to the mountain pine beetle and associated microorganisms, Environ. Entomol., 11:486-492.

Raffa, K. F., and A. A. Berryman, 1982b, Accumulation of monoterpenes and associated volatiles following fungal inoculation of grand fir by the fir engraver, Scolytus ventralis (Coleoptera: Scolytidae), Can. Entomol., 114:797-810.

Raffa, K. F., and A. A. Berryman, 1983a, The role of host resistance in the colonization behavior and ecology of bark beetles, Ecol. Monogr., 53:27-49.

Raffa, K. F., and A. A. Berryman, 1983b, Physiological aspects of lodgepole pine wound responses to a fungal symbiont of the mountain pine beetle, Dendroctonus ponderosae (Coleoptera: Scolytidae), Can. Entomol., 115:723-734.

Ralph, C. P., 1976, Natural food requirement of the large milkweed bug, Oncopeltus fasciatus (Hemiptera: Lygaeidae), and their relations to gregariousness and host plant morphology, Oecologia, 26:157-175.

Reid, R. W., H. S. Whitney, and J. A. Watson, 1967, Reactions of lodgepole pine to attack by Dendroctonus ponderosae Hopkins and blue stain fungi, Can. J. Bot., 45:1115-1126.

Rhoades, D. F., 1977, The antiherbivore chemistry of Larrea, in: "Creosote Bush", T. J. Mabry, J. H. Hunziker, and D. R. Difeo, Jr., eds., pp. 135-175, Dowden, Hutchinson and Ross, Stroudsburg.

Rhoades, D. F., 1979, Evolution of plant chemical defense against herbivores, in: "Herbivores, Their Interaction With Secondary Plant Metabolites", G. A. Rosenthal and D. H. Janzen, eds., pp. 3-54, Academic Press, New York.

Rhoades, D. F., 1983, Herbivore population dynamics and plant chemistry, in: "Variable Plants and Herbivores in Natural and Managed Systems", R. F. Denno and M. S. McClure, eds, pp. 155-220, Academic Press, New York.

Rhoades, D. F., 1985, Offensive-defensive interactions between herbivores and plants: Their relevance in herbivore population dynamics and ecological theory, Am. Nat., 125:205-238.

Rhoades, D. F., and R. G. Cates, 1976, Toward a general theory of plant antiherbivore chemistry, Rec. Adv. Phytochem., 10:168-213.

Rice, W. R., 1983, Sexual reproduction: An adaptation reducing parent-offspring contagion, Evolution, 37:1317-1320.

Richardson, S. D., 1958, Bud dormancy and root development in Acer saccharinum, in: "The Physiology of Forest Trees", K. V. Thimann, ed., Ronald Press, New York.

Rodman, J. E., and F. S. Chew, 1980, Phytochemical correlates of herbivory in a community of native and naturalized Cruciferae, Biochem. Syst. Ecol., 8:43-50.

Rosenthal, G. A., and D. H. Janzen, eds., 1979, "Herbivores, Their Interaction With Secondary Plant Metabolites", Academic Press, New York.

Ryan, C. A., 1978, Proteinase inhibitors in plant leaves: A biochemical model for pest-induced natural plant protection, Trends Biochem. Sci., 5:148-150.

Ryan, C. A., 1983, Insect-induced chemical signals regulating natural plant protection responses, in: "Variable Plants and Herbivores in Natural and Managed Systems", R. F. Denno and M. S. McClure, eds, pp. 43-60, Academic Press, New York.

Ryan, C. A., P. Bishop, G. Pearce, A. Darvill, M. McNeil, and P. Albersheim, 1982, A sycamore cell wall polysaccharide and a chemically related tomato leaf polysaccharide possess similar proteinase inhibitor-inducing activities, Plant Physiol., 68:616-618.

Safranyik, L., D. M. Shrimpton, and H. S. Whitney, 1975, An interpretation of the interaction between lodgepole pine, the mountain pine beetle and its associated blue stain fungi in western Canada, in:

"Management of Lodgepole Pine Ecosystems", D. M. Baumgartner, ed., pp. 406-428, Washington State University Cooperative Extension Service, Pullman.

Schoonhoven, L. M., 1968, Chemosensory bases of host plant selection, Annu. Rev. Ent., 13:115-136.

Schoonhoven, L. M. 1972, Secondary plant substances and insects, Rec. Adv. Phytochem., 5:197-224.

Schroeder, A. C., 1984, Induction of snap bean resistance to the Mexican bean beetle, Masters thesis, University of Delaware, Newark.

Schultz, J. C., 1983a, Impact of variable plant defensive chemistry on susceptibility of insects to natural enemies, in: "Plant Resistance to Insects", P. A. Hedin, ed., pp. 37-54, Symp. Ser. No. 208, Amer. Chem. Soc. Washington.

Schultz, J. C., 1983b, Habitat selection and foraging tactics of caterpillars in heterogeneous trees, in: "Variable Plants and Herbivores in Natural and Managed Systems", R. F. Denno and M. S. McClure, eds., pp. 61-90, Academic Press, New York.

Schultz, J. C., I. T. Baldwin, and P. J. Nothnagle, 1981, Hemoglobin as a binding substrate in the quantitative analysis of plant tannins, J. Agric. Food Chem., 29:823-826.

Schultz, J. C., and I. T. Baldwin, 1982, Oak leaf quality declines in response to defoliation by gypsy moth larvae, Science, 217:149-150.

Schultz, J. C., P. J. Nothnagle, and I. T. Baldwin, 1982, Individual and seasonal variation in leaf quality of two northern hardwood tree species, Am. J. Bot., 69:753-759.

Scriber, J. M., and F. Slansky, Jr., 1981, The nutritional ecology of immature insects, Annu. Rev. Entomol., 26:183-211.

Sharik, T. L., and B. V. Barnes, 1976, Phenology of shoot growth among diverse populations of yellow birch (Betula alleghenienisis) and sweet birch (B. lenta), Can. J. Bot., 54:2122-2129.

Sickerman, S. L., and J. K. Wangberg, 1983, Behavioral responses of the cactus bug, Chelinidea vittiger Uhler, to fire damaged host plants, Southwest Entomol., 8:263-267.

Smith, N. G., 1982, Periodic migrations and population fluctuations by the neotropical day-flying moth Urania fulgens through the isthmus of Panama, in: "The Ecology of a Tropical Forest: Seasonal Rhythms and Long-Term Changes", E. Leigh, A. S. Rand, and D. Windsor, eds., pp. 331-344, Smithsonian Institute Press, Washington.

Smith, N. G., 1983, Host plant toxicity and migration in the dayflying moth Urania, Fla. Entomol., 66:76-85.

Soloman, M. J., and F. A. Crane, 1970, Influences of heredity and environment on alkaloidal phenotypes in Solanaceae, J. Pharm. Sci., 59:1670-1672.

Stewart, R. N., and H. Dermen, 1979, Ontogeny in monocotyledons as revealed by studies of the developmental anatomy of periclinal chloroplast chimeras, Am. J. Bot., 66:47-58.

Swain, T., 1979, Tannins and lignins, in: "Herbivores, Their Interaction With Plant Secondary Metabolites", G.A. Rosenthal and D. H. Janzen, eds., pp. 657-682, Academic Press, New York.

Tallamy, D. W., 1985, Squash beetle feeding behavior: An adaptation against induced cucurbit defenses, Ecology, 66:1574-1579.

Tallamy, D. W., and R. F. Denno, 1981, Maternal care in *Gargaphia solani* (Hemiptera: Tingidae), Anim. Behav., 29:771-778.

Tester, D. F., 1977, Constituents of soybean cultivars differing in insect resistance, Phytochemistry, 16:1899-1901.

Thalenhorst, W., 1958, Grundzuge der Populationsdynamik des grossen Fichtenborkenkafers *Ips typographos* L., Schriftenreihe der Forstlichen Fakultat der Universitat, Gottingen, 21:1-126.

Thielges, B. A., 1968, Altered polyphenol metabolism in the foliage of *Pinus sylvestris* associated with European pine sawfly attack, Can. J. Bot., 46:724-726.

Tostawaryk, W., 1972, The effect of prey defense on the functional response of *Podisus modestus* (Hemiptera: Pentatomidae) to densities of the sawflies *Neodiprion swainei* and *N. pratti banksianae* (Hymenoptera: Neodiprionidae), Can. Entomol., 104:61-69.

Towers, G. H. N., 1980, Photosensitizers from plants and their photodynamic action, Prog. in Phytochem., 6:183-202.

Van Emden, H. F., ed., 1972, "Insect/Plant Relationships", Royal Entomol. Soc. Lond. Symp. 6, 215 pp.

Vité, J. P., A. Bakke, and J. A. A. Renwick, 1972, Pheromones in *Ips* (Coleoptera: Scolytidae): Occurrence and production, Can. Entomol., 104:1967-1975.

Wallace, J. W., and R. L. Mansell (eds.), 1976, Biochemical interaction between plants and insects, Rec. Adv. Phytochem., 10:1-425.

Wallner, W. E. and G. S. Walton, 1979, Host defoliation: A possible determinant of gypsy moth population quality. Ann. Entomol. Soc. Am., 72:62-67.

Waring, R. H., and G. B. Pitman, 1983, Physiological stress in lodgepole pine as a precursor for mountain pine beetle attack, Z. Angew. Entomol., 96:265-270.

Way, M. J., and M. Cammell, 1970, Aggregation behaviour in relation to food utilization by aphids, in: "Animal Populations in Relation to Their Food Resources", A. Watson, ed., pp. 229-247, Symp. British Ecol. Soc. No. 10, Blackwell, Oxford.

Whitham, T. G., 1981, Individual trees as heterogeneous environments: Adaptation to herbivory or epigenetic noise?, in: "Insect Life History Patterns: Habitat and Geographic Variation", R. F. Denno and H. Dingle, eds., pp. 9-27, Springer-Verlag, Berlin.

Wilson, E. O., 1971, "The Insect Societies", Belknap Press of Harvard University Press, Cambridge.

Wood, D. L., 1982, The role of pheromones, kairomones and allomones in the host selection and colonization behavior of bark beetles, Annu. Rev. Entomol., 27:411-446.

Wood, T. K., 1975, Defense in two pre-social membracids (Homoptera: Membracidae), Can. Entomol., 107:1227-1231.

Wood, T. K., 1977, Defense in *Umbonia crassicornis*: Role of the pronotum and adult aggregations (Homoptera:Membracidae), Ann. Entomol. Soc. Am., 70:524-528.

Wright, L. C., A. A. Berryman, and S. Gurusiddaiah, 1979, Host resistance to the fir engraver beetle, Can. Entomol., 111:1255-1262.

Young, A. M., 1978, The biology of the butterfly *Aeria eurimedea agna* (Nymphalidae: Ithomiinae: Oleriini) in Costa Rica, J. Kans. Entomol. Soc., 51:1-10.

Zimmerman, M. H., and C. L. Brown, 1971, "Trees: Structure and Function", Springer-Verlag, New York.

Zucker, W. F., 1983, Tannins: Does structure determine function? An ecological perspective, Am. Nat., 121:335-365.

Zweigelt, F., 1915, Beitrage zur Kenntnis des Saugphanomens der Blattlause und der Reactionen der Pflanzengallen, Zentr. Bakter. Parasitenk., 42:265-335.

DEVISING PEST MANAGEMENT TACTICS BASED ON PLANT DEFENSE MECHANISMS, THEORETICAL AND PRACTICAL CONSIDERATIONS

Kenneth F. Raffa[1]

E. I. Du Pont de Nemours & Co., Inc.
Agricultural Chemicals Department
Experimental Station, Bldg. 402
Wilmington, DE 19898

1. INTRODUCTION

Among naturally occurring plant-insect systems, large-scale mortality or even severe damage to host plants is uncommon. Yet the innate reproductive capacities of the insect species involved are clearly sufficient to cause eruptive population growth. This implies that certain regulatory mechanisms are keeping the activities of insects and insect populations below the point at which plant reproduction is severely impaired. The most critical processes are those which intensify their effects as the insect population grows.

Early attempts to define these regulatory mechanisms concentrated on natural enemies, primarily because their effects are so obvious. While natural enemies certainly play a role in insect population regulation, attempts to model their activities based on well-defined biological parameters have offered some unanticipated insights. The first is that not all mortality contributes to population control. Insect populations exhibit a number of density-dependent and compensatory features, and in some systems natural enemies simply exert a modulating rather than regulating force. Secondly, not all factors that do regulate populations necessarily involve strict mortality. Reduced fecundity, delayed development, and inability to find suitable food can profoundly affect population behavior.

Concurrent with these developments has been a vastly increased knowledge of how plants can deleteriously affect insects. We now know that the morphology, physiology, chemistry, and ecology of plants include many traits that interfere with nearly all aspects of insect life (e.g., Denno and McClure, 1983; Hedin, 1983). This has led to a view of population biology that more strongly emphasizes regulation by lower trophic levels, i.e. the availability and suitability of host plants (Benz, 1974; White, 1974; Berryman, 1978; Cates, 1980; Rhoades, 1983). This seems to explain many naturally occurring epidemics when they do arise. Often the outbreak is preceded by some sort of environmental or demographic change that lowers the general vigor of the host or increases its availability, there-

---

[1] Present address: University of Wisconsin, Department of Entomology, 237 Russell Laboratories, 1630 Linden Drive, Madison, WI 53706.

by altering a previously stable interaction (Rudinsky, 1962; Cates et al., 1983).

Previous authors in this text have discussed specific features of plant-insect interactions in detail. A common theme is that the end product of these interactions is an ecologically and evolutionarily stable relationship. The purpose of this paper is to consider how these plant defense processes can be utilized in pest management. Unlike their occurrence in natural systems, insect outbreaks in agricultural fields are the norm rather than the exception. This has been attributed to a number of causes, but the most commonly accepted is that compared to naturally occurring environments, agroecosystems are characterized by a reduction in habitat heterogeneity, host genetic diversity, and natural enemies (Price, 1976; Southwood, 1977). The agricultural ecosystems are less stable because they are specifically intended to yield some form of unexploited biomass accumulation, i.e. the crop. This problem is further aggravated by the fact that even when a balance can be achieved between host and pest populations, it may occur at an economically unacceptable injury level. Consequently, control measures must be applied to guarantee a profitable harvest. Often these control measures have not provided stability, but instead, have simply been overcome by genetic adaptations within a few pest generations. We need to consider whether those aspects of plant-insect interactions that provide stability under natural conditions can also be manipulated in agricultural fields and commercial forests to enhance yield. Equally important is the question "Can we apply these methods within practical economic, political, and environmental constraints?"

2. GENERAL STRATEGIES TO DEVISING PEST MANAGEMENT SCHEMES

We can divide the role of plant-insect interactions in agriculture into two general approaches. These can be considered as either directed research on crop protection or basic research on plant-insect interactions, respectively, Table 1. These approaches represent different philosophies, research methods and areas of expertise. Historically they have developed quite separately, but ultimately they will have to complement each other if we are to make any major gains (Gould, 1983). We cannot rely on one approach alone.

The directed approach has as its objective the development of practices and tools that can economically protect crops. The strict criteria of being able to compete effectively and economically with established grower practices are applied early in the discovery process. There are three major tactics for manipulating the plant-insect interaction: genetic, chemical and cultural. These tactics are separated for organizational reasons only, and are best applied in an integrated fashion.

The basic approach has as its objective the development of new knowledge about plant-insect interactions. The contribution of this approach is that it provides an evolutionary and ecological context to the patterns we see around us, the primary goal of pure biology. But also, this approach seeks to identify key processes associated with insect or plant success. This can provide some rationale for devising pest management schemes, because it suggests that these key processes can be manipulated. These basic studies are best classified according to major biological components of the interaction (e.g., insect behavior and ecology, plant physiology, and insect toxicology), rather than as control tools.

The directed approach is fairly well established, and there are already some major successes associated with each tactic, Table 2. These

Table 1. Directed and Basic Approaches to Manipulating Plant-Insect Interactions.

Approach	Objective	Strategies
Directed	Develop practices that protect crops.	Identify resistant lines - screening, native range collections.
		Extract natural products - apply or use as molecular models for insecticides, antifeedants, oviposition repellents, IGR's.
		Devise cultural practices correlated with reduced damage.
Basic	Describe plant/insect interactions in ecological, physiological and evolutionary terms; Identify key processes associated with insect or plant success.	Describe environmental and genetic factors affecting host location and acceptance by insects.
		Quantify plant production, allocation, storage, and transport of chemicals that reduce insect success.
		Analyze insect detoxification, sequestration, excretion, avoidance of or interference with plant defense chemicals.

successes provide precedents and suggest the potential for utilizing plant-insect interactions in agricultural systems. In applying genetic breeding, phytochemistry, and altered cultural practices to agriculture, however, there are some key points that we need to consider.

The first is that the directed approach has all the strengths and weaknesses of any empirical method. There is a high likelihood of finding something efficacious if you look hard enough. But without an understanding of how the method is working, we have little chance of achieving its full potential. Because many different mechanisms can give the same result, it is difficult to correlate a specific treatment (whether genetic, chemical, or cultural) with efficacy. Likewise, once a "lead" strain or molecule is detected there is no way of judging the significance or potential of any particular level of insect response. Only continued screening will show whether this activity represents the tip of the iceberg or the point of diminishing returns.

Secondly, there are some unique considerations to the use of exogenously applied plant extracts and their synthetic analogues. Although phytochemicals offer some novel modes of actions, and have a high degree of potential (Whitehead and Bowers, 1983), judicious considerations

Table 2. Examples of Successful Pest Management Tactics Derived from Plant-Insect Interactions.

Tactic	Example
Plant Breeding	Wheat/Hessian Fly
Plant Extract/Analogue Synthesis	Pyrethroids/Lepidoptera
Cultural	Cotton/Boll Weevil

must be made on how best to utilize them for crop protection. Depending on the mode of action, this could either be because they require special properties to fit well in an agricultural setting, or because they are divorced from their ecological context. Some of these considerations are listed in Table 3. For example, the synthetic pyrethroids are very effective control agents, but like other conventional insecticides they are also deleterious to natural enemies. In their natural context, the biologically occurring pyrethrins are not nearly as toxic, but on the other hand, they do not come into contact with and therefore do not adversely affect insect predators. It is only the herbivore that is exposed. A highly desirable property of exogenously applied toxins is that they are taken up rapidly by the plant before eventually breaking down. For example, avermectin, a pentacyclic lactone of microbial origin, has a half-life on the surface of cotton leaves of less than one day because of this property. The result is good residual mite control with low effects on beneficial species (Bull et al., 1984; Dybas and Green, 1984).

Likewise a number of growth regulating compounds are found in plants, and the widespread interpretation that this represents coevolutionary adaptation is probably correct. Yet high activity alone does not assure crop protection. Certain chitinase inhibitors, for example, give 100% Heliothis virescens (F.) (Lepidoptera: Noctuidae) mortality at 10 ppm. Yet they do not economically protect cotton buds from the same insect at 1000 ppm. The problem is that they act slowly, because the insect must first arrive at a susceptible stage, i.e. a molt, before the material is effective. In a natural situation, such defense systems may be adequate because a fairly high level of damage can be tolerated before the plant's reproductive fitness is reduced. From an agricultural standpoint, this is often intolerable. So here, relying on the natural model is not enough. In the case of a chitinase inhibitor, we need to consider its prospects for commercial usage with and without a fast acting additive such as an insecticide or antifeedant.

A third type of problem is illustrated with the behavior modifying compounds such as antifeedants and oviposition repellents. These types of chemicals play a major role in plant defense in nature. But if such materials derived from one plant are to be applied to another crop plant, they would have to provide continuous protection from living insects (Bernays, 1983). This implies that the compounds must either be translocated systemically to the newly growing, unprotected portions of the plant, or that they irreversibly affect the insect, such that it can no longer recognize even untreated tissue as suitable substrate. Such materials do exist, such as, azadirachtin and formamidine which systemically and irreversibly prevent feeding, respectively. Still, this approach differs from the normal phytochemical context, in which allelochemicals can be continuously synthesized, transported, and sometimes

Table 3. Advantages and Disadvantages of Exogenously Applied Plant Extracts and Their Analogues.

Chemical Action	Advantage	Disadvantage
Toxin	Broad Spectrum, Active at Low Rates	Natural Enemy Reduction
Insect Growth Regulators	Active at Low Rates	Slow action - often fail to protect plant
Oviposition Deterrent	Rapid Acting, Prevent Infestation	Must be systemic. Often operate in nature in combination with other constituents
Antifeedant	Rapid Acting, Prevent Feeding	Must be either systemic or irreversible. Often operate in nature in combination with other constituents
Synomones	Enhance Predation, Parasitism	Relies on stable, high density natural enemy population or augmentative releases. Often volatile
Synergists	Prevent Detoxification and Resistance	Compounded discovery, development, registration, manufacture, and formulation costs. Highly specific. Problems with photostability and formulation

actively accumulated at the site of attack.

Likewise, the attraction of insect natural enemies by plant semiochemicals has theoretical potential, but practical utilization can be difficult. Some of the problems are illustrated by recent attempts with a similar tactic, application of kairomones derived from the host insect. For example, indiscriminate foliar application can actually hinder the ability of parasites to concentrate on certain plant structures for which the host insect may have a preference (Chiri and Legner, 1983). Also, the effectiveness of this strategy is related to host density. Lewis et al. (1979) found that kairomone applications were effective at very high pest densities only. From the grower's viewpoint this is rarely acceptable. Finally, we need to consider for each crop system the validity of the underlying assumptions that entomophagous insect populations are continually present, and stable at high densities. Many agroecosystems do not meet these criteria, even under non-sprayed conditions (Stehr, 1982). Also the breadth of the pest species range affected is usually quite narrow, and innate spacing mechanisms that improve efficiency could well be negated. This potential control strategy gives a good illustration of how mere identification of an active compound is not enough to promote crop protection. Extensive research on the integration of plant- and insect-derived chemical messengers, augmentative releases, predator and prey distribution patterns, and available means for controlling coexisting pest species will be necessary before any practical gains can be realized.

Finally, there has been increased appreciation in recent years that plants rely largely on chemical combinations for defense (Kubo and Klocke, 1983). These different compounds may function independently, thereby delaying the evolution of resistance because of the multiple changes required to confer selective advantage, or they may synergistically inhibit metabolic detoxification, again achieving the same effect. Phytochemical applications could mimic this strategy, either by directly applying mixtures of independently active plant-derived materials, or by applying compounds that function synergistically with synthetic insecticides. Although this is a physiologically very sound strategy, it may not be economically competitive. For example, classical insecticide synergists are potentially of very high value for counteracting or delaying insecticide resistance, but they have not yet proven practical for widespread agricultural use (Raffa and Priester, 1985). Their advantages have not yet overcome the added costs of their manufacture, formulation, application, and registration. Again, the natural model has not proven sufficient.

The key point, then, is that the agronomic, ecological, and physiological context of the plant-insect interaction needs to be considered in directed research programs. The more thoroughly we consider the attributes essential for a particular control tactic, the greater the chance of success. Only on such a foundation, can novel control approaches be effectively integrated with other pest management practices, such as trap crops, natural enemy conservation, insecticide mixtures, and resistant plant varieties. The alternative is to repeat more boom-and-bust projects that falsely raise expectations, only to disappoint everyone, and further increase the resistance of growers to new ideas. Because growers are looking for the simplest, most reliable method possible of insuring crop protection, innovative approaches will only be appreciated if they work and are cost-effective.

While the advantage of the directed approach is that it helps focus our attention on potential obstacles, the advantage of the basic approach is to suggest possibilities. Indeed, these differences are often reflected in the people doing either type of research. As applied biologists we wonder if any method will ever approach all of our ideal criteria for insect control. Yet, anyone who has studied insects in their natural environment has marveled that any of them survive at all.

The main advantage of the basic approach is that by asking such questions as how do plants defend themselves, and how do insects overcome these mechanisms, we can identify vital links in the sequence of events that lead to successful exploitation of the host. This approach is potentially more productive than strictly empirical studies, but the obstacles are also more formidable.

A major problem is that we really know very little about the dynamics of plant-insect interactions. Many of the individual, static components have been isolated and identified, such as what toxins and deterrents occur in which plants, and what insect species are found on them. But we do not have a clear idea of how these components behave in interactive processes. For example, coevolutionary theory provides that plants produce a variety of compounds that protect them from many species of herbivores. Because some species of herbivores are able to successfully exploit the host despite these "defenses", the animals are termed

Table 4. Comparison of Constitutive Monoterpene Content (mg/gm) of Resistant and Susceptible Lodgepole Pines.

Resistant	12.7 ± 2,8	no significant difference
Susceptible	11.2 ± 3,7	

Data from Raffa and Berryman, 1982b.
Mean ± 95% C.I.

"adapted". But from a pest management standpoint this does not offer any starting point. We need to know the ways in which plants could potentially resist those insects that are adapted to them. These after all, are our "pest" species.

3. KEY PROCESSES IN PLANT-INSECT ENCOUNTERS

We can approach this problem by first disassembling the various plant and insect feedback processes involved in their relationships. These various processes can then be manipulated so that we can determine how the overall outcome of the interaction is affected. Certain plant-insect systems lend themselves more readily to this approach than others and where generalized models can be developed, they suggest theoretical control strategies that can subsequently be evaluated in practical terms.

3.1. Examples from a Model System

One naturally occurring relationship that serves as a very useful model system is that of bark beetles (Coleoptera:Scolytidae) infesting living conifers. First, the effects on the plant and insect are quite clear-cut. Reproduction by the insect is contingent on host mortality (Raffa and Berryman, 1983a). If the beetles are unable to kill the tree, the brood is eliminated and the beetles are either killed or repelled. Secondly, many of the compounds found in host plants are toxic to the beetles at rates that normally occur in the host (Smith 1965; Raffa and Berryman, 1983b; Raffa et al., 1985). Yet colonization may or may not proceed successfully. Third, there are very high between-plant differences in susceptibility within a single host species. Consequently, the colonizing insects must locate and detect not only the proper species, but also the more favorable individuals within the host population. And finally, this is a system where insect mortality due to natural enemies and interspecific competition is relatively low and where insect population dynamics are strongly influenced by host availability and suitability, i.e. regulated by the lower trophic level (Berryman, 1976, 1978). Some of our findings from this system have a direct bearing on devising tactics for altering plant defense effectiveness.

The occurrence of compounds in plants that are toxic or repellent to insects is not necessarily correlated with successful defense, Table 4. For example, the monoterpenes present in host trees are toxic and repellent to the brood and adults. It ought to follow that resistant trees have higher monoterpene levels, different relative proportions of the more toxic materials, or at least more frequent occurrence of certain groups. We find that this is not the case (Raffa and Berryman, 1982a,b,c, 1983b). None of these parameters are correlated with resistance either in lodgepole pine (_Pinus contorta_ Douglas var _latifolia_ Englemann) - mountain pine beetle (_Dendroctonus ponderosae_ Hopkins) or in grand fir (_Abies_ grandis Douglas Lindley) - fir engraver (Scolytus ventralis Le Conte) interactions.

On closer investigation, we see that this lack of correlation is because constitutive host chemistry tells us very little about the environment actually confronted within the host. Very rapid biochemical changes occur once an attack has commenced, Fig. 1a. There is both an increased concentration of materials already present in the tissue, and the appearance of phytoalexins not previously available. We can simulate these changes by inoculating trees with fungi that are vectored by the beetle. These inoculations are conducted just prior to the flight season. The defensive lesions are removed after a fixed interval, and analyzed by gas-liquid chromatography. The trees are then classified as either

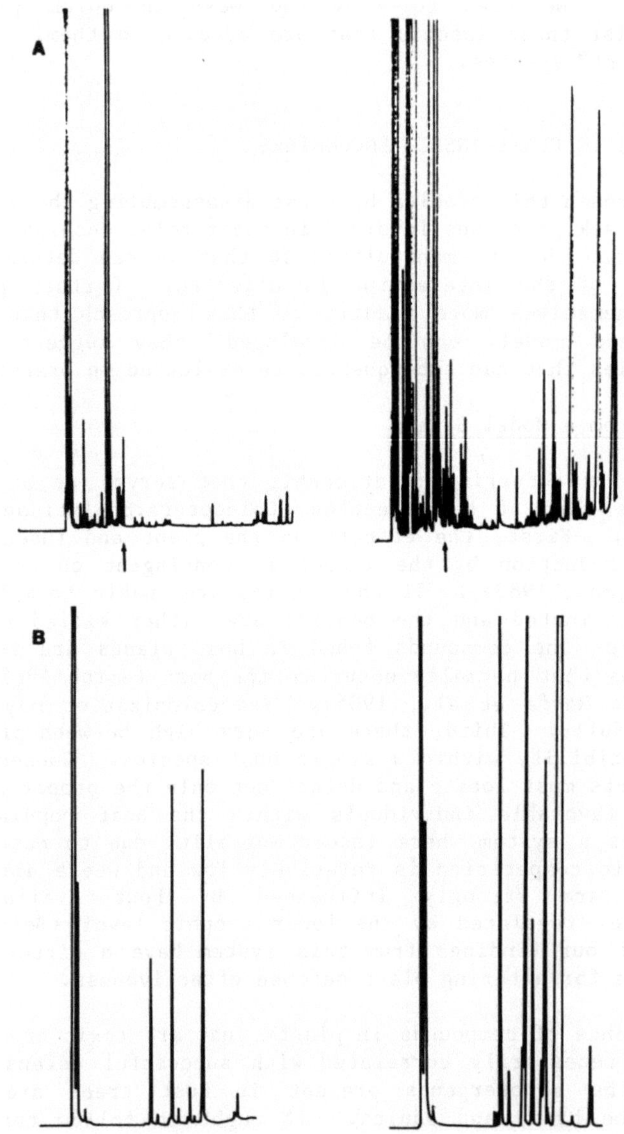

Fig. 1. Comparison of constitutive (left) and reaction (right) tissue of lodgepole pine phloem. Reaction tissue was induced by inoculating trees with fungi vectored by mountain pine beetles. A: Entire acetone extract; B: Monoterpene fraction in pentane extract.

Table 5. Comparison of Induced Monoterpene Accumulation (mg/gm) of Resistant and Susceptible Lodgepole Pines.

Resistant	91.2 ± 63.9	$p < 0.05$
Susceptible	13.4 ± 8.3	

Data from Raffa and Berryman, 1982b.
Mean ± C.I.

resistant or susceptible depending on whether they are subsequently killed by the naturally occurring beetle population or survive. Following controlled inoculation, there is a strong difference between resistant and susceptible trees in the concentration of monoterpenes in induced reaction tissue, Table 5. Resistant trees not only undergo a more extensive response, they also exhibit a more variable response, Table 6. During these compositional changes, there is a disproportionate increase in those compounds that are most toxic (Fig. 2) and repellent. Those biochemical pathways that result in allelochemicals most deleterious to the insect are most pronounced following invasion. This relationship is true for both the lodgepole pine - mountain pine beetle and grand fir - fir engraver systems.

So the key point here is that induced responses can explain a major portion of the difference between resistant and susceptible plants. In this system, all host plants are capable of undergoing an induced response. It is the rate and extent that are critical.

In addition, insect activities can completely eliminate the ability of a host plant to exhibit an induced response that it would otherwise conduct. In the case of the bark beetles, they do this by mass attack. Host monoterpenes are enzymatically oxidized by the beetles and their associated microorganisms into aggregation pheromones (Hughes, 1973; Conn et al., 1984). The combined activities of the rapidly assembled beetles mechanically disrupt and drain the resin ducts. This elimination of potential defense capacity can be demonstrated in several ways. One is by assaying trees for their defensive capacity just prior to the flight season and, if they are mass attacked, repeating the same test during the third day of aggregation. The actual response by the tree during mass

Table 6. Comparison of Compositional (Relative Proportion) Changes in Monoterpene Fraction of Lodgepole Pines Resistant and Susceptible to Mountain Pine Beetles Following Controlled Inoculation with Beetle-Vectored Fungi.

	% Change	
Monoterpene	Resistant	Susceptible
α-pinene	+60	None
Δ-3-carene	−46.2	−17.3
d-limonene	+289	None

Data from Raffa and Berryman, 1982b.

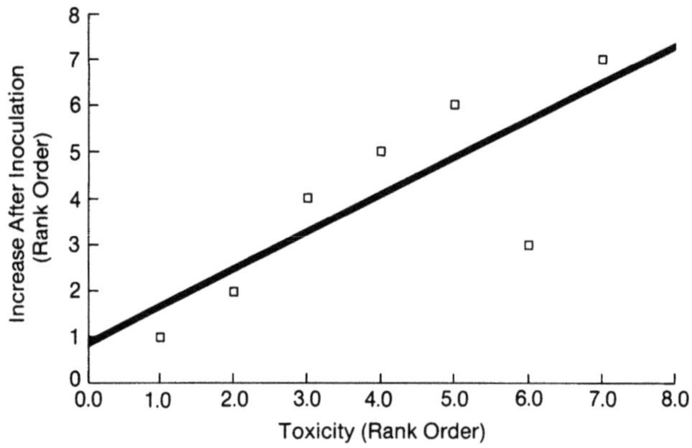

Fig. 2. Correlation of relative increase during induced response of seven monoterpenes present in grand fir phloem with toxicity to adult fir engravers (from Raffa et al., 1985). Response induced by artificial inoculation with fungi vectored by fir engravers (Raffa and Berryman, 1982c).

attack is far below its actual capability, Table 7. Another way of looking at this, and a more quantitative way of measuring it, is to apply varying densities of controlled inoculations. The induced response, expressed in terms of monoterpene content, declines with density in a typically sigmoid dose-response relationship, Fig. 3a. This relationship translates directly into insect success. The deleterious effect of the host on beetle reproduction is nearly total at low insect densities, Fig. 3b. But it declines with increased insect density until no effect at all is manifested above a critical threshold. That is, phytoalexin concentrations fall below those required to kill or repel the insect.

3.2. A Generalized Model

Based on this work, and that of others, we can depict the interaction in a very generalized scheme. The emphasis is on the rate dynamics of

Table 7. Comparison of Defensive Capacity of Lodgepole Pines Prior to and During Mass Attack by Mountain Pine Beetles.

	Pre-Attack Response	Response during 3rd day of attack
Lesion formation to controlled inoculation (mm)	26.5**	15.1
Resin exudation from wound (ml)	1.10*	0.39

* $p < 0.01$, ** $p < 0.001$
Data from Raffa and Berryman, 1983a.
Trees were sampled for 2 parameters related to resistance, lesion formation following controlled fungal inoculation, and resin exudation from a controlled wound, prior to attack. These assays were repeated on the same trees after they were attacked. Statistical tests refer to paired t comparisons on each tree before and during attack. Trees which were not attacked showed no change in these parameters over the same time interval.

several processes. They can be illustrated as acting sequentially, but in many cases they are simultaneous. They can be cooperative or conflicting. Initial feeding or oviposition by the insect results in a response by the plant, Fig. 4, path a. This response induces a localized and/or systemic series of chemical, physiological and morphological changes, Fig. 4, paths b,c. These changes can make the tissue less suitable for the insect, resulting in toxicity, repellency, etc., Fig. 4, path d. Until recently these responses were viewed as either rare events or as being of minor consequence. But we now know they are almost universal (e.g. Hedin, 1983; Denno and McClure, 1983). They have been demonstrated in corn, cotton, pines, firs, tomatoes, potatoes, wheat and just about everywhere anyone has looked. They can be induced by artificial simulation of feeding (Ryan, 1974, 1979; Tallamy, Chapter 8 in this text), inoculation with fungal symbionts (Raffa and Berryman, 1982 b,c), or treatment with insect secretions or extracts (Kile et al., 1974; Sato et al., 1981, 1982). The problem is that these responses are not always of sufficient

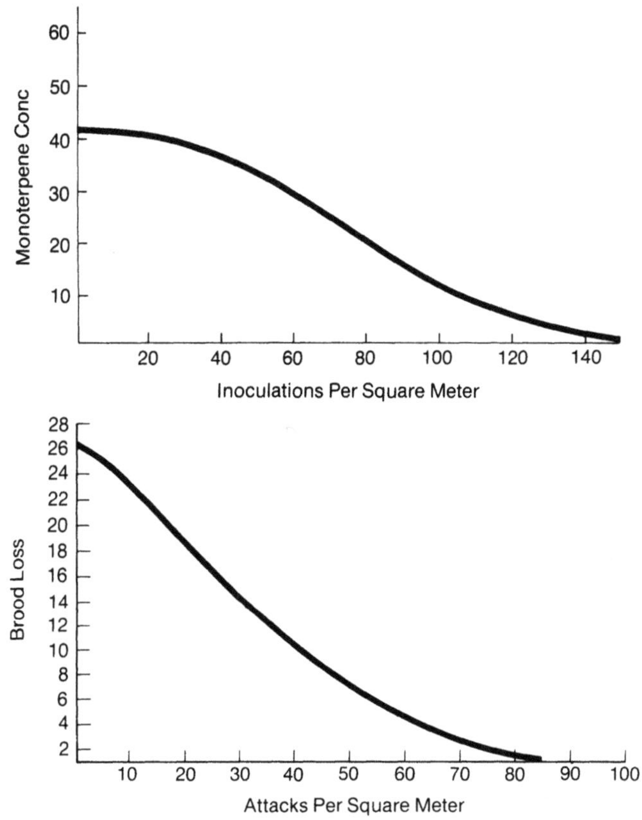

Fig. 3. Ability of bark beetles to eliminate the potential response (Fig. 1) of lodgepole pines to invasion.

Top: Accumulation of monoterpenes at a single inoculation site as a function of inoculum density.

$$Y = 41.6e^{-0.0000139X^{2.47}},\ r^2 = 0.97$$

Bottom: Reduced ability of trees to intoxify brood as a function of attack density.

$$Y = 26.3e^{-0.0015X^{1.5}},\ r^2 = 0.99$$

rate or magnitude to prevent insect behaviors that ultimately interfere with the plant's response, or the effectiveness of the response, Fig. 4, paths f,g,h. In the bark beetle example, host terpenes are converted by cytochrome P-450-catalyzed oxidations into aggregation pheromones, and the combined damage caused by the attracted beetles finally exhausts the defensive reaction Fig. 4, paths e,g. Other examples of direct interference are the insertion of fungal spores near the ovipositional site by a wood wasp, Sirex noctilio F. (Hymenoptera:Siricidae) (Coutts and Dolezal, 1969), and the trenching behavior by Epilachna borealis F. (Coleoptera: Coccinellidae) which prevents translocation of cucurbitacins to the wounded tissue (Tallamy, Chapter 8 in this text). The increased suitability of the host plant for insect development may also be related to factors other than defense per se, such as general stress-related changes in nitrogen availability, moisture content, and photosynthesis that occur during defoliation (White, 1978, 1984; Haukioja and Niemala, 1979; Hirchel and Turner, 1983).

Changes in the insect, Fig. 4, paths e,f can reduce the efficacy of the plant response, without affecting the response itself. For example, the induction of insect detoxification enzyme systems by plant compounds has been demonstrated for a number of plant-insect interactions (Yu et al., 1979; Yu, 1983; Yu, Chapter 4 in this text). These enzyme systems increase the ability of the insect to detoxify plant allelochemicals. The result is that the insect can continue to feed, thereby exhausting the defensive machinery of the plant.

This model also allows for the insect to prevent the defensive response from being initiated, Fig. 4, path h. This is a commonly seen strategy of virulent fungal strains, whereby the sporulating pathogen interferes with detection, and so fails to elicit a host response (e.g. Horsfall and Cowling, 1980; Wood, 1982). We do not have any clearcut examples of this with insects, probably because nearly all insects cause some mechanical damage to the plant and all insects are contaminated with microorganisms. Both of these factors can induce a plant response. Still, it seems likely that this type of host exploitation will be found among sucking insects.

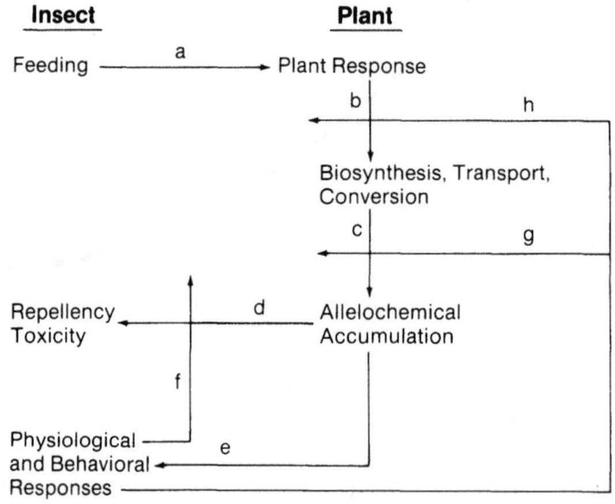

Fig. 4. Generalized description of feedback processes operating during plant-insect encounters. Insect feeding or oviposition induces plant responses (a) that trigger a series of events (b, c) that ultimately make the plant less suitable for the insect (d). Plant defensive responses may be ineffective if corresponding responses by the insect (e) increase insect tolerance to host compounds (f) or inhibit the plant's reaction (g, h).

For the purpose of pest management, this model suggests that regardless of population phenomena, each individual encounter between an insect and its host plant is a very precarious balance. If so, then just looking at the end product of the interaction such as defoliation, insect flight, etc. does not tell us very much. These interactions can generate a wide range of behaviors. In the bark beetle example, an observer could easily conclude either that conifers do not respond to beetle attack or that beetles cannot overcome conifer defenses, both of which are untrue, depending on the encounters observed. These processes need to be disassembled into their components, and these individual components then manipulated, to reveal their potential impact on the system.

Some examples from the bark beetle system can illustrate the effects of such alterations. Each of these components were actually broken down into much more specific sub-processes (Raffa and Berryman, 1982a,b,c, 1983a,b) but for simplicity, they are discussed in general terms. In the case of the bark beetles, the insect response → plant exhaustion pathway, Fig. 4, paths e,g, is regulated by aggregation. We manipulated the arrival rate by constructing screens around trees, and as beetles landed on the screens, we collected beetles and allowed only a fixed percentage of them onto the tree. That is, we reduced the rate of path "g" in Figure 4. We measured the number of landing females by using plexiglass window traps. By controlling the number of entered females we could also calculate the average number of landing females per entered female. This experimental approach provides an index of the attractiveness, Fig. 4, path e, of those beetles inside the tree. When we let all of these naturally arriving beetles in, we saw the typical bark beetle aggregation pattern, Fig. 5, top. Beetle attractiveness reached a peak within four days, until there were enough beetles to kill the tree, and then declined. On screened trees, the aggregation pattern was quite different, Fig. 5, bottom. The chemical attractiveness of the beetles increased more slowly, and never reached even 25% of the peak attractiveness of the controls. This was because a heavy resin flow accumulated around the entry sites, and prevented pheromone emission or some other aspect of pheromone communication (Raffa and Berryman, 1983a). Consequently, the critical threshold of beetles required to overcome resistance was not achieved, and the attacking beetles were entombed within defensive lesions. By quantitatively, but not qualitatively changing the processes involved, we altered the attack dynamics, and therefore the final outcome of the interaction. We gave the trees a chance to "catch up".

This is obviously a crude and impractical way of interfering with the insect's feedback, but it does illustrate the point. The outcome of the interaction can be changed by altering one or more of the processes in this generalized model. A more sophisticated way of achieving the same effect would be to treat defoliating insects so that their detoxification or behavioral systems were inhibited or not induced (Raffa and Priester, 1985). This would render them susceptible to plant compounds, and indeed this is what may occur in some natural systems. For example, natural pyrethrin and compounds closely related to piperonyl butoxide, an inhibitor of pyrethroid metabolism, co-occur in the same plants (Doskotch and El-Feraly, 1969). Also, black pepper produces both insecticides and corresponding synergists that amplify activity (Miyakado et al., 1983).

We can also demonstrate the opposite effect, i.e. the response to reducing pathway b-c in Figure 4. Girdling fir trees inhibits their ability to translocate and/or synthesize monoterpenes involved in defense by about 94% (Raffa and Berryman, 1982c). As a consequence, the likelihood of these trees being attacked successfully increases by about 11 times. Here again, a quantitative change in one of the key rate processes qualitatively affects the outcome of the interaction: trees which otherwise would have survived were killed by the naturally occurring beetle population.

4. MANIPULATING PLANT-INSECT INTERACTIONS

These illustrations show that the outcome of plant-insect interactions can be artificially altered using drastic measures. But what sort of challenges will we face if we try to incorporate these ideas into a practical insect control tactic? There are some major hurdles to overcome, both operational and theoretical. From an operational standpoint, the key question is "How can we do it?"; from a theoretical standpoint, the key question is "What would it accomplish if we could?". The outcome of plant-insect encounters can be modified in several ways. These methods can either be direct, by specifically favoring resistant properties of the plant, or they can be indirect by reducing any means by which insects increase their tolerance of plant defense substances, Table 8. These are analogous to the cultural, chemical, and genetic methods of the directed research approach, but are applied more specifically to key target processes that affect the physiology and behavior of the plants and insects involved. Our knowledge of these specifics will surely increase

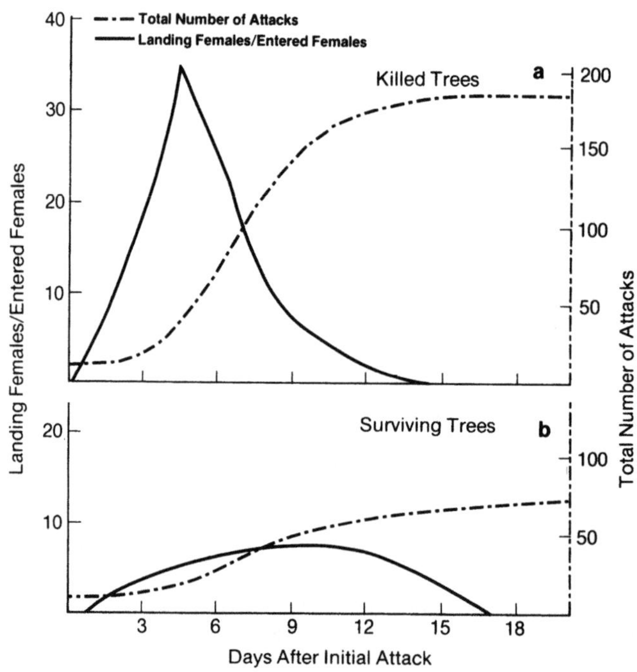

Fig. 5. Altering the outcome of a plant-insect encounter by reducing the rate of a key feedback process in the generalized model (Fig. 4). In control trees (above), the attractiveness (solid line) of individual entered females increased rapidly as more females arrived. This attractiveness continued until there were sufficient beetles (dotted and dashed line) to kill the trees. Test trees (below) were confined within screens and only fixed proportions of those beetles responding to the original entrees were allowed in. Under these conditions, the original beetles could not generate high attractiveness (solid line) because a heavy resin flow covered their entry sites. Eventually, females ceased to arrive before there were sufficient beetles to kill these trees. Raffa and Berryman, 1983a; reprinted with permission.

Table 8. Methods of Enhancing Plant Resistance to Insects.

Method	Examples: Host	Examples: Insect	Examples: Reference
CULTURAL			
1. Increase vigor and/or lower suitability - spacing of plants, altered (increased, decreased, appropriately timed) irrigation	Cotton	Pink Bollworm	Slosser, 1980
2. Ecological - phenological asynchrony, heterogeneity	Collards	Imported Cabbageworm	Root, 1973 Karieva, 1983
CHEMICAL			
1. Increase vigor and/or lower suitability - altered (increased, decreased, appropriately timed) fertilization	White Pine	White Pine Weevil	Xydias and Leaf, 1964
2. Induce Resistance - elicitors, bioregulators	Rice	Rice Stem Borer	Premila and Dale, 1984; Campbell et al., 1984
3. Indirect (increase susceptibility of insect to plant chemicals: inhibit detoxification capacity, disrupt perineurium, etc.)	Cotton	Egyptian Armyworm	Meisner et al., 1977, 1982
GENETIC			
1. Breeding	Wheat	Hessian Fly	Hatchett and Gallun, 1970
2. Genetic engineering	--	--	--
BIOLOGICAL			
1. Induce resistance (fundamentally by chemical signals)	Potato	Colorado Potato Beetle	Ryan, 1978, 1979
	Larch	Larch Bud Moth	Benz, 1974
	Cotton	Two-Spotted Spider Mite	Karban and Carey, 1984

with time, but that alone will not be enough to make these approaches practical. The final decisions made by people involved in pest management will be based on economics.

We must, therefore, consider early in the basic research approach how these processes could fit in crop production. Each crop system is unique, but for purposes of simplicity I have broken it down into two major categories of crop production: agriculture and forestry, Table 9. As in the case of the general plant-insect interaction model, crop production systems need to be further delineated into cotton agroecosystems, pear orchids, loblolly pine plantations, etc. to specifically apply these tactics. However, the agricultural vs. forestry categories illustrate the point that some major constraints exist for every approach considered.

Host suitability can be lowered in agricultural systems by altering such factors as planting dates, crop homogeneity, and spacing. But cultural control measures will only be accepted if the gains in insect control offset the losses due to inefficiency, inconvenience, or both. Another problem is that practices which provide escape from one pest species may benefit another. For example, fertilizing plants clearly enhances their resistance to some insects, but increases susceptibility to others. This is because the relationship between host nitrogen availability and insect utilization is extremely complex, and is quite distinct for every insect-plant system (Mattson, 1980; White, 1984). The same types of examples can be given for planting dates and irrigation (Johnson, 1968; Leigh et al., 1970). In many cases there are physiological limits to the extent to which plant phenology can be altered, and economic limits

Table 9. Practical Considerations in Enhancing Plant Resistance to Protect Crops.

	CROP SYSTEM	
Method	Agriculture	Forestry
Cultural	Suggested methods frequently do not fit well with established practices. Increased yield from insect protection may not offset reduced yield from altered spacing, planting dates, etc. High level of research will be required to integrate all factors of production.	Highly compatible. Major insect pests are less successful in vigorous than nonvigorous hosts. Growing schedules are intended to include long-term objectives. Strong basic research foundation in silviculture already exists.
Chemical	Potentially compatible approach. But, increased allocation of plant resources to defense may require additional fertilization, etc. This increased cost could be offset by broad pest control.	Low per-acre value of forests often makes chemical application costs against insects difficult to justify. Response would have to be long-lasting. Must be compatible with multiple use philosophy in some regions.
Genetic	Proven successes; Major problems are biotype evolution, loss of desirable agronomic characteristics, and narrow species range.	Potentially major problems with biotype evolution. Reproductive cycle of pests may be several hundred times faster than that of host. Heterogeneous introductions will probably be required.

to changes in harvesting and land usage (Hare, 1983). A considerable research effort is needed to integrate such approaches with other crop production practices. In the case of forestry, this may be easier to achieve because there are fewer pest species, and thus less chance for interfering effects. Further, some of the major pest species, such as bark beetles and spruce budworm, show preference for weakened trees. Therefore, practices which promote overall tree vigor can be effective. Finally, a strong body of basic silvicultural research already exists, and economic goals are based on long-term growing schedules.

The chemical enhancement of resistance to insects holds a good deal of promise for agricultural crops (Kogan, 1982; Kogan and Paxton, 1983; Hare, 1983). Improved crop protection from insects and mites has been achieved in response to a variety of plant growth regulators, herbicides, fungicides, and metal-containing compounds, usually by fortuitous observation. More recent efforts have examined the underlying biochemical mechanisms involved, and this is currently an area of intense directed research. Among the agronomically desirable features of the approach are that systemic induction does not require broad coverage as do most insecticides, insect growth regulators, synergists, and antifeedants. Also, the effect can be quite long-lasting (Benz, 1974; Schultz and Baldwin, 1982). On the other hand, chemical enhancement of resistance will necessitate extensive investigations into the effects of the treatment on yield and quality. It seems quite likely that increased allocation to defense could be at the expense of other physiological processes (Smedegaard-Petersen, 1982; Mooney et al., 1983). The best approach may be to enhance the responsiveness of plants, rather than their constitutive resistance (Mansfield and Bailey, 1982). Perhaps additional and/or different fertilization regimes will be needed. One positive feature that could offset this cost in agricultural systems is the broad spectrum of pest control provided by the many plant allelochemicals. Often the same allelochemical is effective against a broad range of insects, fungi, and bacteria (Marini-Bettolo, 1983). Plants probably could not afford to evolve a specific insecticide or fungicide for every pest they encounter, and rely instead on rather generally acting materials. Until recently, the pros and cons of exploiting the plants' defensive abilities were too complex to evaluate, even in theory. This has changed in the past few years, primarily because of the development of vastly improved computer models. These more recent models, such as "COTCROP" for cotton, are based on much sounder biological data than previous attempts. This is particularly true for aspects of plant compensatory growth and the economic value of plant parts during different periods of the growing season. The key feature of these newer models is that they are built from the plant up, rather than the pest down. We are really not far from the point where very accurate insect control choices can be based on all aspects of crop production. To achieve utility in commercial forestry, on the other hand, chemical induction of tree resistance would have to be very inexpensive given the low per-acre value of the crop.

Some of the major advances and problems with genetic resistance have already been discussed. In general, there have been some dramatic successes, but there are often nagging problems with increased susceptibility to other insects or microorganisms, loss of desirable agronomic characteristics, and insect biotype evolution, Table 9. Each of these problems can occur with any method of enhancing plant resistance, however, and as breeding efforts become more focused on specific defense processes, the time and expense required for new plant variety development could go down. The evolution of insect biotypes poses a special problem for forest pest management. Here the difference in reproductive cycles between host and insect is dangerously high with hundreds or even thousands of insect generations during a single host generation time, i.e. harvesting interval. Among naturally coevolved systems, it does not seem wise to reduce the genetic heterogeneity that plays a major role in limiting forest insect populations. Where foreign insect introductions have caused outbreaks among native trees, however, the selection for resistant genes among these nonadapted hosts may be judicious.

The problem of insect biotype evolution should be of concern with any pest management tactic based on plant defenses, not just genetic approaches. The same principles that apply to resistance to synthetic insecticides, for example, are also important to novel approaches. The most important lesson is that control strategies have to be integrated. We have to constantly consider the effects of alternate host and weed species, natural enemies, insecticide treatments, and economic injury levels on the selection pressures to which insects are exposed.

5. CAN PEST MANAGEMENT TACTICS BASED ON PLANT DEFENSES FAVOR STABILITY?

The underlying theme of the previous chapters is that coevolved plant-insect systems have undergone a series of adaptations and counter-adaptations, the result of which is ecological and evolutionary stability. I have approached the subject of plant defenses from the standpoint of efficacy and feasibility first, before considering their effects on system stability. This is because a control strategy must first provide consistent economical plant protection to ever be adopted. This sequence is just as true for natural plant defense mechanisms as for grower practices. The selection pressure on both plants and growers is for immediate competitive fitness. Without success in the short run, no strategy even has the opportunity to become stable. Yet practices that are successful in the short run can also promote instability. Can we avoid this in devising pest management schemes based on plant-insect interactions?

We must first consider in more detail what is meant by "stability". A system is considered stable if it occurs at equilibrium, and will return to the original state variables when disturbed. These properties require that negative feedback pathways predominate over positive feedback pathways (Berryman, 1981). In nature, this often occurs when insect activities cause biotic changes, such as resource depletion, natural enemy reproduction, and host plant alterations, that render their habitat less suitable. Agricultural systems do not generally have this property. Resource depletion is not an acceptable form of insect control and economic considerations dictate that the insect's ability to damage the host is curtailed immediately.

The traditionally applied control measures have successfully provided this immediate control, but these perturbations have not resulted in a return to the original stage. Each new chemical, genetic, or cultural control strategy has placed an intense and more importantly, unidirectional selection pressure on the insect. The objective, then, is to incorporate factors into the insect's environment that reduce the fitness of genetic adaptations to the control strategy.

Is this theoretically possible? Can resistance to the control strategy be at least partially nonadaptive? The conifer-bark beetle system provides a natural, coevolved model for considering this question. The defensive response of grand fir is much more extensive than that of lodgepole pine (Raffa and Berryman, 1982b,c). Firs undergo much more pronounced induced transformations. These differences relate to ecological differences between grand fir and lodgepole pine, particularly in the consequence of host mortality to host reproductive success (Raffa, 1981). According to classical resistance theory then, fir engravers should evolve greater physiological tolerance than mountain pine beetles. We have tested this idea and find that this simply is not the case: the fir engravers are actually less resistant (Raffa et al., 1985). The reason is that insects have several ways of dealing with plant defense chemicals, each of which has a reproductive cost (Raffa and Berryman, 1983a; Berryman et al., 1985; Tallamy, Chapter 8 in this text). Bark beetles can avoid high concentrations of host toxins by either overwhelming trees with mass attacks or by colonizing severely weakened trees with low defenses. The advantage to mass attack is that the number and nutrient quality of suitable trees within the host range are increased. The disadvantages to attacking healthy trees are the increased chance of attack failure and high intraspecific competition. As the resistance of the potential host increases, the density of beetles required to overcome these defenses rises, and the available food and space per colonizer decline (Raffa and Berryman, 1983a). In the case of the fir engraver, beetles which only enter unusually weakened trees outcompete those individuals which invade more toxic environments. First, insects that attack healthy fir trees experience a low likelihood of success, because the tree may entomb them in necrotic lesions before other beetles arrive. Secondly, if aggregation were successful, more beetles would be required to kill the tree than could be produced within the available substrate. The average beetle would fail to replace itself. Among pine beetles, however, the advantages of entering relatively healthy trees offset the disadvantages because of the host species' lower commitment to defense. There is a reasonable chance of survival within pine resin by partially tolerant individuals, at least until additional colonizers arrive. These trees can be overcome at attack densities that allow brood-to-parent ratios far above 1.0 (Berryman, 1976; Raffa and Berryman, 1983a). Those insects which accept a broad portion of the host population are most fit. Under these conditions, physiological resistance has evolved. This difference in host-mediated selective pressures can be determined by comparing the relative defensive capacity of colonized (prior to colonization) and uncolonized trees. In the case of fir engravers, the colonized trees had a resistance capacity that was only 6.7% of that of the avoided trees, Table 10. These trees are clearly distinct from the rest of the host population, in that some severe stress factor such as disease or age had lowered their resistance. In the case of the mountain pine beetle, the colonized trees had a resistance capacity that was 14.7% of that of the avoided trees, Table 10. This beetle can therefore include a broader physiological condition of lodgepole pine within its host range.

Even though pest management is concerned with population phenomena, selection operates at the level of individual insects. Each insect must compete with other insects, and the optimal behavioral - physiological - and ecological strategy will prevail. This optimal strategy may or may not result in large scale exploitation of the host. Fir engraver - grand fir relationships are quite stable, and the insect culls off those few trees which due to age, disease, or environmental stress fall well below average vigor. Mountain pine beetle - lodgepole pine relationships, on the other hand, show eruptive outbreaks often resulting in over 95% host mortality and completely changing the nature of the stand. This naturally occurring example provides support for the views of Kogan and Ortman (1978) and Gould (1983, 1984) who have argued that insect aversion behavior and plant antibiotic constituents need to be considered in an inclusive fashion. Gould (1983, 1984) has proposed that systems involving both properties are less likely to promote resistant biotype evolution, because there would be no advantage in possessing either reduced nonpreference or increased detoxification capacity in the absence of the other trait. Those individuals that were only resistant to a plant's or

Table 10. Comparison of Host Selection Behavior by Mountain Pine Beetles and Fir Engravers. Defensive Capacity Determined by Monoterpene Accumulation in Response to Controlled Inoculation.

Insect	Average defensive capacity of avoided trees / Average defensive capacity of colonized trees
Mountain Pine Beetle	6.8
Fir Engraver	15.0

chemical's antifeedant properties would be poisoned, and those that were only metabolically resistant would never exploit the host. A similar combination of traits keeps the fir engraver - grand fir relationship stable, with mortality to the host population equilibrating at low rates. In the case of mountain pine beetles, aversion to host toxins may be either disadvantageous or advantageous, depending on beetle population density (Raffa and Berryman, 1983a). There are probably corresponding changes in gene frequencies governing host acceptance between endemic and epidemic populations (Raffa, 1981).

In the bark beetle example, the key to stability is the availability of these weakened hosts. In an agricultural system, it could be susceptible varieties, trap crops or wild hosts. One attribute that facilitates this approach is that many of the major agricultural pest species are polyphagous. Another possibility is to systemically provide greater protection for certain plant parts than others. For example, we could theoretically provide better protection of cotton buds than leaves. Under these conditions, Heliothis larvae that chose leaves, a much less damaging behavior than bud entry, could still reproduce, and perhaps have a selective advantage. Physiological resistance to the chemicals protecting the bud would not be a prerequisite for survival. This strategy may be applicable to a number of pests where the less valuable plant parts can sustain the insect to maturity. It could conceivably apply to both novel control methods and traditional systemic insecticides. Such an idea would obviously have to be thoroughly studied and computer-simulated, from both an insect genetics and crop production standpoint before actually being attempted. But again, natural systems do provide examples of just this sort of relationship.

Recent findings on plant-insect relationships provide some similar ideas on how such stabilizing elements could be introduced into agricultural systems. An example is the production of allomones that mimic aphid alarm pheromones by wild potatoes (Gibson and Pickett, 1983). These pheromone mimics act as natural repellents. This is an interesting defense strategy because it places the insect in a position of conflicting selective pressures. Those aphids that become resistant to the alarm pheromone, either by physiological habituation or genetic selection, can now more fully exploit the host. On the other hand, their very mechanism of resistance renders them more susceptible to predators. Breeding for or spraying alarm pheromones is not necessarily an economically sound control tactic, but this example does provide an illustration of systems with strong negative feedback components. Whitham (1981) and Schultz (1983) have provided similar examples of how natural enemy populations can stabilize plant-insect interactions. Within such a feedback system, enhancement of plant chemical and physical attractiveness to natural enemies could play a valuable role, even if a strict reliance on semiochemicals is not sufficient.

Enhancing plant resistance can be a stabilizing force even in those cases that do not provide complete control. For example, because the rate of insecticide resistance development is more rapid among insects with short generation times (Georghiou, 1983), Leeper et al. (1985) have argued that we can capitalize on cases where induced plant resistance causes delayed development. The integration of traditional insecticides and induced plant resistance could theoretically delay insecticide resistance, a major source of instability among pest populations.

Yet in all likelihood, there are limits to the stabilizing influences we can incorporate into agriculture. The problem is that many of the negative feedback processes operating in nature are either agronomically unacceptable or beyond our control. For example, coevolutionary adaptations between plants and insects can be stabilizing when both the plant and the insect are successful at reproducing. A plant defensive characteristic may cause slower growth, smaller size and lower fecundity on the part of the insect, and provide some reduction in defoliation to the host. The plant need not intensify its commitment to defense because there have not been selective pressures for 100% protection. In an agricultural field, on the other hand, growers are not willing to accept appreciable damage, and the total commitment to defense, whether by plants or human control measures must always be high. Further, defensive processes evolve within the context of many selective pressures, both biotic and abiotic. The ecological circumstances under which stable coevolutionary patterns arose are quite different from those in which these plants are now cultivated. Factors such as escape in place and time, habitat heterogeneity, and multifaceted natural enemy complexes are not likely to be operating. There is a natural tension between the desire of the grower for simplicity and uniformity on the one hand, and the complex ecological basis for non-outbreak insect behavior on the other. This is the point at which the gap between the knowledge from basic biological studies and direct application for crop protection is greatest.

6. SUMMARY

In summary, many of the plant defense processes we see in nature have very strong relevance to insect control. They provide some well-founded ideas for devising pest management tactics. These tactics could potentially exert the constraints on insect populations that are imposed by host plants in naturally occurring systems. But agricultural tactics will have the same constraints on them as evolutionary processes: they must be efficacious and economical in the short-run to be adopted, regardless of their potential for conferring long-term benefit or stability. Secondly, plant protection tactics need to be integrated, particularly in such a way as to avoid unidirectional selection pressures, to lessen or delay genetic resistance among our target species. Natural systems provide some examples where conflicting selective pressures imposed on the insect population can maintain stability, and suggest some theoretical approaches for applying them to pest management. Finally, the chances of success are greatest where applied and theoretical problems are addressed in collaborative fashion. From a theoretical standpoint we need to better understand how the various components of coadapted plant-insect associations interact. We can best develop this understanding by emphasizing the dynamic aspects of these relationships under natural conditions, and by manipulating key processes under controlled conditions. From an applied standpoint, we need to identify the relative advantages and disadvantages of each control tactic in each crop system, and direct our research toward control methods that are most compatible with all aspects of economical crop production. Effective integration of basic and directed research ideas poses the greatest challenge to devising pest management practices based on plant defense.

7. ACKNOWLEDGEMENT

The critical review of J. M. Scriber, Department of Entomology, University of Wisconsin, Madison, is greatly appreciated.

8. REFERENCES

Bernays, E. A, 1983. Antifeedants in crop pest management, in "Natural Products for Innovative Pest Management", D. L. Whitehead and W. S. Bowers, eds., pp. 259-271, Pergamon Press, New York.

Benz, G., 1984. Negative Rückkoppelung durch Raum- und Nahrungskonkurrenz sowie zyklische Veränderung der Nahrungsgrundlage als Regelprinzip in der Populationsdynamik des Grauen Larchenwicklers, Zeiraphera diniana (Guenee) (Lep., Tortricidae), Z. angew Entomol., 76:196-228.

Berryman, A. A., 1976. Theoretical explanation of mountain pine beetle dynamics in lodgepole pine forests, Environ. Entomol., 5:1225-1233.

Berryman, A. A., 1978. Towards a theory of insect epidemiology, Res. popul. Ecol., 19:181-196.

Berryman, A. A., 1981. "Population systems. A general introduction", Plenum Publ. Corp., New York.

Berryman, A. A., B. Dennis, K. F. Raffa and N. C. Stenseth, 1985. Evolution of optimal group sizes in small predators attacking large prey, with particular reference to bark beetles (Coleoptera: Scolytidae), Ecology, 66:898-903.

Bull, D. L., W. G. Ivie, J. G. Mac Connel, V. F. Gruber, C. C. Ku, B. H. Arison, J. M. Stevenson, and W. J. A. Vanden Hewel, 1984. Fate of avermectin B in soil and plants, J. Agric. Food Chem. 32:94-102.

Campbell, B. C., B. G. Chan, L. L. Creasy, D. L. Dreyer, L. B. Rabin and A. C. Waise, Jr., 1984. Bioregulation of host plant resistance to insects, in "Bioregulators: Chemistry and Uses", R. L. Ory and F. R. Rittig, eds., pp. 193-204, ACS Symp., Ser. No. 257, Amer. Chem. Soc., Washington.

Cates, R. G., 1980. Feeding patterns of monophagous, oligophagous, and polyphagous insect herbivores: The effect of resource abundance and plant cemistry, Oecologia, 46:22-31.

Cates, R. G., R. A. Redak, and C. B. Henderson, 1983. Patterns in defensive natural product chemistry: Douglas fir and western spruce budworm interactions, in: "Plant Resistance to Insects", P. A. Hedin, ed., pp. 3-20, ACS Symp., Ser. No. 208, Amer. Chem. Soc., Washington.

Chiri, A. A., and E. F. Legner, 1983. Field applications of host searching kairomones to enhance parasitization of the pink bollworm (Lepidoptera: Gelechiidae), J. Econ. Entomol., 76:254-255.

Conn, J. E., J. H. Borden, D. W. A. Hunt, J. Holman, H. S. Whitney, O. J.

Spanier, H. D. Pierce, Jr., and A. C. Oehlschlager, 1984. Pheromone production by axenically reared Dendroctonus ponderosae and Ips paraconfusus (Coleoptera:Scolytidae), J. Chem. Ecol., 10:281-290.

Coutts, M. P., and J. E. Dolezal, 1969. Emplacement of fungal spores by the woodwasp, Sirex noctilio, during oviposition, For. Sci., 15:412-416.

Denno, R. F., and M. S. McClure, eds., 1983. "Variable Plants and Herbivores in Natural and Managed Systems", Academic Press, New York. 717 pp.

Doskotch, R. W., and F. S. El-Feraly, 1969. Isolation and characterization of (+)-sesamin cyclopyrethrosin from pyrethrum flowers, Can. J. Chem., 47:1139-1142.

Dybas, R. A., and A. St. J. Green, 1984. Avermectins: Their chemistry and pesticidal activity, in: "Proc. British Crop Protection Conference, Pests and Diseases", pp. 947-954, Lavenham Press LTD, Suffolk.

Georghiou, G. P., 1983. Management of resistance in arthropods, in: "Pest Resistance to Pesticides", G. P. Georghiou and T. Saito, eds., 769-792, Plenum Publ. Corp., New York.

Gibson, R. W., and J. A. Pickett, 1983. Wild potato repels aphids by release of aphid alarm pheromone, Nature, 302:608-609.

Gould, F., 1983. Genetics of plant-herbivore systems: interaction between applied and basic study, in: "Variable Plants and Herbivores in Natural and Managed Systems", R. F. Denno and M. S. McClure, eds., pp. 599-653, Academic Press, New York.

Gould, F., 1984. Role of behavior in the evolution of insect adaptation to insecticides and resistant host plants, Bull. Ent. Soc. Am., 30:34-41.

Hare, J. D., 1983. Manipulation of host suitability for herbivore pest management, in: "Variable Plants and Herbivores in Natural and Managed Systems", R. F. Denno, and M. S. McClure, eds., pp. 655-680, Academic Press, New York.

Hatchett, J. H., and R. L. Gallun, 1970. Genetics of the ability of the Hessian fly, Mayetiola destructor, to survive on wheats having different genes for resistance, Ann. Entomol. Soc. Amer. 63:1400-1407.

Haukioja, E., and P. Niemela, 1979. Birch leaves as a resource for herbivores; Seasonal occurrence of increased resistance in foliage after mechanical damage of adjacent leaves, Oecologia, 39:151-159.

Hedin, P. A., 1983. "Plant resistance to insects", Symp. Ser. 208. American Chemical Society. Washington.

Heichel, G. H., and N. C. Turner, 1983. CO_2 assimilation of primary and regrowth foliage of red maple (Acer rubrum L.) and red oak (Quercus rubra L.): response to defoliation, Oecologia, 57:14-19.

Horsfall, J. G., and E. B. Cowling, 1980. "How Plants Defend Themselves", Vol. V in: "Plant Disease: An Advanced Treatise", J. G. Horsfall and E. B. Cowling, eds., Academic Press, New York.

Hughes, P. R., 1973. Dendroctonus: production of pheromones and related compounds in response to host monoterpenes, Z. Angew. Entomol., 73:294-312.

Johnson, N. E., 1968. Insect attack in relation to the physiological condition of the host tree, Entomol. and Limnol. Mimeogr. Rev. 1:1-46, New York State College of Agriculture.

Karban, R., and J. R. Carey, 1984. Induced resistance of cotton seedlings to mites, Science, 225:53-54.

Kareiva, P., 1983. Influence of vegetation texture on herbivore populations: resource concentration and herbivore movement, in: "Variable Plants and Herbivores in Natural and Managed Systems", R. F. Denno and M. S. McClure, eds., pp. 259-189, Academic Press, New York.

Kile, G. A., P. J. Bowling, J. E. Dolezal and T. Bird, 1974. The reaction of Pinus radiata twigs to the mucus of Sirex noctilio in relation to resistance to Sirex attack, Aust. For. Res. 6:25-34.

Kogan, M., 1982. Plant resistance in pest management, in: "Introduction to Insect Pest Management", R. L. Metcalf and W. H. Luckmann, eds., pp. 93-134, John Wiley and Sons, New York.

Kogan, M., and E. F. Ortman, 1978. Antixenosis - a new term proposed to define Painter's "nonpreference" modality of resistance, Bull. Entomol. Soc. Am., 24:175-176.

Kogan, M., and J. Paxton, 1983. Natural inducers of plant resistance to insects, in: "Plant Resistance to Insects", P. A. Hedin, ed., pp. 153-172, Symp. Ser. No. 208, Amer. Chem. Soc., Washington.

Kubo, F., and J. A. Klocke, 1983. Isolation of phytoecdysones as insect ecdysis inhibitors and feeding deterrents, in: "Plant Resistance to Insects", P. A. Hedin, ed., pp. 329-346, Symp. Ser. No. 208, Amer. Chem. Soc., Washington.

Leeper, J. R., R. T. Roush, and H. T. Reynolds, 1986. Tactics for preventing or managing resistance in arthropods, in: "Pesticide Resistance; Strategies and Tactics for Management", pp. 335-346, NRC Board on Agriculture, National Academy Press.

Leigh, T. F., D. W. Grimes, H. Yamada, D. Bassett, and J. R. Stockman, 1970. Insects in cotton as affected by irrigation and fertilization practices, Calif. Agric., 24:12-14.

Lewis, W. J., M. Beevers, D. A. Nordlund, H. R. Gross, Jr., and K. S. Hagen, 1979. Kairomones and their use for management of entomophagous insects. II, Investigations of various kairomone-treatment patterns for Trichogramma spp., J. Chem. Ecol., 5:673-680.

Mansfield, J. W., and J. A. Bailey, 1982. Phytoalexins: current problems and future prospects, in: "Phytoalexins", J. A. Bailey and J. W. Mansfield, eds., pp. 319-324, John Wiley and Sons, New York.

Marini-Bettolo, G. B., 1983. The role of natural products in plant-insects and plant-fungi interaction, in: "Natural Products for Innovative Pest Management", D. L. Whitehead and W. S. Barnes, eds., pp. 167-186, Pergamon Press, New York.

Mattson, W. J., 1980. Herbivory in relation to plant nitrogen content, Annu. Rev. Ecol. Syst., 11:119-161.

Meisner, J, K. R. S. Ascher, M. Zur, and C. Eizik, 1977. Synergistic and antagonistic interactions of gossypol with some OP-compounds demonstrated in Spodoptera littoralis larvae by topical application, Zeitschr. Pflanzenkrankh. Pflanzensch., 89:571-574.

Meisner, J., K. R. S. Ascher, M. Zur, and E. Kabonci, 1982. Synergistic and antagonistic effects of gossypol for phosfolan in Spodoptera littoralis larvae on cotton leaves, J. Econ. Entomol., 70:717-719.

Miyakado, M., I, Nakayama, N. Ohno, and H. Yoshioka, 1983. Structure, chemistry, and actions of the Piperaceae amides: new insecticidal constituents isolated from the pepper plant, in: "Natural Products for Innovative Pest Management", D. L. Whitehead and W. S. Bowers, eds., pp. 369-382, Pergamon Press, New York.

Mooney, H. A., S. L. Gulman, and N. D. Johnson, 1983. Physiological constraints on plant chemical defenses, in: "Plant Resistance to Insects", P. A. Hedin, ed., pp. 21-36, Symp. Ser. No. 208, Amer. Chem. Soc., Washington.

Premila, K. S., and D. Dale, 1984. Induction of resistance in rice plants to insect pests by the application of chelated metal complexes, Crop Prot., 3:187-192.

Price, P. W., 1976. Colonization of crops by arthropods: non-equilibrium communities in soybean fields, Environ. Entomol., 5:605-611.

Raffa, K. F., 1981. "The role of host resistance in the colonization behavior, ecology, and evolution of bark beetles", Ph.D. Diss., Washington State Univ., Pullman.

Raffa, K. F., and A. A. Berryman, 1982a. Gustatory cues in the orientation of Dendroctonus ponderosae (Coleoptera:Scolytidae) to host trees, Can. Entomol., 114:97-104.

Raffa, K. F., and A. A. Berryman, 1982b. Physiological differences between lodgepole pines resistant and susceptible to the mountain pine beetle and associated microorganisms, Environ. Entomol., 11:486-492.

Raffa, K. F., and A. A. Berryman, 1982c. Accumulation of monoterpenes and associated volatiles following fungal inoculation of grand fir with a fungus vectored by the fir engraver Scolytus ventralis (Coleoptera: Scolytidae), Can. Entomol., 114:797-810.

Raffa, K. F., and A. A. Berryman, 1983a. The role of host plant resistance in the colonization behavior and ecology of bark beetles (Coleoptera: Scolytidae), Ecol. Mongr., 53:27-49.

Raffa, K. F., and A. A. Berryman, 1983b. Physiological aspects of lodgepole pine wound responses to a fungal symbiont of the mountain pine beetle, Can. Entomol., 115:723-734.

Raffa, K. F., A. A. Berryman, J. Simasko, W. Teal, and B. L. Wong, 1985. Effects of grand fir monoterpenes on the fir engraver beetle (Coleoptera:Scolytidae) and its symbiotic fungi, Environ. Entomol., 14:552-556.

Raffa, K. F., and T. M. Priester, 1985. Synergists as research tools and in agriculture, J. Agric. Entomol., 2:27-45.

Rhoades, D. F., 1983. Herbivore population dynamics and plant chemistry, in: "Variable Plants and Herbivores in Natural and Managed Systems", R. F. Denno and M. S. McClure, eds., pp. 155-220, Academic Press, New York.

Root, R. B., 1973. Organization of a plant-arthropod association in simple and diverse habitats: the fauna and flora of collards (Brassica oleracea), Ecol. Monogr., 43:95-124.

Rudinsky, J. A., 1962. Ecology of Scolytidae, Annu. Rev. Entomol., 7:327-348.

Ryan, C. A., 1974. Assay and biochemical properties of the proteinase inhibitor inducing factor, a wound hormone, Plant Physiol., 54:328-332.

Ryan, C. A., 1979. Proteinase inhibitors, in: "Herbivores: Their Interaction with Secondary Plant Metabolites", G. A. Rosenthal and D. H. Janzen, eds., pp. 599-618, Academic Press, New York.

Sato, K., I. Uritani, and T. Saito, 1981. Characterization of the terpene-inducing factor isolated from the larvae of the sweet potato weevil, Cylas formicarius Fabricius (Coleoptera:Brenthidae), Appl. Entomol. Zool., 16:103-112.

Sato, K., I. Uritani, and T. Saito, 1982. Properties of terpene-inducing factor extracted from adults of the sweet potato weevil, Cylas formicarius Fabricius (Coleoptera: Brenthidae), Appl. Entomol. Zool., 17:368-374.

Schultz, J. C., 1983. Impact of variable plant defensive chemistry on susceptibility of insects to natural enemies, in: "Plant Resistance to Insects", P. A. Hedin, ed., pp. 37-55, Symp. Ser. No. 208, Amer. Chem. Soc., Washington.

Schultz, J. C., and I. T. Baldwin, 1982. Oak leaf quality declines in response to defoliation by gypsy moth larvae, Science, 217:149-151.

Slosser, J. E., 1980. Irrigation timing for bollworm management in cotton, J. Econ. Entomol., 73:346-349.

Smedegaard-Petersen, V., 1982. The effect of defense reactions on the energy balance of resistant plants, in: "Active Defense Mechanisms in Plants", R. K. S. Wood, ed., pp. 299-316, Plenum Publ. Corp., New York.

Smith, R. H., 1965. Effect of monoterpene vapors on the western pine beetle, J. Econ. Entomol., 58:509-510.

Southwood, T. R. E., 1977. The relevance of population dynamic theory to pest status, in: "Origins of Pest, Parasite, Disease and Weed Problems", J. M. Cherrett and G. R. Sagas, eds., pp. 35-54, Blackwell Scientific Publications, Oxford.

Stehr, F. W., 1982. Parasitoids and predators in pest management, in: "Introduction to Insect Pest Management", R. L. Metcalf and W. H. Luckmann, eds., pp. 135-174, John Wiley and Sons, New York.

White, T. C. R., 1974. A hypothesis to explain outbreaks of looper caterpillars, with special reference to populations of Selidosema suavis in a plantation of Pinus radiata in New Zealand, Oecologia, 16:279-301.

White, T. C. R., 1978. The importance of a relative shortage of food in animal ecology, Oecologia, 33:71-86.

White, T. C. R., 1984. The abundance of invertebrate herbivores in relation to the availability of nitrogen in stressed food plants, Oecologia, 63:90-105.

Whitehead, D. L., and W. S. Bowers, 1983. "Natural Products for Innovative Pest Management", Pergamon Press, New York.

Whitham, T. G., 1981. Individual trees as heterogeneous environments: adaptation to herbivory or epigenetic noise? in: "Insect and Life History Patterns: Habitat and Geographic Variations", R. F. Denno and H. Dingle, eds., pp. 9-27, Springer-Verlag, New York.

Wood, R. D. S., 1982. "Active Defense Mechanisms in Plants", Plenum Publ. Corp., New York.

Xydias, G. K., and A. L. Leaf, 1964. Weevil infestation in relation to fertilization of white pine, For. Sci., 10:428-431.

Yu, S. J., 1983. Induction of detoxifying enzymes by allelochemicals and host plants in the fall armyworm, Pestic. Biochem. Physiol., 19:330-336.

Yu, S. J., R. E. Berry, and L. C. Terriere, 1979. Host plant stimulation of detoxifying enzymes in a phytophagous insect, Pestic. Biochem. Physiol., 12:280-284.

INDEX

Abies grandis, 278, 307, 309, 310, 318-320
Acalymma vittatum, 17, 21, 219
Accessory cells, 8, 9
Acephate, 161
Acetaldehyde, 168
Acetylation, 119
N-Acetylation, 124
Acetylcholine, 235, 259
Acetylcholinesterase, 88, 162, 259, 263
  carbamylation, 259
  inhibition, 88, 162, 259
  phosphorylation, 88, 162, 259
Acetyl transferases, 124
Acheta domesticus, 78, 214
Acrolepiopsis assectella, 78
Acrythosiphon pisum, 104, 179, 181
Actinomycetes, 126
Action potential, 258, 259
Acyl-coA:glycine N-acyltransferase, 123
Acyl-coA synthase, 123
Acylthioesterase, 182
Acyl transferase, 123
A.D., 48, 49, 59, 264, 265
Adaptations, 303
  defensive, 73
  genetic basis, 44, 48, 302
  heterosis, 43, 48, 51, 53, 57-59
  physiological basis, 43, 50, 183-187
Adenosine, 4
Adriamycin, 99
Adynerin, 99, 184
Adzuki bean weevil (Callosobruchus chinensis), 95
A-esterases, 74, 100, 101
Aflatoxin, 99
Ageratum houstonianum, 178, 235
Aglycones, 123, 178, 184, 215
Aggregation pheromones, 212-214, 281, 287, 309, 312
Agrilus, 283
Agroecosystem, 302, 305, 316
Ajugarin I, 24
AK-reductases (carbonyl reductases), 74, 99, 100, 224, 264
L-Alanine, 14,
Alarm pheromones, 182, 186, 320
Alcohol dehydrogenases, 98, 99
Alcohol sulfotransferase, 125
Aldehyde dehydrogenase, 237
Aldehyde reductases, 99, 100
Aldrin, 76-78, 82, 92, 93, 154-158, 165, 189-194, 227
Alfalfa, 155, 161
Alfalfa looper (Autographa californica), 155, 161
Aliesterases, 100
Aliphatic hydroxylation, 84
Alkaloids, 15-18, 28, 29, 77, 89, 99, 166, 168, 169, 179, 180, 235-240, 266, 275
  atropine, 32, 168, 236
  benzylisoquinoline alkaloids, 49
  caffeine, 4-6, 30, 32, 168, 235
  chaconine, 28
  cotinine, 98, 236, 237, 264
  DMDP, 18, 29
  isoboldine, 15
  isoquinoline alkaloids, 235
  lycorine, 16
  N-methyl flindersine, 15
  monocrotaline, 237-239
  morphine, 236
  nicotine, 18, 28, 32, 33, 89, 98, 154, 161, 166, 179, 186, 187, 195, 235-237, 263, 264
  Papaveraceae alkaloids, 179
  pyrrolizidine alkaloids, 77, 96, 166, 237-239
  quinine, 18, 28, 32, 160
  quinolizidine alkaloids, 179, 184
  senkirkine, 17
Solanum alkaloids, 28
  solanine, 28
  sparteine, 28, 32
  strychnine, 168
  tobacco alkaloids, 154, 235-237
  tomatine, 16, 28
  trewiasine, 17

Alliaria petiolata, 235
Allium porrum, 78
Allomones, 175, 320
Allyl glucosinolate (sinigrin), 14, 154, 159, 160, 168, 234
Allyl isothiocyanate, 234
Alsophila pometaria, 275
Amblyseius fallacis, 117
American cockroach (Periplaneta americana), 115, 116, 119, 121, 124, 178, 180, 229, 236, 237, 263
Amiben, 181
Amidase, 100
Amine oxide reductases, 100
Amino acids,
  alanine, 14
  γ-aminobutyric acid, 14, 29
  arginine, 230, 231, 262
  aspartic acid, 233
  asparagine, 233
  canaline, 117, 230
  canavanine, 117, 230, 231, 262
  β-cyanoalanine, 126, 233
  cysteine, 31
  glutamic acid, 117, 124, 231
  glutamine, 117, 231
  glycine, 119, 123, 124
  homoserine, 230
  isoleucine, 232
  phenylalanine, 228, 232
  ricinine, 179
  serine, 124
  tryptophan, 188
  tyrosine, 184
  valine, 232
Amino acid conjugation, 119, 123-124, 231
γ-Aminobutyric acid, 14, 29
Amiton, 195
Ammonia, 230, 231
Amylostereum areolatum, 281
Anacridium melanorhodon, 117, 228
trans-Anethole, 166
Aniline, 92, 94, 99
Anopheles albimanus, 110
A. arabiensis, 110
Anoxia tolerance, 262
Antheraea pernyi, 107
A. polyphemus, 114
Anthocaris cardamines, 235
Anthocyanins, 228
Anthonomus grandis, 79, 108, 114, 119, 213, 218, 278, 304
Antibiosis, 48, 49, 58, 59, 319
Anticarsia gemmatalis, 158
Antifeedants, 13, 49, 215, 217, 218, 302, 304, 305, 317, 319
Antixenosis, 49, 54, 58
Aonidiella aurantii, 262
Aphids, 175-196, 279
Aphins, 182
Aphis cytisorum, 184
A. fabae, 115, 178
A. nasturtii, 178
A. nerii, 104, 184, 185, 187, 189, 190, 193
A. pisum, 184, 189, 194
A. pomi, 181
Apigenin, 185
Apis mellifera, 109, 117, 119, 179
Aplysia, 3
Apples, 177, 277
Approximate digestability (A.D.), 48, 49, 59, 264, 265
Arene oxides, 86
Arene oxide hydrolases (epoxide hydrolases), 74, 75, 80, 81, 85-87, 116, 117, 127, 153, 154, 160-162, 166, 167, 189-194
Arginase, 117, 230, 231
Arginine, 230, 231, 262
Arginyl tRNA-synthetase, 230, 262
Argyrotaenia citrana, 77, 107, 117, 119
A. velutiana, 120
Aristolochic acid, 30
Armored scales, 261
Armyworms, 276
Aromatic hydroxylation, 85
Arrhenius activation energy, 82
Arthrospores, 281
Aryl amine acetyl transferase, 124
Aryl esterases, 74, 100, 101
Aryl sulfatases, 223
Arylphorin, 226, 227
Asclepias syriaca, 99, 185, 187, 193, 265, 282, 284
Asparagine, 233
Aspartic acid, 233
ATPases, 224, 261, 262
ATPase pump,
  calcium-dependent, 228
  potassium-dependent, 9, 28
Atropine, 32, 168, 236
Atta texana, 110
Attagenus piceus, 109
Aucubin, 216
Autographa californica, 155, 161
A. precationis, 287
Avermectins, 178, 304
Avoidance behavior, 273
  to triggering defenses, 279, 280
  spatial, 277-279
  temporal, 274-276
Azadirachta indica, 29
Azadirachtin, 25, 29, 304
Azinphos methyl, 162
Azoreductases, 99

Baccharis halimifolia, 275

Bacteria, 213, 232, 234, 317
Balsam poplar, 45
Baltimore (Euphydryas phaeton), 215, 216
Banded cucumber beetle (Diabrotica balteata), 214
Barbiturates, 91, 96, 153, 154, 158-162, 168, 238
Bark beetles, 84, 165, 212-214, 279-281, 286, 307-314, 317-320
  axenic, 213
  pheromones, 84, 212-214, 281, 287, 309, 312
Basiconic sensilla, 5
Bean aphid (Aphis fabae), 115, 178
1,2-Benzenediol (pyrocatechol), 49, 50, 58
Benzo(a)pyrene, 92
Benzylisoquinoline alkaloids, 49
Bermudagrass, 155, 159
Besseya, 216
B-esterases, 74, 100, 101
Big-toothed aspen (Populus grandidentata), 44, 45
Biochanin A, 181
Bioenergetics, 43
Biotypes, 57, 59, 316, 317, 319
Biphenyl, 92, 155-158, 169
Blaberus giganteus, 103
Black blow fly (Phormia regina), 3, 9, 29, 31, 33, 79, 113, 154
Black cherry (Prunus serotina), 45-47, 50-59
Black fire beetle (Melanophila acuminata), 281
Black pepper (Piper nigrum), 95, 313
Black swallowtail (Papilio polyxenes), 77, 227, 228, 263
Blattella germanica, 102, 229, 237
Blow fly (Calliphora erythrocephala), 116, 236
Boettcheria peregrina, 28
Boll weevil (Anthonomus grandis), 79, 108, 114, 119, 213, 218, 278, 304
Bombyx mori, 5, 228
π-Bond oxygenations, 83, 85-88
Boophilus microplus, 101
Botanical insecticides,
  nicotine, 18, 28, 32, 89, 98, 154, 161, 166, 179, 186, 187, 195, 235-237, 263, 264,
  pyrethrins, 86, 95, 166, 178, 214, 215, 258, 259, 264, 265, 304, 312
  rotenone, 188, 229, 230
Brevicoryne brassicae, 115, 178, 189, 234
Broccoli, 159, 161
Bromosulfophthalein, 121
Brown planthopper (Nilaparvata lugens), 104, 181
Buckeye butterfly (Junonia coenia), 215
Buckthorn aphid (Aphis nasturtii), 178
2-Butanone, 122
Butyl iodide, 120
Butylated hydroxytoluene, 153

Cabbage, 155, 159
Cabbage aphid (Brevicoryne brassicae), 115, 178, 189, 234
Cabbage looper (Trichoplusia ni), 29, 77, 79, 101, 105, 114, 124, 155, 159-161, 218, 236, 238
Cabbage white (Pieris napi), 235
Caffeic acid, 185
Caffeine, 4-6, 30, 32, 168, 235
Calabar bean, 260
Calactin, 99, 225
Calcium-dependent ATPase pump, 228
California red scale (Aonidiella aurantii), 262
Calliphora erythrocephala, 116, 236
Callosobruchus chinensis, 95
C. maculatus, 108, 115, 233
Calotropin, 99, 225
Campesterol, 220
Camphene, 167
Camphor, 167, 213
Campoletis sonorensis, 76
Canaline, 117, 230
Canavalia, 230, 262
L-Canavanine, 117, 230, 231, 262
Cantharis consors, 76
Carbamate insecticides, 77, 161, 169, 177, 191-193, 259
Carbamoyloximes, 178
Carbaryl, 86, 159, 161, 178, 179, 260
Carbofuran, 191-193
Carbonyl reductases, 74, 99, 100, 224, 264
Carbophenothion, 162
Carboxylesterases, 74, 75, 100-114, 160, 189-191, 194, 214, 216-218, 237
Cardamine cordifolia, 278
Cardenolides (cardiac glycosides), 99, 166, 168, 169, 184, 185, 188, 223, 224, 235, 261, 262, 265
  adynerin, 99, 184
  calactin, 99, 224
  calotropin, 99, 224
  digitoxin, 166, 168, 224
  ouabain, 224, 261, 262,

uscharidine, 99, 224
Cardiac glycosides (cardenolides), 99, 166, 168, 169, 184, 185, 188, 223, 224, 235, 261, 262, 265
Δ-3-Carene, 309
β-Carotene, 155, 167
Carotenoids, 160
Carrot, 155, 159, 160, 193, 194, 227
Carvone, 167
Caryedes brasiliensis, 117, 230, 231, 262
Caryophyllene, 114
Castilleja, 216
Catalpa speciosa, 215
Catalpa sphinx (Ceratomia catalpae), 215
Catalpol, 216
Catechins, 181
Catharanthus roseus, 215
Cattle tick (Boophilus microplus), 101
Ceanothis, 45
Cecropia (Hyalophora cecropia), 82, 107, 261
Celery, 159, 160, 193, 194
Cellulose, 115
Ceratitis capitata, 112, 119
Ceratomia catalpae, 215
CH-hydroxylations, 83-85
  N-dealkylation, 85
  O-dealkylation, 85
  S-dealkylation, 85
Chaconine, 28
Chalcodermus aeneus, 178
Chalcone oxides, 117
  phloretin, 185
  phlorizin, 19, 181, 185
Chaparrinone, 26
Checkerspot butterflies (Euphydryas), 215, 216
Chelone glabra, 216
Chemoreceptive sensilla structure, 7-9
Chemoreceptive sensilla function, 9-13
Chemosensory cells, 3-14, 28-34
Cherry, 277
Chinese cabbage, 159
Chinese mantid (Tenodera aridifolia), 220
Chitin, 115
Chitinase inhibitor, 304
Chlordimeform, 31
Chlorinated hydrocarbon insecticides, 159
Chlorobenzene, 85
1-Chloro-2,4-dinitrobenzene, 120, 121, 159
Chlorogenic acid, 21, 188
p-Chloromercuribenzoate (PCMB), 11, 31
p-Chloro N-methylaniline, 82, 85, 92, 93, 155, 227, 239
5,22,24-Cholestatrienol, 221
Cholesterol, 87, 188, 220-222
Cholinesterases, 88, 89, 100
Choristoneura fumiferana, 17, 21, 276, 279, 317
Chromenes, 80, 167, 178, 185, 188, 225-227, 260
Chrysanthemum cinerariaefolium, 178, 214, 259
Chrysin, 185
Chrysolina brunsvicensis, 13
Chrysomelid beetles, 98, 177, 178, 191, 228, 232
Chrysomyia putoria, 112
Chrysopa carnea, 76, 105
Cicindela oregona, 76
Cigarette beetle (Lasioderma serricorne), 236
trans-Cinnamaldehyde, 120
Cinnamic acid, 19, 185
Cinnamonitrile, 19
Cis-epoxide hydrolase, 87, 116, 117, 190, 193, 194
Cis-stilbene oxide, 116, 189
Clerodin, 23, 31
Clofibrate, 178
Clover, 181
Coadaptation, 59
Coccinella septempunctata, 221
Cockroaches, 78, 79, 99, 115, 116, 119, 121, 122, 124, 229, 261
Coevolution, 59, 263, 265, 306, 318, 321
Colchicine, 160
Colias eurytheme, 107
Collards, 159, 315
Colorado potato beetle (Leptinotarsa decemlineata), 12, 13, 31, 109, 176, 178, 179, 315
Common skipper (Poanes hobomok), 215
Conchosin B, 218
Confused flour beetle (Tribolium confusum), 221
Conifers, 48, 318
Conoderus vespertinus, 236
Contact chemoreception, 7-13
  sensitivity, 9, 10
  specificity, 9, 11
Coremata, 238, 239
Coriander, 155
Corn, 155, 156, 158-161, 169, 176, 184, 186, 191-193, 311
Corn earworm (Heliothis zea), 29, 95, 101, 106, 179, 226, 264, 265
Corn-leaf aphid (Rhopalosiphum maidis), 176
Coronopilin, 218

Corpora allata, 216, 217
Costelytra zaelandica, 20, 119
Cotinine, 98, 236, 237, 264
Cotton, 155, 156, 158-163, 176, 177, 218, 304, 311, 315, 317, 320
Cotton stainer (Dysdercus), 119, 189
Cottonwood, 45
p-Coumaric acid, 185
Coumarins, 166, 168, 181
Coumesterol, 176
Cowpea, 115, 155, 156, 158-161, 169
Cowpea curculio (Chalcodermus aeneus), 178
Cowpea weevil (Callosobruchus maculatus), 108, 115, 233
Creatonotos gangis, 239
Crotalaria, 237, 239
Cucumber, 159, 218
Cucurbita andreana, 288
C. pepo, 284, 285, 288
Cucurbitacins, 26, 166, 218-220, 284-288, 312
  cuc B, 219, 285
  cuc D, 219, 285
  cuc E, 26
Cucurbits, 275, 284
Culex pipiens complex, 111
C. quinquefasciatus, 110
C. tarsalis, 110
C. triaenorhynchus, 111
Curari, 259
Cutin, 191
Cutworms, 283
Cyanide, 125, 126, 231-233, 262, 267
β-Cyanoalanine, 126, 233
β-Cyanoalanine hydrase, 233
β-Cyanoalanine synthase, 126, 233
Cyanogenesis, 179, 232
Cyanogenic glycosides, 49, 179, 180, 231-233, 262, 267
  dhurrin, 179, 232
  linamarin, 123, 232
  lotaustralin, 123, 232
  prunasin, 14
Cyanogens, 122, 232, 275
Cyanohydrins, 123
Cycad, 122
Cycasin, 122
Cyclodiene insecticides, 153, 161
Cypermethrin, 161
Cysteine, 31
Cytochrome oxidase, 232, 262
Cytochrome P-450, 49, 74-97, 153-158, 161, 165, 167-169, 188-193, 195, 312
  microsomal oxidases, 81
  mixed-function oxidases, 82
  polysubstrate monooxygenases (PMSOs), 74, 81
  PSMOs, 74, 81
  age/activity, 77-79
  biochemical characteristics, 80-83
  electron transport, 81
  induction, 96, 154-158, 193, 214, 237, 312
  inhibition, 92, 95, 96, 227, 239, 264
  isoenzymes, 81, 154
  measurement of, 89-92
  mitochondrial, 79
  occurrence, 75-77
  reactions catalyzed, 83-89
  regulation, 96, 97
  synergists, 95, 123, 227, 238, 264, 265, 305, 306, 313, 317
  tissue distribution, 79, 80
Cytochrome P-450 inhibitors, 92, 95, 96, 227, 238, 239, 264, 265, 313
  dihydropipercide, 95
  lignans, 95,
  myristicin, 95, 264, 265
  piperonyl butoxide, 87, 88, 91, 95, 96, 165, 227, 264, 265, 313
  pipercide, 95
  safrole, 166
  SKF 525-A, 238
  tropital, 123
Cytochrome P-420, 91

2,4-D, 181
Dactynotus ambrosiae, 104
Dalapon, 181
Danaidone, 238
Danaus plexippus, 99, 224, 261, 264, 265, 266, 282
Dark-sided cutworm (Euxoa messoria), 6
Daunorubicin, 99
DDE, 120
DDT, 84, 153, 177-179, 193, 257, 258, 260, 264, 266
DDT-dehydrochlorinase, 120, 258
Dealkylation, 79, 84, 85, 93, 94, 155
Defenses,
  blocking, 283-288
  constitutive, 273, 304, 308, 317
  expenditures, 280
  facultative, 273
  inducible, 273, 275, 279, 280, 283-287, 304, 309-312, 317, 321
  overwhelming of, 280-283, 287
  qualitative, 275
  quantitative, 274
Dehydrochlorination, 120, 258
7-Dehydrocholesterol, 221
Dehydrogenases, 98
  alcohol, 98
  aldehyde, 237

Delia brassicae, 13, 162
Demeton, 162
Dendroctonus brevicomis, 213
D. frontalis, 213
D. ponderosae, 213, 281, 287, 307-310, 319, 320
D. terebrans, 165, 214
Descurainia richardsonii, 235
Desert locust (Schistocerca gregaria), 16, 24, 29, 79, 103, 119
Desmosterol, 221
Desulfuration,
oxidative, 87, 88
of parathion, 155-157
Deterrent, deterrency, 1, 13, 180, 182, 189
Deterrent chemosensory cell, 14, 28, 29, 32
Dhurrin, 179, 232
Diabrotica balteata, 214
D. barberi, 191, 192
D. cristata, 219
D. undecimpunctata, 218, 219
D. virgifera, 184, 186, 191, 192, 218, 219
Diapause, 46, 57
Diataraxia oleracea, 108
Diatraea grandiosella, 101, 107
D. saccharalis, 107
Diazinon, 77, 161, 162
Diazoxon, 162
1,2-Dichloro-4-nitrobenzene, 119, 121, 159, 194
Dicofol, 84
Dicoumarol, 99
Dieldrin, 178, 189, 191, 195
Difference spectrum, 90, 91, 154
carbon monoxide, 91
type I, 90, 91, 214
type II, 90, 91
Differential grasshopper (Melanoplus differentialis), 178, 179, 236
Digitonin, 29
Digitoxin, 166, 168, 224
Dihydroisodrin, 92
Dihydropipercide, 95
Dihydroquercetin, 180
2,4-dihydroxy-7-methoxy-1,4-benzoxazin-3-one (DIMBOA), 176
2,5-dihydroxymethyl-3,4-dihydroxypyrrolidone (DMDP), 18, 29
DIMBOA, 176
Dimethoate, 178, 179
6,7-Dimethoxy-2,2-dimethylchromene, 80
N,N-dimethyl p-nitrophenyl carbamate, 92, 94
Dinoseb, 193
Dioclea, 230, 262
Dioxins, 96
Dioxygenases, 82, 97, 193
Diploptera punctata, 79
Disparlure, 87
Disulfoton, 162
Diterpenes, 23, 24, 167, 218
ajugarin I, 24
clerodin, 23, 31
gibberellic acid, 167
gossypol, 114, 155, 160, 176, 218
heliosides, 176, 218
kalmitoxin I, 24
phytol, 167
Dithiothreitol (DTT), 31
DMDP, 18, 29
DNA-adduct, 228, 266
Dorsolateral spaces, 224, 232
Drosophila, 58, 98, 99, 239
D. hydrei, 112
D. melanogaster, 3, 112, 154
DTT, 31
Dysdercus, 119, 189
Dysdercus cingulatus, 114
D. koenigi, 230

Eastern tent caterpillar (Malacosoma americanum), 28, 277
E.C.D., 48, 49, 57, 59, 265
Ecdysones, 80, 153, 167, 168, 220, 222, 223
Ecdysone 20-monooxygenase, 80, 222
Ecdysonoic acid, 223
Ecdysteroids, 188
ecdysone, 80, 220, 222, 223
20-hydroxyecdysone, 220, 222, 223
makisterone A, 188
Efficiency of food conversion (E.C.D.), 48, 49, 57, 59, 265
Egyptian armyworm (Spodoptera littoralis), 15, 23, 27, 29, 80, 105, 114, 125, 258, 315
Electromyogram (EMG), 4
Empoasca fabae, 176
Endoplasmic reticulum, 79, 80, 82, 97
Endosulfan, 178
Endosymbionts, 184
English oak (Quercus robur), 274, 279
Entomoscelis americana, 12, 28
Ephestia kuhniella, 108
Epilachna borealis, 218, 283-288, 312
E. tredecimnotata, 284, 287
E. varivestis, 109, 178, 221, 278, 287
Epoxidation, 189, 192
of aldrin, 76, 77, 78, 82, 93,

154–158, 165, 189, 190, 193, 194
of fucosterol, 80, 221
of heptachlor, 92, 93, 154, 158, 169
Epoxide hydratases, 74, 116
Epoxide hydrolases (arene oxide hydrolases, epoxide hydratases), 74, 75, 80, 81, 85–87, 116, 117, 127, 153, 154, 160–162, 166, 167, 189–194
cis-specific, 87, 116, 117, 190, 193, 194
induction, 117, 160–162
isoenzymes, 116
juvenile hormone-degrading, 190, 191, 217
microsomal, 80, 116
soluble, 116, 117
substrates, 116, 160–162
trans-specific, 87, 116, 117, 189, 190, 193
10,11-Epoxyfarnesoic acid, 225
1,2-epoxy-3-(p-nitrophenoxy) propane, 120
Z-9,14-Epoxytricosane, 87
Ergosterol, 167
Eryodictyol, 180
Essential oil, 49, 155, 159, 161, 166, 265
Esterases, 74, 100–114, 159, 160, 166 189–193, 237
aryl esterases, 74, 100
carboxylesterases, 74, 100–114, 189–193
induction, 114, 159, 160, 193
isoenzymic forms, 101, 114
juvenile hormone, 114, 160, 217
general esterase, 101
paraoxonases, 100
pheromone-degrading, 9
specificity, 114
substrates, 102–113
Estragole, 166
Estrogens, 230
Ethacrynic acid, 121
Ether hydrolases, 116, 117
N-Ethylmaleimide, 31
β-Ethylstyrene diol, 116
trans-β-Ethylstyrene oxide, 116, 189
β-Ethylstyrene diol, 116
Euchaetes egle, 284
Eugenol, 168
Euphydryas anicia, 216
E. phaeton, 215, 216
European corn borer (*Ostrinia nubilalis*), 124, 176, 226
Europhium clavigerum, 281
Euxoa messoria, 6
Excretion, 185–187, 227, 231, 236
FAD monooxygenases, 89, 97, 98
Fall armyworm (*Spodoptera frugiperda*), 25, 77, 88, 89, 98, 105, 114, 117, 119, 155–169, 193, 194, 218, 227
Fall cankerworm (*Alsophila pometaria*), 275
trans-β-Farnesene, 186
Farnesol, 167
Farnesyl pyrophosphate, 216
Feeding behavior, 2–7, 303
biting, 4–6
chemosensory input, 2, 3, 5–7
deterrent, 1, 13, 178–182, 215, 217– 219, 260, 263
foraging, 277
girdling, 283, 313
gregarious, 215, 282
incitant, 1
information flow, 7
inhibition, 13–33
mechanosensory input, 2, 3
opportunistic, 279, 280
regulation of, 1
restrictions, 44, 260, 261
solitary, 279
specializations, 44, 260, 261
stealthy, 279
stimulant, 1, 181, 182
trenching, 284–288, 312
vein snipping, 284
visual input, 2, 3, 175
Ferulic acid, 185
Festuca rubra, 229
Firs, 311, 313
Fir engraver (*Scolytus ventralis*), 278, 308–310, 319, 320
Fisetin, 185
Flavobacterium, 116
Flavone, 155, 157–160, 162, 166
Flavonoids, 19, 20, 99, 153, 169, 178–182, 184–188, 194, 228–230
apigenin, 185
anthocyanins, 228
biochanin A, 181
catechins, 181
chlorogenic acid, 21, 188
chrysin, 185
dicoumarol, 99
dihydroquercetin, 180
eryodictyol, 180
fisetin, 185
hesperetin, 185
kaempferol, 185
liquiritigenin, 181
luteolin, 180
morin, 179, 180
myricetin, 185
β-naphthoflavone, 154, 158, 160
naringenin, 179–181, 185

naringin, 180
neohesperidin, 180
phloretin, 185
phlorizin, 19, 181, 185
proanthocyanidines, 181
procyanidin, 20, 179
quercetin, 99, 155, 160, 180, 185, 229
quercitrin, 180
rutin, 99, 180, 188, 229
taxifolin, 185
tricin, 229
Flavoprotein, 81, 262
Flavoprotein reductase, 81
Flesh fly (Sarcophaga bullata), 79, 113, 154
Flour beetle (Tribolium castaneum), 109
Foraging behavior, 277
Forest self-rejuvenation, 214
Forest tent caterpillar (Malacosoma disstria), 286
Forestry, 316, 317
Formamidine, 31
Free radicals, 82, 95, 99, 266
Fructofuranose, 11, 29
Fructose, 29
Fruitfly, 116
Fucosterol, 80, 87, 220-222
Fucosterol epoxide lyase, 80, 222
Fungal symbiont, 281
Fungicides, 317
Furanocoumarins, 95, 159, 160, 167, 227, 228, 265, 266
isopimpinellin, 19
isopsoralen, 228
psoralen, 228
xanthotoxin, 95, 158-160, 166, 168, 227, 228, 265,
Furanose subsite, 11

GABA, 14, 29
GABA-sensitive cell, 29
Galeal α-sensilla, 13
Galeal styloconic sensilla, 6
Galleria mellonella, 108, 114
Gallic acid, 118, 185, 228
Galling aphids, 181
Gap junction, 8
Gene amplification, 51
Gentisic acid, 167
Geraniol, 167
German cockroach (Blattella germanica), 102, 229, 237
Gibberellic acid, 167
Gibberellins, 181
Girdling, 283, 313
Glaucolide A, 217
Glucose, 4, 6, 12, 114
Glucose conjugation, 119, 122, 123, 189
Glucose-sensitive cell, 31
Glucose transferases, 80, 122, 123, 127, 189
α-Glucosidases, 194
β-Glucosidases, 189, 193, 194
Glucosidation, 119, 122, 123, 189
Glucosides, 122, 123
N-glucosides, 122, 123
O-glucosides, 122, 123
S-glucosides, 123
Glucosinolates (mustard oil glycosides), 115, 160, 166, 168, 180, 233-235, 278
sinigrin (allylglucosinolate), 14, 160, 168
Glucosinolate-sensitive cell, 28
Glucosinolase, 115, 189, 234
β-d-S-glucosyl-5-mercaptouracil, 122
β-d-S-glucosyl thiophenol, 122
Glucuronidases, 115
Glucuronidation, 118, 119, 122
Glutamic acid, 117, 124, 231
Glutamic acid dehydrogenase, 231
Glutamine, 117, 231
Glutamine synthetase, 231
Glutathione, 75, 85, 97, 118,
Glutathione transferases, 74, 75, 81, 118-121, 127, 153, 154, 158, 159, 161, 166, 168, 189, 193, 194, 234, 235
activity measurement, 119, 120
induction, 120, 158, 159, 194
isoenzymic forms, 120, 121
occurrence, 119
substrates, 121
Glycine, 119, 123, 124
Glycine acyl transferase, 123
Glycoalkaloids, 176
Glycoprotein, 258
α-Glycosidases, 115
β-Glycosidases, 115, 122, 219, 233
Glycosidases, 9, 100, 114-115, 189, 193
Glycosides, 178, 183, 188, 195, 223
Glycosyl hydrolases, 115
Glyphosate, 181
Gossypium, 218
Gossypol, 114, 155, 160, 176, 218
Granary weevil (Sitophilus granarius), 109, 233
Grand fir (Abies grandis), 278, 307-310, 318-320
Greater wax moth (Galleria mellonella), 108, 114
Green peach aphid (Myzus persicae), 114, 178, 181, 189-191, 193, 236
Green rice leafhopper (Nephotettix cincticeps), 104, 259
Gregarious feeder, 282
Gromphadorhina portentosa, 78, 99,

103, 116, 124, 214
Groundsel tree (Baccharis halimifolia), 275
Group transferases, 115, 118-126
amino acid transferases, 123, 124
glutathione transferases, 74, 75, 81, 118-121, 127, 153, 154, 158, 161, 166, 168, 189, 193, 194, 234, 235
hexose transferases, 122, 123
N-acetylation, 124
phosphotransferases, 124
sulfotransferases, 124, 125, 223
thiosulfate sulfur transferase, 125, 126, 232, 233
Growth characteristics, 48, 49
Growth performance (R.G.R.), 48, 49, 51, 53-55,
Growth regulators, 153, 303-305, 317
Guineensine, 95
Gymnemic acids, 31
Gypsy moth (Lymantria dispar), 22, 24, 28, 29, 77, 84, 87, 95, 101, 107, 275

Habitat heterogeneity, 302, 321
Hair pencils, 238, 239
Hairy indigo, 158
Harrisonin, 25
Heliconian butterflies, 126, 233, 262,
Heliconius melpomene, 233
Heliocides, 176, 218
Heliothis armigera, 29,
H. virescens, 77, 87, 106, 114, 119, 218, 231, 236, 304, 320
H. zea, 29, 95, 101, 106, 179, 226, 264, 265
Heptachlor, 92, 93, 154, 158, 169
Herbicides, 180, 181, 317
amiben, 181
2,4-D, 181
dalapon, 181
glyphosate, 181
maleic hydrazide, 181
picloram, 181
Hesperetin, 185
Hesperis matronalis, 235
Hessian fly, 304, 315
Heterocampa guttivitta, 277, 283, 286
Heterosis, 43, 48, 51, 53, 57-59
Heterozygotes, 48
trans-2-Hexenal, 154
Hexose transferases, 122, 123
Hill coefficient, 11
Hippodamia convergens, 76, 109
Histamine, 124
Homoserine, 230
Honey bee (Apis mellifera), 109, 117, 119, 179
Honeydew, 181, 183, 185, 188, 194
Hop tree, 45
Hormones, 80, 97, 153, 154, 167, 168
juvenile, 80, 102, 103-110, 112-114, 117, 153, 160, 161, 167, 168, 178, 190, 213, 216, 217
molting, 79, 80, 125, 153, 167, 168, 188, 222, 223
plant, 159, 188
prothoracotropic, 222
Host race formation, 59
Host vigor, 301
House cricket (Acheta domesticus), 78, 214
House fly (Musca domestica), 75, 79, 82, 84, 85, 87, 91, 101, 111-112, 114, 116, 119, 121-124, 127, 154, 159, 188, 214, 236, 257, 258, 264
Hyalophora cecropia, 82, 107, 261
H. euryalis, 48
H. gloveri, 48, 107, 160
Hybrid breakdown, 46, 50
Hybrid vigor, 43, 48, 51, 53, 57-59
Hybrid zone, 43, 48, 58
Hybridization studies, 46-53
Hydrolases, 74, 100-118,
A-esterases, 74, 100, 101
aliesterases, 100
amidase, 100
arginase, 117, 230, 231
arylesterases, 74, 100, 101
ATPases, 223, 261, 262
B-esterases, 74, 100, 101
carboxylesterases, 74, 75, 100-114, 160, 189-192, 194, 214, 216-218, 237
epoxide hydrolases, 74, 75, 80, 81, 85-87, 116, 117, 127, 153, 160-162, 166, 167, 189-194
esterases, 100-114, 189-192, 194
ether hydrolases, 116, 117
glycosidases, 114-116, 189-194
S-glycosyl hydrolases, 115, 116, 189
phosphatases, 189
sulfatases, 223
Hydroprene, 178
Hydroquinone, 123
Hydroxydanaidal, 238
20-Hydroxyecdysone, 80
trans-4-Hydroxy-2-hexenal, 238
Hydroxylations, 83-88
Hydroxypulegone, 84, 165
Hygroreceptors, 8
Hypericin, 14

Imidazoles, 95
Imported cabbageworm (Pieris rapae),

107, 234, 235, 315
Imported willow leaf beetle (Plagiodera versicolora), 275
Incitant, 1
Indoles, 153, 155, 168, 169
Indoleacetic acid, 188
Indole-3-acetaldehyde, 168
Indole-3-acetonitrile, 155, 157-160, 162, 166
Indole-3-carbinol, 155, 157-162, 166-168
Indole alkaloids, 215
Induction,
  cytochrome P-450, 96, 154-158
  enzyme, 58, 194, 231, 264, 265
  epoxide hydrolases, 117, 160-162
  esterases, 114, 159, 160, 193
  glutathione transferases, 120, 158, 159, 194
Iodomethane, 121
Ion channels, 8, 9, 28,
Inositol-sensitive cell, 10, 30, 31
Insect outbreaks, 302
Insect resistance, 176, 189, 265, 266, 308-314, 317
Insecticide resistance, 58, 75, 91, 177, 178, 189, 257-260, 264, 266, 305, 306, 321
Insecticide synergists, 95, 123, 227, 265, 305, 306, 313, 317
Insecticide toxicity, 161-165, 177, 178, 258
Insecticides, 161-165, 321
  acephate, 161
  aldrin, 76-78, 82, 92, 93, 154-158, 165, 189-194, 227
  amiton, 195
  avermectins, 178, 304
  azinphos methyl, 162
  carbaryl, 86, 159, 161, 178, 179, 260
  carbofuran, 191-193
  carbophenothion, 162
  chlordimeform, 31
  cypermethrin, 161
  DDT, 84, 153, 177-179, 193, 257, 258, 260, 264, 266
  demeton, 162
  diazinon, 77, 161, 162
  dieldrin, 178, 189, 191, 195
  dihydroisodrin, 92
  dimethoate, 178, 179
  dinoseb, 193
  disulfoton, 162
  endosulfan, 178
  formamidine, 31, 304
  heptachlor, 92, 93, 154, 158, 169
  hydroprene, 178
  isophenphos, 162
  kinoprene, 178
  lindane, 178
  malathion, 161, 162
  methamidofos, 161
  methomyl, 77, 161
  methyl parathion, 88, 161, 260
  monocrotophos, 161
  nicotine, 18, 28, 32, 89, 98, 161, 166, 179, 186, 187, 195, 235-237, 263, 264
  parathion, 155-157, 162, 178
  permethrin, 77, 102, 105, 106, 112, 124, 161, 258
  phorate, 88, 155-158, 162, 163, 165
  pyrethrins, 86, 95, 166, 178, 214, 215, 258, 259, 264, 265, 304, 312
  rotenone, 188, 229, 230
  schradan, 195
  trichlorfon, 161
Ips paraconfusus, 212, 213
Ipsdienol, 213
Ipsenol, 213
Iridoid glycosides, 215, 216
  aucubin, 216
  catalpol, 216
  specionin, 21
Iridoidal, 215
Iridomyrmex, 215
Isoboldine, 15
Isoflavonoids, 20
  coumesterol, 176
  rotenone, 188, 229
  (-)-vestitol, 20
Isoleucine, 232
Isophenphos, 162
Isopimpinellin, 19
Isopsoralen, 228
Isoquinoline alkaloids, 235
Isothiocyanates, 125, 234

Japanese beetle (Popillia japonica), 76, 79, 158
Juglone, 21
Junonia coenia, 215
Juvenile hormones, 80, 102, 103-110, 112-114, 117, 153, 160, 161, 167, 168, 178, 190, 213, 216, 217
Juvenile hormone esterase, 114, 160, 217

K^+ channel conductance, 28, 32
K^+ specific ATPase pump, 9
Kaempferol, 185
Kairomones, 305
Kalmitoxin I, 24
KDR, 258
Ketoreductases, 99, 100, 224
Kinoprene, 178
Knock down resistance (KDR), 258

Labidomera clivicollis, 284
Laccases, 193
Lace bugs, 282
Laodelphax striatellus, 104
Larch, 315
Larch bud moth (Zeiraphera griseana), 276, 315
Large milkweed bug (Oncopeltus fasciatus), 103, 177, 179, 223-226, 261, 262,
Lasioderma serricorne, 236
Lateral galeal styloconica, 12
Lateral styloconicum, 29
Latex, 282, 284, 287
Leaf folder, 280
Leaf miner, 261, 279
Leaf roller, 280
Leaf tier, 280
Leaf webber, 280
Learning, 3
Leek (Allium porrum), 78
Leptinotarsa decemlineata, 12, 13, 31, 109, 176, 178, 179, 315
Lignans, 95
  dihydropipercide, 95
  guineensine, 95
  myristicin, 227, 265
  pipercide, 95
  safrole, 166
Lima bean, 193
Limonene, 155, 165, 167, 214, 309
Linamarin, 123, 232
Lindane, 178
Lindera benzoin, 44, 45
Lipaphis erysimi, 178
Lipases, 101
Lipid peroxidation, 97
Lipophorin, 226
Lipopolysaccharide, 278
Lippia stoechadifolia, 190
Liquiritigenin, 181
Liriodendron tulipifera, 45-59
Locusta migratoria, 2, 6, 80, 103, 115, 119
L. migratoria cinerascens, 119
L. migratoria migratoroides, 230
Lodgepole pine (Pinus contorta), 280, 281, 307-309, 318, 319
Lotaustralin, 123, 232
Lucilia cuprina, 101, 111
Lycorine, 16
Luteolin, 180
Lygus lineolaris, 2, 178
Lymantria dispar, 22, 24, 28, 29, 77, 84, 87, 95, 101, 107, 275

Macromolecular adducts, 96, 117, 118, 119, 266, 238, 239
Macrosiphum euphorbiae, 104, 184, 185, 187, 189-194
Madagascar cockroach (Gromphadorhina portentosa), 78, 99, 103, 116, 124, 214
Magnolia acuminata, 44, 45
M. virginiana, 44, 45, 57
Makisterone A, 188
Malacosoma americanum, 28, 277
M. disstria, 286
Malathion, 161, 162
Maleic hydrazide, 181
Malus, 181
Mamestra brassicae, 10, 12, 29,
Manduca sexta, 2, 4-6, 10, 12, 30, 32, 79, 80, 107, 124, 187, 220, 221, 230, 231, 236, 261, 263, 264
Manipulating plant-insect interactions, 302, 303, 314-318
Mannosyl transferase, 122
Marbled white butterfly (Melanargia galathea), 229
Marine invertebrates, 3, 48
Maxillary palpi, 12
Maxillary styloconica, 14
Mealworm (Tenebrio molitor), 108, 114-116, 119, 178
Mechanoreception, 7
Meconium, 215
Medial galeal styloconica, 12
Medicago sativa, 181
Mediterranian fruit fly (Ceratitis capitata), 119
Melanargia galathea, 229
Melanophila acuminata, 281
Melanoplus differentialis, 178, 179, 236
M. sanguinipes, 218
Melezitose, 183, 194, 195
Membrane,
  fluidity, 258
  potential, 258
  sterols, 220, 221
Menthol, 155, 158-160, 167
Menthone, 155, 167
Mercaptopyruvate sulfurtransferase, 232
Mercapturic acids, 121, 234
Methamidophos, 161
Methomyl, 77, 161
7-Methoxy-4-methylcoumarin, 154
Methoxyresorufin, 154
Methylation, 118, 119
Methylazoxymethanol, 123
3-Methylcholanthrene, 91, 153
Methyl farnesoate, 80
N-Methyl flindersine, 15
Methyl iodide, 120
Methyl paraoxon, 88
Methyl parathion, 88, 161, 260
4-Methyl umbelliferone

glucosidation, 123
24-Methylenecholesterol, 221
Methylenedioxyphenyl compounds, 95, 227, 265
  dihydropipercide, 95
  myristicin, 95, 227, 265
  pipercide, 95
  piperonyl butoxide, 87, 88, 91, 95, 96, 165, 227, 237, 264, 265, 313
  safrole, 166
  synergists, 95, 123, 227, 265, 305, 306, 313, 317
  tropital, 123
Mexican bean beetle (Epilachna varivestis), 109, 178, 221, 278, 287
Microbes, 193, 213
Microbial symbionts, 183
Microorganisms, 98, 115, 312, 317
Migratory grasshopper (Melanoplus sanguinipes), 218
Migratory locust (Locusta migratoria), 2, 6, 80, 103, 115, 119
Milkweed (Asclepias), 99, 185, 187, 193, 224, 265, 282, 284
Millet, 155, 159
Millipedes, 232, 262
Mite control, 304
Mixed-function oxidases, 74, 82
Mollusk, 3
Molting hormones, 79, 80, 125, 153, 167, 168, 188, 222, 223
Monarch (Danaus plexippus), 99, 223, 224, 261, 264, 265, 266, 282
Monocrotaline, 237, 239
Monocrotophos, 161
Monooxygenases, 74-98
Monoterpenes, 21, 84, 155-160, 278, 280, 307-313
  camphene, 167
  camphor, 167, 213
  carene, 309
  carvone, 167
  geraniol, 167
  limonene, 155, 165, 167, 214, 309
  menthol,155, 158-160, 167
  menthone, 155, 167
  myrcene, 167, 213
  pinene, 114, 154, 155, 159-161, 165, 167, 212-214, 237, 309
  pulegone, 84, 160, 165-167
  pyrethrins, 86, 95, 166, 178, 214, 258, 259, 264, 265, 304, 312
  specionin, 21
  as pheromone precursors, 212-214
  as inducers, 154, 155, 159, 160, 165, 214
Morin, 179, 180
Morphine, 236
Motivation, 3
Mountain magnolia (Magnolia acuminata), 44, 45
Mountain pine beetle (Dendroctonus ponderosae), 213, 281, 287, 307-310, 319, 320
Musca domestica, 75, 79, 82, 84, 85, 87, 91, 101, 111, 112, 114, 116, 119, 121-124, 127, 154, 159, 188, 214, 236, 257, 258, 264
Mustard aphid (Lipaphis erysimi), 178
Mustard oils, 115-116, 160
Mustard oil glycosides (glucosinolates), 115, 160, 166, 168, 180, 233-235, 278
Myoinositol, 4, 5, 6, 31
Myrcene, 167, 213
Myricetin, 185
Myristate, 182
Myristicin, 95, 227, 265
Myrosinase, 116, 189, 234
Myrtenol, 212
Myzus persicae, 114, 178, 181, 189-191, 193, 236

N-dealkylation, 83, 85
N-demethylation, 83, 85
N-oxidation, 83, 89
N-oxide reductase, 99
Na^+,K^+ ATPases, 223, 224, 261, 262
Na^+ channel conductance, 9, 32, 258
NADPH cytochrome c (P-450) reductase, 81
NADPH cytochrome P-450 reductase, 76, 81
NADPH hemoprotein reductase, 81
NADPH oxidation, 84, 89, 90, 167, 168, 214
Naphthalene, 154
β-Naphthoflavone, 154, 158, 160
Naphthoquinone, 30
1-Naphthyl acetate, 101-113, 159, 160, 189
2-Naphthyl acetate, 101, 104, 105, 110, 111, 112
Naringenin, 179-181, 185
Naringin, 180
Natural enemies, 301, 302, 304-307, 318, 320, 321
Neem extract, 29
Neem tree (Azadirachta indica), 29
Neohesperidin, 180
Neonate larvae, 44-46, 50-52, 54, 55, 57-59
Nephotettix cincticeps, 104, 259
Nerium oleander, 184
Nerolidol, 167
Nerve poison, 232, 260
Neuroethology, 3-7

Nicotiana, 179
Nicotine, 18, 28, 32, 33, 89, 98, 154, 161, 166, 179, 186, 187, 195, 235-237, 263, 264
metabolism, 89, 98, 236, 237, 263, 264
toxicity, 154, 263
Nilaparvata lugens, 104, 181
Nitriles, 234
p-Nitroanisole, 94, 155
Nitrobenzene, 99
4-Nitrophenol, 123, 125
Nitrophenyl acetate, 102-106, 108, 109, 112
p-Nitrophenyl butyrate, 101-106
p-Nitrophenyl sulfate, 125
Nitroreductases, 99
Non-protein amino acids, 230, 231, 262
Northern corn rootworm (Diabrotica barberi), 191, 192
Notonecta, 115

Oak, 95
Octopaminergic receptors, 33
Oleander aphid (Aphis nerii), 104, 184, 185, 187, 189, 190, 193
Olfaction, 2, 3
Omphalea, 276
Oncideres, 283
Oncopeltus fasciatus, 103, 177, 179, 224-226, 261, 262
Oncophanes americanus, 110, 117, 119
Operophtera brumata, 274
Opportunistic feeder, 279
Orange tortrix (Argyrotaenia citrana), 77, 107, 117, 119
Ordeal bean, 260
Organophosphates, 87, 88, 178, 189, 259, 264
Ostrinia nubilalis, 124, 176, 226
Ouabain (G-strophanthin), 224, 261, 262
Oxidases, 74-98,
S-Oxidation, 83, 88, 89, 98
N-Oxidation, 83, 89, 98
Oxidative desulfuration, 83, 87, 88
Oxygenation at unshared electron pair, 83, 88, 89

Pacific beetle cockroach (Diploptera punctata), 79
Papaveraceae alkaloids, 179
Paper birch, 45
Papilio eurymedon, 44-46, 48
P. glaucus australis, 44-47, 52, 55-57, 59
P. g. canadensis, 44-47, 51-59
P. g. glaucus, 43-59
P. g. rutilus, 44-47, 55-57, 59
P. multicaudatus, 44
P. palamedes, 44
P. polyxenes, 77, 227, 228, 263
P. troilus, 44, 45
Paraoxon, 100, 101, 264
Parapatry, 46
Paraphlepsius irroratus, 105
Parathion, 155-157, 162, 178
Parsley, 155, 159, 160, 162-165, 193, 194, 227
Parsnip (Pastinaca sativa), 95, 159, 168, 266
Parthenium, 218
Pastinaca sativa, 95, 159, 168, 266
PCMB, 11, 31
Pea aphid (Aphis pisum), 184, 189, 194
Peanut, 155, 156, 158-161
Pectinases, 188
Pectinophora gossypiella, 218, 315
Pediobius foveolatus, 76, 110
Pentamethylbenzene (PMB), 91, 153, 154, 264, 265
Peppermint, 154, 155, 159, 161, 166
Peppermint oil, 155, 159, 161, 166
Peridroma saucia, 154
Periplaneta americana, 115, 116, 119, 121, 124, 178, 180, 229, 236, 237, 263
Permethrin, 77, 102, 105, 106, 112, 124, 161, 258
Peroxidases, 188
Persea borborna, 44
Pest management tactics, 302
Phaedon, 119
Phase I reactions, 73-75, 118, 120, 126, 127
Phase II reactions, 73-75, 118, 120, 127
Phaseolus vulgaris, 158, 159, 161, 278
Phenobarbital, 91, 153, 154, 158-162, 168, 238
Phenolics, 49, 178, 181, 182, 184-188, 228, 277, 279
Phenolic acids, 185-187
caffeic acid, 185
trans-cinnamic acid, 19, 185
p-coumaric acid, 185
ferulic acid, 185
gallic acid, 118, 185, 228
gentisic acid, 185
protocatechuic acid, 122, 185
syringic acid, 185
Phenol oxidases, 182, 188
Phenol sulfotransferase, 125
Phenotypes, 49
Phenylalanine, 228, 232
trans-4-Phenyl-3-buten-2-one, 121
2-Phenylethyl isothiocyanate, 155, 160, 168
Phenylheptatriyne, 21

Phenylpropanoids, 19-21, 181, 224-228
trans-anethole, 166
cinnamonitrile, 19
coumarins, 166, 168, 181
chromenes, 167, 178, 185, 188, 225-227, 260, 280
estragol, 166
eugenol, 168
isopimpinellin, 19
myristicin, 95, 227, 265
pipercide, 95
psoralen, 228
safrole, 166
tropital, 123
umbelliferone, 114, 168
xanthotoxin, 95, 158-160, 166, 168, 227, 228, 265
Phenylpropenes, 166-169, 265
Pheromones,
aggregation, 212-214, 281, 287, 309, 312
alarm, 182, 186, 320
bark beetle, 84, 212-214, 281, 287, 309, 312
danaidone, 238
disparlure, 87
emission, 313
hydroxydanaidal, 238
ipsdienol, 213
ipsenol, 213
recruitment, 277
trail, 277
trans-β-farnesene, 182, 186
tricosenes, 87
Phloem feeders, 175-196, 234, 236
Phloem loading, 175
Phloem-mobile herbicides, 181
Phloem toxicology, 180-182
Phloem translocation, 180-182, 283
Phloretin, 185
Phlorizin, 19, 181, 185
Phorate, 88, 155-158, 162, 163, 165
Phormia regina, 3, 9, 29, 31, 33, 79, 113, 154
Phosphatase, 100
Phosphate conjugation, 124
Phosphatidylcholine, 82, 83
Phosphatidylethanolamine, 82
Phosphodiesterase, 32
Phospholipid, 82, 258
Phosphorothionate, 88, 162, 163
Phosphotransferases, 124
Photinus pyralis, 76
Phototoxin, 228, 266, 280
Phyllotreta nemorum, 26
Physicochemical properties of toxicants, 177
Physostigma venenosum, 260
Physostigmine, 33, 100, 260
Phytoalexins, 308, 310
Phytol, 167
Phytosteroids, 87, 123, 167, 181, 188, 220-222
ergosterol, 167
fucosterol, 80, 87, 220-222
sitosterol, 155, 157, 167, 188, 220
stigmasterol, 87, 155, 167, 188, 220, 221
Pi-bond oxygenation, 83, 85-88
Picloram, 181
Pieris brassicae, 5, 10, 12, 14, 16, 18, 19, 21, 24, 29, 32, 236
P. napi, 235
P. rapae, 107, 234, 235, 315
α-Pinene, 114, 154, 155, 159-161, 165, 167, 212-214, 237, 309
β-Pinene, 155, 159, 167, 212
Pink bollworm (Pectinophora gossypiella), 218, 315
Pinus contorta, 280, 281, 307-309, 318, 319
P. radiata, 281
Piper nigrum, 95
Pipercide, 95
Piperonyl butoxide, 87, 88, 91, 95, 96, 165, 227, 237, 264, 265, 313
Pith tunnelers, 279
Plagiodera versicolora, 275
Plant hormones, 159, 188
Plantain looper (Autographa precationis), 287
Plant resistance, 230, 231, 235-237, 260-263
Platynota ideusalis, 119
Plodia interpunctella, 108
Plutella maculipennis, 108
PMB, 91, 153, 154, 264, 265
Poanes hobomok, 215
Poikilocerus bufonius, 224
Polyacetylenes, 266
Polymorphism, 46
Polyphenisms, 59
Polyphenols, 274, 281
Polyphenol oxidases, 188, 282
Polysubstrate monooxygenases (PSMO), 74-97
Pomace fly (Drosophila melanogaster), 3, 112, 154
Ponasterone, 223
Popillia japonica, 76, 79, 158
Population dynamics, 279, 301, 307
Populus grandidentata, 44, 45
P. tremuloides, 44-59
Porina moth (Wiseana cervinata), 106, 119
Potatoes, 159-161, 176, 193, 310, 315, 320
Potato aphid (Macrosiphum euphorbiae), 104, 184, 185,

187, 189-194
Potato leafhopper (Empoasca fabae), 176
Precocenes, 80, 167, 178, 185, 188, 225-227, 260
Primary hybrids, 47
Proanthocyanidins, 181
Procyanidin, 20, 179
Prodenia (Spodoptera) litura, 15
Proinsecticides, 162
Protease, 95
Proteinase inhibitor, 279
Prothoracotropic hormones, 222
Protocatechuic acid, 122, 185
Prunasin, 14
Prunus serotina, 45-47, 50-59
Pseudaletia unipuncta, 106
Pseudoplusia includens, 106, 159, 161
PSMOs, 74-97
Psoralen, 228
Pulegone, 84, 160, 165-167
Pulegone-1,2-oxide, 190
Pyranose subsite, 11
Pyrethrin I, 86, 215
Pyrethrins, 86, 95, 166, 178, 214, 215, 258, 259, 264, 265, 304, 312
Pyrethroids, 33, 161, 169, 177, 178, 189, 195, 214, 258, 304
Pyrethrum, 165
Pyrocatechol, 49, 50, 58
Pyrrhocoris apterus, 103, 114
Pyrrolizidine alkaloids, 77, 96, 166, 237-239
  monocrotaline, 237, 239
  senkirkine, 17
  usaramine, 239

Quaking aspen (Populus tremuloides), 44-59
Quercetin, 99, 155, 160, 180, 185, 229
Quercitrin, 180
Quercitrinase, 115
Quercus robur, 274, 279
Quinine, 18, 28, 32, 160
Quinolizidine alkaloids, 179, 184
Quinones, 181, 182, 194

Radish, 159, 160, 234
R.C.R., 48, 49, 59
Receptors, 219, 257
  acethylcholine, 263
  infrared-sensitive, 281
  nicotinic, 259
Receptor protein, 8, 96
Recruitment pheromone, 277
Red bay (Persea borborna), 44
Red turnip beetle (Entomoscelis americana), 12, 28
Redbanded leafroller (Argyrotaenia velutiana), 120
Reductases, 74, 98-100, 264
  amine oxide, 99
  azo, 99
  carbonyl, 74, 99, 100, 224, 264
  NADPH cytochrome c (P-450), 76, 81
  nitro, 99
  N-oxide, 99
Reductions, 98-100
Regulatory genes, 58, 97
Relative consumption rate (R.C.R.), 48, 49, 59
Relative growth rate (R.G.R.), 48, 49, 51, 53-55
Reserpine, 33
Resin, 280, 281, 287, 309, 310, 313, 314
Resistance to insecticides, 58, 75, 99, 177, 178, 189, 257-260, 264, 266, 305, 306
Resistance to insects, 176, 189, 265, 266, 308-314, 317
Resistance to plants, 230, 231, 235-237, 260-263,
R.G.R., 48, 49, 51, 53-55
Rhagium inquisitor, 115
Rhagoletis pomonella, 12
Rhamnus, 45
Rhodanese, 125, 126, 232, 233
Rhodnius prolixus, 236
Rhopalosiphum maydis, 176
Rice, 315
Rice stem borer, 315
Ricinine, 179
Rotenone, 188, 229, 230
Rutin, 99, 180, 188, 229

Saddleback caterpillar (Sibine stimulea), 218
Saddled prominent (Heterocampa guttivitta), 277, 283, 286
Safrole, 166
Salamanders, 48
Salicin, 49, 98, 228
Salicylaldehyde, 98, 228
Salt-sensitive cells, 14, 31
Samia cynthia, 108
Santonin, 159, 160, 162
Saponin, 29, 31, 180, 266
Sarcophaga bullata, 79, 113, 154
Sawflies, 282, 283
Schistocerca gregaria, 16, 24, 29, 79, 103, 119
S. vaga, 217
Schizaphis graminum, 19, 20, 104, 178, 179, 188
Schradan, 195
Scolytus multistriatus, 21
S. ventralis, 278, 308-310, 319, 320
Scopoletin, 114, 168

Seed borers, 261, 262
Seirarctia echo, 115
Selection pressure, 260, 261, 264-266
Semiochemicals, 305, 320
Senecio, 237
Senkirkine, 17
Sensilla, 6, 7-13
Sensory modalities, 1-7
Septate junction, 8
Sequestration, 183-185, 215, 216, 219, 223, 224, 229, 232, 239, 263
Serine, 124
Serotonine, 124
Sesquiterpene lactones, 49, 151, 217, 218
Sesquiterpenes, 22, 23, 167, 216-218
  farnesol, 167
  glaucolide A, 217
  nerolidol, 167
  santonin, 159, 160, 167
  shiromodiol diacetate, 22
  tulirinol, 22
  warburganal, 22, 30, 31, 217
Sheath cells, 8
Sheep blow fly (Lucilia cuprina), 101, 111
Shiromodiol diacetate, 22
Sibine stimulea, 218
Silk moth (Antheraea polyphemus), 114
Silk worm (Bombyx mori), 5, 228
Sinigrin (allyl glucosinolate), 14, 154, 159, 160, 168, 234
Sirex noctilio, 281, 312
Sitophilus granarius, 109, 233
Sitosterol, 87, 155, 167, 188, 220
SKF 525-A, 238
Skunk cabbage, 232
Snap bean (Phaseolus vulgaris), 158, 159, 161, 278
Sodium channel, 9, 32, 214, 258
Sodium gate, 258
Solanine, 28
Solanum alkaloids, 28
Solenopsis invicta, 110
Solitary feeders, 279
Sorbic acid, 182
Sorbitol, 14
Sorghum, 155, 159, 179, 232
Southern armyworm (Spodoptera eridania), 49, 77-79, 82, 84, 87, 90-92, 101, 105, 116, 119, 126, 154, 155, 159, 161, 165, 178, 193, 194, 214, 217, 218, 223, 227, 234, 237, 239
Southern corn rootworm (Diabrotica undecimpunctata), 218, 219
Southwestern cornborer (Diatraea grandiosella), 101, 107
Soybean looper (Pseudoplusia includens), 106, 159, 161
Soybeans, 155, 156, 158, 193
Sparteine, 28, 32
Specionin, 21
Spectral perturbation, 84, 90, 91
Spicebush (Lindera benzoin), 44, 45
Spilosamia virginica, 218
Spodoptera eridania, 49, 77-79, 82, 84, 87, 90-92, 101, 105, 116, 119, 126, 154, 155, 159, 161, 165, 178, 193, 194, 214, 217, 218, 223, 227, 234, 237, 239
S. exempta, 14, 15, 23, 24, 25, 29, 30,
S. frugiperda, 25, 77, 88, 89, 98, 105, 114, 117, 119, 155-169, 193, 194, 218, 227
S. littoralis, 15, 23, 27, 29, 80, 105, 114, 125, 258, 315
S. litura, 19, 23,
S. ornithogalli, 218
Spruce budworm (Choristoneura fumiferana), 17, 21, 276, 279, 317
Squalene, 167
Squash, 218, 220
Squash beetle (Epilachna borealis), 218, 283-288, 312
Stable fly (Stomoxys calcitrans), 76, 122
Stealthy feeders, 279
Sternechus tuberculatus, 230
Steroids, 87, 167, 220-223
Steroid hormones, 80, 153, 167, 168, 220, 222, 223, 230
Steroid sulfotransferases, 125
Steroidal glycoalkaloids, 28
  chaconine, 28
  solanine, 28
  tomatine, 16, 28
Stilbene glycol, 116
cis-Stilbene oxide, 116, 189
Stilbenes, 116, 181
Stigmasterol, 155, 167, 220, 221
Stimulant, 1
Stink bugs, 282
Stomoxys calcitrans, 76, 122
Streptomyces avermitilis, 178
S. peuceticus, 99
Streptomycin, 213
Striped cucumber beetle (Acalymma vittatum), 17, 21, 219
Strophanthus gratus, 261
Strychnine, 168
Styloconica, 6, 7, 12, 14, 29, 30, 31
Styrene glycol, 162
Styrene oxide, 161

Sucrose, 194
Sugar maple, 277
Sugar-sensitive cells, 9, 10, 11, 12, 13, 28, 29, 31, 33
Sulfane sulfur, 125, 232
Sulfotransferases, 124-125, 223
Sulfoxidation, 155-158, 163, 165
Swallowtail butterflies (Papilio), 43-59
Sweet bay (Magnolia virginiana), 44, 45, 57
Sweet potato, 160
Sycamore, 45
Symbiont, 183, 188, 194, 281, 308-312
Synapse, 259
Synergists, 95, 123, 227, 265, 305, 306, 313, 317
Synomone, 305
Syringic acid, 185

Tanacetum (Chrysanthemum) cinerariaefolium, 259
Tannins, 117, 273, 275
Target site insensitivity, 195, 223, 228, 230, 232, 236, 257-267
Tarnished plant bug (Lygus lineolaris), 2, 178
Tarsal hair, 12
Taxifolin, 185
Teleogryllus commodus, 102
Tenebrio molitor, 108, 114-116, 119, 178
Tenodera aridifolia, 220
Tent caterpillar (Malacosoma), 275
Terpenes, 29-31, 153, 176, 180, 181, 212-220
Tetraneurin A, 218
Tetranychus urticae, 117, 158, 159, 161, 193, 194, 259, 263, 315
Tetraopes, 284
Tetraterpenes, 166, 167
Tetrodotoxin, 9
Thermobia domestica, 102
Thermoreception, 8
Thin layer chromatography (TLC), 184
Thiocyanate, 120, 234
Thioglucosidases, 115, 234
Thiosulfate sulfurtransferase, 125, 126, 232, 233
Thlaspi arvense, 235
Tobacco alkaloids, 154, 235-237
Tobacco budworm (Heliothis virescens), 77, 87, 106, 114, 119, 218, 231, 236, 304, 320
Tobacco hornworm (Manduca sexta), 2, 4-6, 10, 12, 30, 32, 79, 80, 107, 124, 187, 220, 221, 230, 231, 236, 261, 263, 264
Tobacco wireworm (Conoderus vespertinus), 236
Tomatine, 16, 28
Tomatoes, 160, 176, 279
Trail pheromone, 277
Trans-anethole, 166
Trans-cinnamaldehyde, 120
Trans-epoxide hydrolase, 116, 117
Transferases, 74, 118-126,
Trans-2-hexenal, 154
Trans-4-hydroxy-2-hexenal, 238
Trans-4-phenyl-3-buten-2-one, 121
Trap crops, 306, 320
Tree locust (Anacridium melanorhodon), 117, 228
Trenching, 284-288, 312
Trewiasine, 17
Trialeurodes vaporarium, 104
Triatoma infestans, 101, 103, 153
Tribolium castaneum, 109
T. confusum, 221
Trichlorfon, 161
Trichoplusia ni, 29, 77, 79, 101, 105, 114, 124, 155, 159-161, 218, 236, 238
Tricin, 229
Z-9-Tricosene, 87
Z-14-Tricosen-10-one, 87
Z-14-Tricosen-10-ol, 87
2-Tridecanone, 176, 260
Trimeresia miranda, 22
2,4,6-Trinitrobenzene sulfonic acid (TNBS), 11
Triphenyl phosphate, 153
Trirhabda baccharidis, 275
Triterpenes, 25-27, 166, 167, 218-220, 284-287
  azadirachtin, 25, 29, 304
  chaparrinone, 26
  cucurbitacins, 26, 166, 218-220, 284-288, 312
  gymnemic acids, 31
  harrisonin, 25
  squalene, 167
  withanolide E, 27
  ziziphins, 31
Tropital, 123
Tryptophan, 188
Tufted apple bud moth (Platynota ideusalis), 119
Tuliptree (Liriodendron tulipifera), 45-59
Tulirinol, 22
Turnip beetle (Phaedon cocleariae), 119
Two-spotted spider mite (Tetranychus urticae), 117, 158, 159, 161, 193, 194, 259, 263, 315
Tyramine, 124
Tyrosinases, 188, 193
Tyrosine, 184

UDP-glucosyl transferases, 74, 122, 123
Umbelliferone, 114, 168
Umbellifers, 95, 193, 265, 266, 280
Urania moth, 276
Urea, 230
Urease, 117, 230, 231
Usaramine, 239
Uscharidine, 99, 224
Utetheisa bella, 77
U. ornatrix, 239

Vagrant grasshopper (Schistocerca vaga), 217
Valine, 232
Variegated cutworm (Peridroma saucia), 154
Vein snipping, 284
Velvetbean caterpillar (Anticarsia gemmatalis), 158
Verbenol, 212, 213
Verbenone, 213
Vernonia, 217, 218
Vespula flavopilosa, 76
(-)-Vestitol, 20
Voltinism, 46

Warburganal, 23, 30, 31, 217
Water boatman (Notonecta), 115
Watercress, 234
Weevils, 178
Western corn rootworm (Diabrotica virgifera), 184, 186, 191, 192, 218, 219
Wiseana cervinata, 106, 119
Withanolide E, 27
Wheat, 304, 310, 315
White pine, 315
White pine weevil, 315
Whiteflies, 179
Willow, 228
Winter moth (Operophtera brumata), 274
Wood wasp (Sirex noctilio), 281, 312

Xanthommatin, 95
Xanthotoxin, 95, 158-160, 166, 168, 227, 228, 265

Yellow-striped armyworm (Spodoptera ornithogallii), 218
Yellow woollybear (Spilosoma virginica), 218
Yponomeuta evonymellus, 14

Zeiraphera griseana, 276, 315
Ziziphins, 31
Zucchini, 284, 287
Zygaena trifolii, 232
Zygaenid moths, 122, 232, 262

FSC
www.fsc.org
MIX
Papier aus verantwortungsvollen Quellen
Paper from responsible sources
FSC® C105338

If you have any concerns about our products,
you can contact us on
ProductSafety@springernature.com

In case Publisher is established outside the EU,
the EU authorized representative is:
Springer Nature Customer Service Center GmbH
Europaplatz 3, 69115 Heidelberg, Germany

Printed by Libri Plureos GmbH
in Hamburg, Germany